STRONGER

STR**O**NGER

The Untold Story of Muscle in Our Lives

MICHAEL JOSEPH GROSS

DUTTON

DUTTON

An imprint of Penguin Random House LLC
penguinrandomhouse.com

Copyright © 2025 by Michael Joseph Gross

Penguin Random House values and supports copyright. Copyright fuels creativity, encourages diverse voices, promotes free speech, and creates a vibrant culture. Thank you for buying an authorized edition of this book and for complying with copyright laws by not reproducing, scanning, or distributing any part of it in any form without permission. You are supporting writers and allowing Penguin Random House to continue to publish books for every reader. Please note that no part of this book may be used or reproduced in any manner for the purpose of training artificial intelligence technologies or systems.

DUTTON and the D colophon are registered trademarks of Penguin Random House LLC.

Permissions appear on page 433 and constitute an extension of the copyright page.

LIBRARY OF CONGRESS CATALOGING-IN-PUBLICATION DATA
has been applied for.

ISBN: 9780525955238 (hardcover)
ISBN: 9781101986707 (ebook)

Printed in the United States of America
1st Printing

BOOK DESIGN BY SILVERGLASS STUDIO

This book aims to provide useful information that serves as a starting point for your exercise/strength building program but is not intended to replace the medical advice of your doctor. Please consult with your doctor before beginning a new exercise program or making significant health decisions, particularly if you believe you have any medical conditions that may require treatment. Publisher and author specifically disclaim responsibility for any loss or damage that may result from the use of information contained in this book.

For Steve—my fighter and my heart

. . . In one moment of time
the winds are variable, blowing in different directions.

—Pindar, Seventh Olympian Ode

Contents

Prologue   xiii
Introduction   xv

Part I Mark the Field
How words and work make muscle and mind   **1**

CHAPTER 1 Give and Receive   3

CHAPTER 2 Break and Build   31

CHAPTER 3 Live and Die   75

Part II Run the Risk
How strength shapes identity   **109**

CHAPTER 4 Born and Made   111

CHAPTER 5 Big and Small   137

CHAPTER 6 Old and New   179

Part III Gain the Prize
How muscle is a matter of life and death   **217**

CHAPTER 7 Heavy and Light   219

CHAPTER 8 Push and Pull   261

CHAPTER 9 Fall and Rise   305

Conclusion	347
Epilogue	359
Acknowledgments	363
Source Notes	369
List of Illustrations	433
Index	437

STRONGER

Prologue

Croton, on the southeastern coast of present-day Italy. Sixth century BC.

In the grassy field, the man walked toward the grazing calf—and picked it up.

Small hooves swung, slightly, in the air. Then came back down flat on the dirt.

The rest of that day, and again the next, the calf kept grazing, filling up with grass. And the man walked across the field again, wrapped his arms around the calf again, hoisted it, and set it down again.

Every day he lifted the calf until the calf became a bull, and the man became so strong that his tale still survives: the myth of Milo, the strongman of Croton.

Milo was a real person, a six-time Olympic victor in wrestling, the most celebrated athlete of antiquity. The tale about him and the calf is fiction—but in modern times, some have made the story true.

Or as true as it can be.

One who makes the story true, in the mid-1930s AD, is a teenager in Birmingham, Alabama.

The young man, Thomas Lanier DeLorme, falls ill. Doctors diagnose him with rheumatic fever, an inflammatory disease believed to weaken the heart, and they order him to rest. While spending four whole months in bed, to pass the time he reads a magazine, *Strength and Health*, full of facts about exercise and fables of strongmen such as Milo of Croton. Inspired, he gets well enough to wander local junkyards searching for machine parts; and from the junk, he builds himself a barbell; and by lifting it, he gradually builds up his strength.

The young man who grew strong by lifting weights becomes a doctor in the United States Army Medical Corps, and in 1944 he goes to work in a Chicago military hospital full of wounded soldiers. Orthopedic surgeons fix the soldiers' injured legs, but after surgeries, healing is slow. Can Dr. DeLorme do something to help them heal faster? He has an idea. But it's risky.

At a time when most mainstream doctors say weak muscles should never be challenged to work near their limit of strength, Thomas DeLorme bucks the system. For rehab after surgery, he prescribes strenuous weight training, based on a technique of testing "maximum exertion" or effort. And he prescribes weight-lifting exercises in a standard dosage, which after several years of experiment he will refine to this protocol: three sets of ten repetitions, four times a week, lifting weights heavy relative to a person's maximal strength.

The prescription works. It works so well that *strength*, in much scientific literature, comes to be defined as maximal force exertion. Years later, after the whole U.S. Army hospital system has adopted his technique; after his prescription has been used to rehabilitate polio patients, and it proves to be safe, feasible, and effective for women and men, adolescents and adults; after the doctor has retired from the Army, he reflects on his revolutionary treatment, giving credit where it's due.

Glancing back through time, DeLorme tips his cap to Milo. "For centuries it has been known that if a person lifts progressively larger loads," he writes, "the muscles, in response to the work stimulus, will hypertrophy and increase in strength."

Introduction

Lifting a calf, turning a page, and every other voluntary movement happens by means of muscle.

No matter how you think of yourself—strong or weak, large or small—you are substantially made of muscle. Most adults are made up of at least 30 percent muscle. Many of us are closer to 40 percent muscle, and a few of us exceed 50 percent. Muscle is also one of the body's most plastic tissues, changing its size and properties based on people's habits of diet and care, work and rest. Human skeletal muscle is a primary organ of metabolism, the chemical processes that sustain life. When a child is growing, when an adult is injured or sick, and as every body executes the constant processes of cellular wear and tear and repair, proteins in muscle do the work of generation, healing, and regeneration. Muscle contraction is the basis of an extensive signaling network in the body, too. Working muscles produce secretions called myokines that circulate to the brain, liver, heart, intestines, and other organ systems, regulating biological functions that make for thriving life.

From early adolescence onward, the kind of muscular work shown in the myth of Milo and prescribed by Thomas DeLorme can yield a wealth of benefits. Progressive resistance exercise can build confidence and reduce anxiety; improve bone density, blood pressure, aerobic fitness, body composition, metabolic health, insulin sensitivity, depression, and sleep; prevent and treat type 2 diabetes and cardiovascular disease; reduce the risk of several types of cancer; increase resistance to injury; and decrease the likelihood of falls and of osteoporotic fracture. In old age, muscle increasingly decides who can live independently and who cannot. Your ability to stand and go where you want to go—your independence, autonomy, and agency—your effectiveness in the world—will depend on muscle, to the last day of your life.

In recent decades, we have all witnessed what can happen when people make a practice of lifting weights. The bodies of public figures in many fields have been transformed: Derek Jeter, Serena Williams, and Cristiano Ronaldo; Madonna, Beyoncé, and Taylor Swift; Marc Jacobs, Oprah Winfrey, and Jeff Bezos; Ruth Bader Ginsburg, Michelle Obama, and Volodymyr Zelensky. At the same time, a related change has happened in the general population. Signs of the change emerge from almost thirty years of data collected by the Centers for Disease Control between 1988 and 2017, in surveys that asked American adults what forms of physical activity or exercise they spent the most time doing. Among more than fifty types of activities tracked by these surveys over that whole period—from bowling to fishing to running—the one that grew most popular most quickly was lifting weights. The number of people who said they lifted weights more often than they did other types of exercise increased by more than 34 percent in those years.

But on the other hand, the absolute number of people who made lifting weights their main form of exercise remained low. That number grew barely more than 1 percent in thirty years—from 3.2 percent to 4.3 percent of the population. And national physical activity surveys in many countries find that vast majorities of people do little or no exercise of any kind that would strengthen their muscles.

The truth is, few of us take much active interest in our own muscles, except for the young and athletic, whose interest tends to be tied to relatively short-term payoffs: winning games, or dates, or clicks.

It can be easy to lose sight of muscle's importance in every stage and every function of the widest range of lives because muscle is easy to typecast. Just hearing the word *muscle* can trigger instant thoughts of bodybuilders, or combat-sport athletes like boxers and mixed martial artists, or elite soldiers in actual military combat, or influencers and models and movie stars, exuding erotic privilege. For many people, the word *muscle* can sound inherently sinister, because muscle was glorified in pseudoscientific theories and ideologies that have been used to justify sexism, racism, colonialism, authoritarian rule, and mass atrocities.

Books that investigate muscle's meanings to aesthetics, sex, violence, injustice, and oppression do valuable work. But this book does a different kind of work, focused on the long-term, existential significance of muscles in our lives: the unnerving fact that muscles—and our individual and collaborative abilities to exert muscular strength—modulate our power to act upon the world.

In 1937, Charles Scott Sherrington, the Nobel Prize–winning British neurophysiologist whose lifework was to map the nervous system's amalgamation with the muscular system, reflected that "the importance of muscular contraction to us can be stated by saying that all man can do is to move things, and his muscular contraction is his sole means thereto." If we updated the gendered language, a more accurate one-line summary of muscle's central role in our lives would be hard to devise. And so, with a slight paraphrase—to open that statement wide, to invite everybody in—Sherrington's words bear repeating: . . . *all we can do is to move things, and our muscular contraction is our sole means thereto.*

Try to imagine: How would muscle look different to you, if you had never heard of Arnold Schwarzenegger or The Rock?

Or if that's not possible to imagine, try this instead: Think how the *world* could look different if every time you heard someone say *muscle*, the first person you thought of was not some big guy who had taken steroids, but your grandmother.

Making that shift is one of the best things you can do for yourself and for the people you love. Status quo views of muscle, by contrast, keep people stuck in destructive zero-sum games, pitting aspects of ourselves against each other—the superficial and the serious, brain versus brawn—even though significant evidence shows these conflicts have no legitimate basis in biology.

The brain's posterior cingulate cortex, the seat of empathy, self-awareness, and emotional memory—which is also the first part of the brain to atrophy in Alzheimer's disease, even before people show any signs of memory loss—"actually increases in size when you do weight-lifting exercise," according to one of the researchers who discovered this, at the University of Sydney in Australia.

For ages, though, most people have been raised on mind-body dualism, the notion that experience can be neatly divided into the physical and mental, or spiritual. Getting ourselves out of the rut requires some reflection.

Often attributed to René Descartes, the seventeenth-century French philosopher, mind-body dualism has ancient roots. By the fourth century BC, in Athens, Plato taught that a person's body, or *sōma*, contains and is activated by an incorporeal faculty, a kind of spirit, or *psychē*, that is superior to the body. Plato said the body is like a tomb or prison for the soul.

Mind-body dualism became a tenet of scientific medicine and of the medical dogma that moderate care of each—cultivating a sound mind in a sound body—is the key to health. One of history's most influential doctors, Galen of Pergamon, who lived in the Roman empire during the second century AD, was fanatic about moderation. Galen denounced athletics as bad for health because athletic competition involved striving for excellence, the opposite of moderation, for the sake of winning a prize.

Galen savaged those who disagreed with him, and he saved special venom for big men with lots of meat on their bones. Such athletes "do not even know that they have a soul," he wrote. "For they are so busy accumulating a mass of flesh and blood that their soul is extinguished as if beneath a heap of filth, and they are incapable of thinking about anything clearly; instead they become mindless like the irrational animals."

Ancient medicine's antipathy to athletics developed in part because of muscle, Galen's writings imply, and the prejudice endured. In World War II, it was an obstacle for Thomas DeLorme. When the doctor prescribed weight training to rehabilitate injured soldiers, medical colleagues disapproved. A few years later, DeLorme wrote that "the mere mention of large muscles provokes in most people, and especially those of the medical profession, a decided antipathy" because "almost everyone is bewildered and repulsed by the so-called body builder."

In the 1950s and 1960s, when mainstream science showed that physical activity is imperative for everyone who wants to live a long and healthy life, positive messages about cardiovascular fitness were commonly joined with negative judgments about muscular fitness. One of the twentieth century's most popular books about exercise, published in 1968, was *Aerobics* by Kenneth Cooper, then a thirty-seven-year-old United States military physician. *Aerobics* denigrated muscular fitness and said that lifting weights was "like putting a lovely new coat of paint on an automobile that really needs an engine overhaul."

Cooper articulated a common prejudice against weight training, based on cardiac concerns, that evidence would later refute. By the time he turned seventy years old, the author of *Aerobics* was committed to a regimen of lifting weights.

Individuals can change more rapidly than institutions and cultures, however. Still today, the medical profession shows relatively little interest in muscle. There is no medical specialty for the treatment of muscle, few doctors routinely measure or assess patients' muscle mass or strength, and few medical schools require their students to take any classes about any kind of exercise. Government, health insurance, and hospital policies in most countries make little to no provision for doctors to prescribe exercise to patients, especially in an ongoing way, even for conditions proven to be more effectively prevented or treated by exercise than by drugs or surgeries.

Athletics, on the other hand, especially since the 1970s, has become thoroughly, intensively muscle-conscious, as weight training has helped propel the steady advancements of world-record-breaking performance. In sports from swimming to stock-car racing, lifting can make the difference between winning and losing.

Divergent views of muscle in medicine and athletics describe a contrast of values: steadiness versus striving. Doctors want patients to keep steady regimens, in line with ideals of stable, constant, balanced health.

Coaches want athletes to strive for peak condition on competition day, to help them win the prize of victory.

From muscle's point of view, it's not possible to take sides in these disputes. To stay well, even at a baseline level, all the way through life, muscles need to have regular chances to really shine, to show how hard they're able to work. They also need to rest and recover: Even the strongest muscles can't be excellent all the time.

The mind-body problem is no problem for muscle—it is nonsense—because muscle stops working and fades out of existence without constant interaction with the neurological system. Mind and muscle are not enemies. They're the best of friends.

A clearer view of muscle starts with considering some facts about how muscle works.

When your hand is hanging at your side and then you lift it, opposing muscles on your upper arm shorten and lengthen—contract and relax—to bend the elbow. *Biceps contract, and triceps relax.*

Reversing the motion, when you drop your hand back down, muscles reverse roles. *Triceps contract, and biceps relax.*

Lift your hand or drop it, and muscle shows what it is: a system of symmetries, managed by orchestrated tension.

As limbs rotate around joints, muscles activate and deactivate, contract and relax.

All physical activity is paradoxical, in this sense: Movement depends on what muscles *don't* do, as much as it depends on what they do.

Both are necessary, each in its time—and the same is true of each side of the pairs of concepts shaping how we talk of muscles.

Start with nature versus nurture: Some people are stronger or more muscular than others; is the difference inborn, or does it depend on what people do? To the latter question: What *do* you do? Do you have to lift weights, or is walking enough exercise? Whatever kind of exercise you do, how should you do it? Should you move fast or slowly? Lift heavy weights or lighter ones? How much does the size of muscles matter? Is big always stronger than small?

Put like that, basic questions about muscle may sound like they have one right answer. Oppositions can polarize. But where muscle is concerned, few oppositions are true polarities. Look closely, and most prove to be paradoxes. Born and made, heavy and light, fast and slow, big and small: Those antagonists actually need each other.

This book, structured by such paradoxes, shows people navigating tensions and finding answers to vital questions about muscle, answers grounded in the central fact of muscle's critical, universal importance to life, for individuals and societies.

Cultivating that kind of awareness of muscle can be a constant struggle for the cultural reasons already mentioned, and for a material reason, too: Modern life is designed to marginalize muscles. Driving to school or to work and then sitting in chairs for most of each day eliminate much of the need for many of us to engage large groups of muscles involved in locomotion and posture, including muscles in the hips, back, and trunk. Hunching and slouching aggravate imbalances of tension among muscles, imbalances that can develop into aches and pains, especially in the back, neck, and shoulders.

"You have to redistribute that tension," says Charles Stocking, who spent four years after college moonlighting as a strength and conditioning coach for Olympic athletes, among others, while working toward his PhD in classics—studying ancient Greek language and culture, with an emphasis on religious rituals of sacrifice. The focus of his research includes athletics, because some of the earliest Greek athletic contests were religious rituals.

Now on the faculty of the University of Texas at Austin, with a joint appointment in classics and kinesiology, the study of human bodily movement, Stocking still works out several times a week, mainly by lifting weights and running sprints. He is motivated in large part by a most practical goal: to minimize the damage caused by sitting at his desk all day, so as to prevent occasional back pain from becoming chronic. Relieving and preventing pain are essential to what Stocking calls a "process of continuous self-overcoming" that, in his experience, makes for a good life at home, at work, and with friends.

He lifts heavier weights today, in his forties, than he did in his twenties—and in those earlier days, in California, he set a junior state record in the squat. Stocking understands his own unusual strength to be highly contingent on help and knowledge he has received from friends, coaches, teachers, and others. But the roots of this understanding extend far deeper than his own personal history.

Stocking is an expert on the words that some of the earliest Greek poets used to describe strength. The poets often portrayed warriors and athletes not as having strength but as *receiving* strength. They did not consider strength mainly as an individual accomplishment based on individual effort. Their experience of strength, which depended partly on what a person did, and partly on what help and gifts the person received from the gods, can seem paradoxical to us. Stocking says, "The paradox is that an ancient warrior is defined by his force, but that force is contingent on the gods."

Stocking also studies the rivalry between ancient athletics and medicine. Athletics is older than scientific medicine, and the origins of athletic contests show that trainers and athletes, no less than doctors and patients, understood the ultimate concerns of their respective interactions to be matters of life and death. The belief was not merely philosophical or figurative. It was also physiological, and was in some ways physiologically sound by current scientific standards, according to Stocking's close readings of ancient texts.

In his workouts and in his writing, Charles Stocking shows building muscle and strength in a fresh light. Athletic training marks the field of life's possibilities because it gives people freedom, and nothing less: freedom to do the things we want to do in the world.

"The main thing it has done for me is that I don't have a sense of limitations," Jan Todd said about weight training. It was 1978, and Todd was in her twenties, near the peak of a trailblazing pursuit of muscular strength. For a full decade, she was listed in *The Guinness Book of World Records* as the strongest woman in the world. Then she became a coach in the sport of powerlifting, and she led the United States national men's team and the women's team to their respective world championships. Now in her early seventies, she still lifts weights. She also sees exercise in broad historical perspective, in her work as an academic historian—she, too, is at the University of Texas at Austin, where she is chair of the Department of Kinesiology and Health Education and directs an archive of physical culture and sports, an archive housed in the university's football stadium. The term *physical culture*, though seldom heard today, is a name for the multitude of practices by which people pursue health, strength, endurance, beauty, and athletic victory.

Lifting weights has helped Jan Todd, as an athlete and as a scholar, run the risk of setting goals beyond what few, if any, of her peers dared to aim for, goals she has consistently reached. In addition to her many world-record-setting lifts, Todd helped write the first scientific strength training guidelines for women. She recovered umpteen surprising lost chapters from the history of physical culture, including Victorian-era traditions of women practicing heavy resistance exercise; and she has led her profession as president of one of the world's largest academic associations of sport historians.

Strength training, for her, has always been a practice in satisfying "a fascination in what is difficult," as she once told Johnny Carson on *The Tonight Show*. Defying stereotypes, she has done the difficult work of showing that muscular strength and knowledge of strength training are essential to female health, and access to such knowledge and practice is essential to social equality and opportunity.

"As a woman," Todd says, "if you feel physically stronger, you're going to be less afraid, more willing to try new things, and have more of a sense—I think—of yourself as a whole person."

For all kinds of people, from our first days to our last, muscle mass and muscular strength are crucial for well-being. For older people, it is never too late to start discovering how progressive resistance exercise can make life better, and for adolescents, it is never too early to build strength and muscle, with lifelong positive effects "from head to toe," in the words of Maria Fiatarone Singh, who holds a professorship at Sydney Medical School and a chair of exercise and sport science at the University of Sydney.

Fiatarone Singh and a group of her colleagues were the ones who found that the brain's posterior cingulate cortex, the seat of empathy, grows larger when people do weight-lifting exercise. The discovery amazed her, as did another scientist's finding that the hippocampus, the part of the brain responsible for memory, grows larger when people do aerobic exercise. These two different kinds of physical activity, producing two different sets of effects, illustrate a principle that structures her research: *specificity* of exercise adaptation.

Doing resistance exercise produces one set of results. Doing aerobic exercise produces another set of results. Their Venn diagram shows ample overlap of outcomes, but it also shows that exercise "requires a targeted prescription depending on what is the change that you're looking for in either physiological capacity or disease risk," says Maria Fiatarone Singh.

The principle of specificity means, as she often repeats, "It matters what you do."

For more than thirty-five years, Fiatarone Singh has been one of the world's leading researchers on exercise and health, "trying to figure out," she says, "what is the right dose, what is the right modality, for a particular syndrome or disease that might occur across the lifespan." Most of her research investigates how progressive resistance exercise may be used as medicine. She has shown that weight training can be an alternative treatment for some conditions, such as depression, and can be a supplementary treatment for others, such as type 2 diabetes.

Her work also verifies that progressive resistance exercise produces a spectrum of benefits that aerobic exercise alone does not provide, including increased muscle mass and bone density and reduced risk of falls. For such reasons, she says, "If you have to choose only one" form of exercise, "or if you can only do one—it should be progressive resistance training," and she says this is especially true for older people.

"You can get away without lifting weights when you're young," she adds. "You can't really get away with it when you're older." And when she says "lifting weights," she means lifting weights. "Although body-weight training is appropriate when you are just beginning a program, or for small muscle groups such as the calves," she says, "it is difficult to provide the continuous progression to heavier loads with body weight alone, particularly for the larger leg muscles—and that progression is necessary for optimum adaptation and clinical benefits." Heavy weight training is the *only* type of exercise that can build strength and muscle for the oldest, frailest people, Fiatarone Singh established, when she became the first physician to train nonagenarians and centenarians using Thomas DeLorme's classic strength training prescription.

The caveat, as always with strength training, is that *heavy* is a relative term. If the heaviest weight that you are able to lift is the weight of your own two arms, then you start by raising your arms. By starting where you are, wherever that may be, lifting weights can produce life-changing results.

In weight training, it's absolutely critical to know what you're doing: to learn to do each exercise as safely as possible by following proper forms of movement, along the lines of proven regimens like the one that Thomas DeLorme described. At the same time, lifting weights is a practice of constant experiment, a perpetual adventure of improvisation. For the past few years, Charles Stocking, Jan Todd, and Maria Fiatarone Singh have been my generous guides in this adventure, and they are about to become your guides, too.

This book tells their stories in detail, including some of the hopes, the losses, the fascinations, and the struggles that have shaped their pursuits of strength, because these stories are integral to their important scholarly and scientific work.

Taken together, all the stories in this book—about athletic and medical training techniques for building muscle and its attributes, including strength—show people reaching for one prize above all, a prize almost beyond naming. The prize is freedom to do what we want to do in life—freedom that assures our independence so as to fortify our interdependence. The prize is aptitude to do things that are difficult. The prize is capability to make yourself, and therefore also the world, into something different, and possibly even better, than what might otherwise have been.

The prize is life.

PART I

MARK THE FIELD

How words and work make muscle and mind

CHAPTER 1

Give and Receive

"The number of societies in human history where lots of people in many socioeconomic groups are playing sports is almost zero. We have that experience in the modern world. But historically it's really rare," says Paul Christesen, a professor of ancient Greek history at Dartmouth College. "The Greeks did it before we did," he adds. "If we want to see another society that took that stuff as seriously as we're starting to take it, there are not many places to look." But to see ancient Greek athletics in the context of ancient Greek physical culture is an experience of continual surprise.

For example, ancient Greek athletic training regimens are among the earliest programs of muscular exercise in recorded history. In fact ancient Greek art depicts so many muscular physiques, it may seem safe to assume, as one authoritative book about Greek sculpture claims, that ancient Greek athletes "knew about the muscle-building process which modern weight-trainers term 'progressive resistance.'" Yet few reports of ancient athletic training regimens have survived, and none of them clearly describe techniques of muscle-building, not even the one that's commonly assumed to be the template for modern progressive resistance training—the story of Milo and the calf.

If you could travel back to Milo's time to ask ancient Greek athletes how they trained their muscles, they would probably not understand your question. None of them—not athletes or trainers, and not even doctors—seem to have imagined that muscles had to do with movement until the athletic festival at Olympia had been going on for about five hundred years. Early Greek experiences of muscle—combining ignorance of its purpose with mastery of its development—constitute a paradox. It is one of many paradoxes of ancient Greece, the patriarchal society built on slavery that created democracy.

Mind and body, individual and community, function and beauty: Today, many people consider such elements of experience as distinct, if not opposed. But the elements of each pair are inextricable, according to some ancient athletic traditions.

For Charles Stocking, long hours sitting at the desk are part of the job.

His work as a scholar of ancient Greek and a university professor entails many hours of sitting almost every day. And all his desk work, which may appear to be mainly work of the mind—reading and writing—also trains his muscles. Teaches his muscles how to be, or how not to be.

Sitting in a chair, he teaches groups of muscles on the front of his body, flexor muscles, to tighten up—including upper-body muscles that pull arms and shoulders forward, such as the biceps and pectorals, and lower-body muscles, at the junction of pelvis and legs, that pull the hips forward.

Engagement of those muscles—as well as disengagement of the opposite sets of muscles on the back of the body—helps explain why, after sitting for a while, he feels stiff and starts to ache, even though he's not old, even though a doctor would say he's in good shape. Standing up, moving around a bit, he feels more comfortable again. Ache and stiffness subside when he engages the extensor muscles on the back of his body—including upper-body muscles that retract the shoulders, such as the rhomboids, and lower-body muscles that pull the hips back as he rises from his chair, such as the glutes.

Still, the training effect of time spent sitting persists. Steady, low-level engagement of nerve and muscle stays focused on those flexors, on the front of his body, and when this continues for weeks, months, and years, the body changes.

This is how he sees the situation: "Left to its own devices, your flexors will tighten up and take you back to the fetal position, whence you came—if you don't do something about it."

Charles Stocking is probably the only classics professor who is also a record-setting powerlifter. He broke the junior state record for his weight class in California by squatting 562.1 pounds in 2003—almost triple his body weight at the time—when he was twenty-three years old. He went to graduate school in classics at the University of California, Los Angeles, while at the same time working as a strength and conditioning coach for several UCLA Bruins teams and for individual USA Olympic athletes, and he was a private trainer for high school athletes. Coaching, he likes to say, was his "ethnographic fieldwork in physical culture," though he was, and still is, a participant observer in physical culture, simultaneously inside and outside the phenomena he studies.

Stocking is the author of four books, including *Homer's* Iliad *and the Problem of Force*, a study of eight Greek words for strength in the ancient epic about the Trojan War; and he is on the faculty of two academic departments at the University of Texas at Austin. He teaches both classics and kinesiology. For one person to work in these two very different fields is highly unusual, and it may be unique; but by the time you finish reading this book, Stocking's combination of competencies may make such perfect sense as almost to seem poetic. His example may even make you decide that one of the best ways to understand exercise is as a kind of language—a language that sustains some of the best things in life. With *life*, in Stocking's case, usually meaning another day of walking the dog; helping to care for his family, friends, and students; and getting his work done at the desk.

Back in his coaching days, a big part of Stocking's job was guiding athletes through a sticky wicket he calls "the sport-specific paradox." He explains the concept to me: "The more you do a sport, the better you get. But the better you get at the sport, that leads to overuse injuries, and eventually not being able to do the sport."

As a coach, Stocking identified basic movement patterns in various sports, then designed training programs that emphasized the opposite, contrasting movements, to reduce athletes' risk of overuse injuries. For example, when you extend your leg to kick a ball, the main muscles involved are the hip flexor muscles on the front of the legs, extending from hips to knees. Athletes who do a lot of kicking, Stocking says, are wise to balance that work by training muscles that pull in the opposite direction—the hamstrings and glutes, extensor muscles on the back of the legs, spanning from buttocks to knees.

As a professor, Stocking now applies the same principle to his own body, to manage what could be called the desk-job-specific paradox. When he begins work every morning, he is aware that some flexor muscles on the front of his body will be pulling forward most of the day, while some extensor muscles on the back of his body will spend those same hours fairly starved for attention. So before the sitting starts, Stocking tries to compensate some of his extensor muscles, in advance, for their impending deprivations.

A short routine, he says, "locks the body into proper posture, which sitting will inevitably destroy." It's a simple regimen of two exercises—a lower-body move and an upper-body move, one for the butt and one for the back.

The biggest muscle in the butt, the gluteus maximus, is a priority because it is the biggest, thickest, most powerful muscle in the body. The muscle's enormous size makes it a striking anomaly of human muscular anatomy, compared to all other animals. "Being endowed with prominent rounded buttocks is the unique privilege of humans," wrote the evolutionary biologists Françoise K. Jouffroy and Monique F. Médina in their 2006 study of the glutes, "A Hallmark of Humankind." No other mammal even *has* a gluteus maximus. The analogous muscle in other primates is called the gluteus superficialis.

The gluteus maximus, therefore, ranks high on the list of characteristics that make humans human, in terms of comparative anatomy. The muscle's extraordinary size, according to Jouffroy and Médina, suggests that actions it enables "have been of paramount importance during the course of human evolution." The muscle is most active during powerful movements, including "jogging, running, sprint-starting, leaping, and walking up stairs or a slope," and "with straightening up from stooping or squatting positions." It is "inactive, or quasi-inactive" during normal walking, and when a person stands still, sits, or reclines.

The design of modern life has eliminated much need for engaging the glutes. Standard heights of chairs, beds, tables, and desks make deep squatting unnecessary in most people's daily lives. Elevators and escalators spare the need for stepping up. Even central heating helps people "avoid squatting and crouching for hearth upkeep." In such a world, the evolutionary biologists write, properly stimulating the gluteus maximus often "requires recreating unaccustomed conditions, to be found only in sports and body-building rooms."

If people don't seek out such unaccustomed conditions, making special efforts to challenge the glutes, these muscles can become so estranged from the nervous system, it's almost as if they are forgotten. The sorry state of *gluteal amnesia*—an actual clinical term—can set in.

"Gluteal amnesia," Charles Stocking says, with half a chuckle, "is my favorite term in the world." Even as sarcasm, that's quite a statement, because within the field of classics, Stocking is a philologist. From the Greek *philo*, for "loving," and *logos*, for "word," philology concerns how language is structured and how it develops through history. Philology is a discipline of word-loving, and Stocking loves words in seven languages. To celebrate his fortieth birthday, he read the Bhagavad Gita in Sanskrit.

Stocking's strategies for avoiding gluteal amnesia include an exercise called the hip thrust. With the lower edges of his shoulder blades pressed to the side of a bench, a bed, or a sofa, and with his feet flat on the ground in front of him, his knees bent, and his trunk muscles braced—to keep his spine in neutral position—Stocking extends his hips and contracts his glutes.

Hip thrusts teach the contractile tissue of the giant complex of muscles in the rear to pull, faintly and constantly, against the force of the hip flexors, in a balanced tug-of-war. That way, Stocking says, "you can sit for a certain amount of time and not be crippled afterward."

Stocking's favorite exercise for the upper body is a rowing motion. Doing rows is a bit like pulling oars in a boat: retracting the shoulders and drawing the elbows back, as if trying to pinch the shoulder blades together.

Sometimes Stocking does rows with dumbbells or on a machine at the gym, but his favorite form of the row is the reverse pull-up. Positioning himself beneath a horizontal bar—which could be a railing at a playground or a bar on a rack at the gym—he raises his arms to grip the bar with his palms at shoulder width. He fully extends his legs, with toes pointed up and heels dug into the ground—or, to make the move more challenging, he digs his heels into a raised surface. Then, while stiffening the muscles of his trunk and contracting his glutes and quads, all to fix the length of his body as a kind of lever, he pulls himself up, touches his chest to the bar, and pinches his shoulder blades together at the top of the movement.

Stocking tries to do three sets of 20 or so repetitions, approximately every 90 minutes or whenever his back starts feeling uncomfortable, each day he spends mostly sitting at his desk. Even with a formula so flexible, Stocking's deskproofing workout is a lot of work. "The degree of volume and intensity that's required for these muscles to stay tight—people underestimate that," he says. "And the amount that has to be done on a daily basis—people underestimate that. When we sit all day, the hip flexors get really tight. The quads get really tight. We're basically just in that fetal position. And so the amount of work you have to do just to be able to sit, and not develop overuse injuries from sitting—it has to be a *lot*, on a daily basis, the other way."

On Charles Stocking's desk, among his papers and next to his computer, he keeps a slender, curved piece of bronze, about the size of a banana.

It's called a strigil—rhymes with *vigil*. Stocking bought his strigil in Greece, in the gift shop at the Archaeological Museum of Olympia. It is a reproduction, a bronze cast of an original that dates to the fourth or fifth century BC.

In ancient Greece, an athlete always applied olive oil to his skin

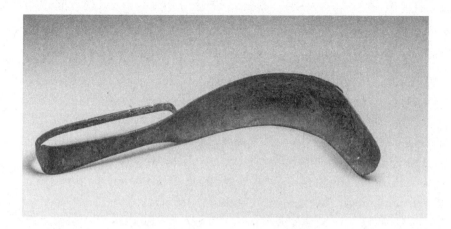

before training or competing. (I say *his*, because almost all surviving evidence of Greek athletics is of men's athletics, though there is also evidence of women's training and competitions, on which more later.) After training or competing, athletes used strigils to scrape their bodies clean of the oil, along with sweat and dust that had accumulated during their exertions. They saved this gooey mixture, called *gloios*, and sometimes they sold it, because people believed it had medicinal properties.

In the art of ancient Greece, when a man is shown with a strigil, even if the depiction shows nothing else about him, it always means he is an athlete.

But what did it mean, in ancient Greece, to be an athlete?

This is a driving question of an academic conference that Stocking runs at Olympia in Greece every summer, a symposium on ancient and modern athletics at the International Olympic Academy, an educational center near Olympia's archaeological site.

The language we use when we talk about athletics shapes our experience of athletics, and more generally, that language shapes everyone's bodily experience, Stocking contends. "Every one of us has one thing in common with ancient Greeks," he says on the first day of the conference. "We all have bodies."

Then he asks with a faint smile, "But are they the same bodies?"

At the border of the Altis, the sanctuary and sacred olive grove of Zeus, where the ancient competitions are believed to have begun, a few dozen students, mostly from universities in Greece and the United States, huddle around Charles Stocking and his wife, Catherine Pratt, an archaeologist who studies the Aegean Bronze Age. The couple are leading a tour of the small site, which occupies a little more than one square kilometer in the hollows of the valleys of two rivers, the Kladeios and Alpheios, in the northwest Peloponnese region of Greece. It's the middle of July. Even in the shade it's hot, and without a breeze, odors have no place to go, so they stagnate, pressed together, smells of bloom and rot. Our group stands amid a scattering of ruins, stones shifted and eroded by centuries of earthquakes, armies, and floods. In our ears, the insect-songs of cicadas sound so loud, Stocking and Pratt almost have to shout to be heard.

One of the oldest parts of the site is the "black ash layer," Catherine Pratt says. In this thick layer of soot, archaeologists found broken bronze tripods, cooking vessels on three-legged stands. Some of the earliest events here, scholars believe, were gatherings of local chieftains for ritual feasts, beginning after 900 BC. The feast gatherings may have involved some kind of athletic competition, too—if only a simple footrace or horse race—and the prize for winning may have been the cooking pot. "And that same cooking pot would have been left, in turn, as a thank offering to Zeus," Stocking says.

The whole experience was probably kind of like going to a barbecue, he adds, and he specifies: "a barbecue shared between humans and gods."

Over time, these early feasts evolved, encompassing more elaborate religious rituals and offerings to the gods involving animal sacrifice. Ancient Greek sacrifice was typically a social experience of people coming together to enjoy abundance and plenty. "At one point," Stocking says, "as many as one hundred bulls were sacrificed at Olympia for participants and spectators alike, both human and divine." Athletics became so central to these ritual events that the Greek word for "assembly," *agōn*, became the name for "contest."

But to start with, ritual athletic contests at Olympia began around 700 BC, many archaeologists believe. The earliest such gatherings are said to have involved a single event, a footrace, as the linchpin of a complex of rituals honoring a hero and a god.

In the night, the black ram was sacrificed.

It happened inside a circle of stones, on the altar of a hero, a man of superhuman abilities who lived long ago and was descended from a god. Inside this stone circle, the ram's throat was cut. The ram's blood ran down into the earth, through a hole in the ground.

Then the carcass of the ram was lit on fire on the altar, and its fur and flesh and bones were burned up completely, until nothing but ashes were left.

The next morning, athletes gathered at the hero's altar. It was late summer, when even the morning air was muggy. At the altar, the smell of the previous night's ritual—blood and smoke and ash—was still thick.

Standing by the altar of the hero, looking straight ahead, the athletes saw the altar of the god—the altar of Zeus. The altar of the god was situated a distance of roughly 200 modern meters away.

On this morning, for Zeus, priests had sacrificed a bull. Men from all over Greece were present—not only the athletes but their fathers and their trainers, and others representing many cities. They all watched as the best parts of the bull, the fleshy thighs, were consecrated and laid out on the god's altar to be burned.

Then one priest of Zeus stood at the altar to the god, holding a torch. This priest was the center of attention for everyone. Especially for the athletes, who watched him from their place at the altar of the hero.

The priest of Zeus raised his arm—lifting the torch—and the race was on.

The distance between altars was called a *stade*. The race across that distance was the *stadion*.

The purpose of the race was to complete the sacrifice to Zeus.

The sprinting athletes strove to win, to be first to reach the priest holding the torch; and the first one there took that torch from the priest's hand and held it in his own, lowering it to the altar and lighting the offering—and then everyone erupted with joy, the sounds of celebration so loud they drowned out the screams of the cicadas.

Because the victor won for everyone.

And everyone would feast that night—from the strong young winner to the oldest, feeblest man—and they would feast beyond feasting, sharing this most precious offering to the greatest god. Enjoying the rare, expensive, hearty, strong fine flesh of bulls, at the highest banquet table. First the athletes, who had the best parts of the meat. And then everyone.

Scholars argue about what sacrifice meant to ancient Greeks, and what it meant specifically in early competitions at Olympia. Some say the rituals were expiations of guilt for acts of violence that filled the days in times of constant, unremitting battle and war. Early athletic contests evolved from "rituals of compensation for the catastrophe of death," in the words of one respected scholar.

The ritual footrace was also a chance for people to serve a meal—and share a meal—with the gods, Charles Stocking argues. "For me, the point of sacrifice is the consumption and honor you give the gods," he says. "Athletes participate in a prestige economy shared with the gods."

A prestige economy is an exchange of prestige goods. To archaeologists, prestige goods are fine things that, no matter how useful, were not strictly necessary for subsistence-level survival. (Bronze was a prestige good; grain was a subsistence good.) In ancient Greece, "Meat is a prestige good. It's very expensive," Stocking says. "Ancient Greeks did not see meat as nourishment alone." Even athletes did not routinely eat meat, as a normal part of their training diets, until the Olympic contests had been going on for about three hundred years.

When the victor of the footrace lit the sacrificial fire, then, it was more than a symbolic gesture. Victory made him the crucial mediator among the priests who prepared the sacrifice and the god who received it and the people who shared in it. By winning, the victorious athlete actually stepped into the realm of exchange with the god. On behalf of the priests and all the people, the victor *gave* to the god. The victor gave meat to the god in the same concrete, immediate sense that I might pass you the potatoes at the dinner table.

"The gods are *there*," Stocking says, trying to convey the sense of immediate contact with the divine.

He says this often: "The gods are *there*." It's a line from *The Religious Beliefs of the Greeks*, an influential work of history by a nineteenth-century German philologist. "This has to be understood, to understand ancient Greece. *The gods are there.* They're *physically* there. *Everywhere*"—and the gods have bodies, Stocking adds, extending one arm, as if to say, *Like this*.

No one knows exactly where the altars stood, the markers of the starting line and finish of Olympia's early stadion; but in the general vicinity of where they probably were, Charles Stocking's older brother, Damian Stocking, also a trained classicist—he teaches at Occidental College in Los Angeles—stands in the shade beneath a tree, imagining the event for my benefit. "It would have been *reeking*. Reeking with blood and smoke," he says. "Also human shit. There was no sanitation." Adding layers of sensation—"And loud. Crazed!"—he remembers one ancient writer's description of sensory overload at this festival for Zeus, and he paraphrases a couple of lines—"*You become diseased, you become hot, there's too many people.*" Then he interrupts himself—"It's like Coachella gone wrong!"—and then he channel-switches back to antiquity: "*But it's worth it, to see the things of beauty we saw.* It's worth it."

From boyhood, the Stocking brothers were raised to be runners. Their father, who was a track coach and a devout Catholic with a penchant for mysticism, tried to teach his children habits of self-sacrifice by forcing them to run. Kids in this family competed in 10K races by the time they were third graders, and if you so much as grumbled, there were consequences. A standard punishment for any kind of back talk was being made to kneel on the hard kitchen floor in front of a print of Leonardo da Vinci's *The Last Supper*, with arms fully outstretched and each hand gripping a heavy Bible, for half an hour.

After Charles Stocking tells me this, it's his wife Catherine who breaks the pause. "*That's* why he has such strong shoulders," she says.

By 1994, when he was fourteen years old, the young Charles was lead singer of a punk band, Vox Pop. Wherever he went, he carried a piece of carpet pad in his backpack, for when he felt like break dancing. He dressed mostly in black, but he was dying to have a pair of white leather sneakers like the ones in the window of a store at the mall; and after he stole them, the store owner called the police, who called the boy's dad, who brought Charles home, pronounced him permanently grounded, and turned him over to his other son Damian, with a terse instruction: "Fix him."

Damian, who is fourteen years older than Charles, was teaching classics at the high school that Charles attended. Damian didn't know how to fix a juvenile delinquent, and he didn't have time to figure that out. Since Homer was what he knew, he decided that the epic poems of Homer, the *Iliad* and the *Odyssey*, would be his way of fixing Charles. He taught his little brother Homer's stories of rage, suffering, trickery, honor, and glory, and he taught Charles Homer's language. Then Damian moved away, went to graduate school for a doctorate in classical literature. The brothers kept up the Greek lessons by telephone, reading Plato together.

Charles applied to college and got into Stanford, where he planned to study classics. Damian thought Charles should prepare for freshman year by studying a standard textbook of Greek grammar, especially the book's appendix, a list of verbs. Damian so emphatically urged Charles to learn those verbs that, to be safe, Charles decided he might as well memorize the appendix—which is organized into three thousand numbered paragraphs.

"Because of that, I went straight into graduate classes," Stocking remembers. "It was like training with a ten-kilo discus and then showing up to competition." (A standard competition discus weighs 2 kilograms.)

Where had he learned Greek? the other students in his classes wanted to know.

"On the phone," he would answer. "1-800-GOT-PLATO."

While we are at Olympia, Charles Stocking and other professors meet with small groups of students for classroom discussions at the International Olympic Academy, a modernist complex on a hillside within walking distance of the archaeological site. Asked to describe the passions or questions that brought them to Olympia, the students open up:

"My passion is paraphrasing," one declares. "I love to translate Euripides. I write all the time new translations."

Another says, "My maternal grandfather ran in the 1896 Olympics."

A third raises existential questions: "I want to know why people believe the things they do."

On the topic of the conference—how the language of athletics shapes the experience of athletics and the body—one group discusses what it means to call someone an athlete, and where that word comes from. The oldest root of *athlete* is the Greek *athlos*, which has a double meaning, Stocking says.

One meaning of *athlos* is "contest for a prize," and the term *athlos* also signifies the prize itself. One of the oldest accounts of Greek athletic competition, near the end of Homer's *Iliad*, is of games played at a funeral, exertions affirming life in the face of death and angling for various prizes—and the prizes include a big cooking pot.

Another meaning of *athlos* is "labor." This, too, comes from the *Iliad*, perhaps the oldest written reference to one of the most popular characters in Greek myth: Herakles, the demigod of strength, now better known as Hercules, his Latin name. The myths of Hercules describe his intensely demanding feats, each a life-and-death struggle: killing and capturing wild animals and monsters, stealing magical objects, and doing atrocious chores. At one point, when he's given a single day to clean out a king's stable where three thousand cattle have been living for thirty years—and if he misses his deadline, he gets killed—strong, fast Hercules completes the impossible task by redirecting two rivers that spray away the muck like giant garden hoses. Then, to celebrate finishing the job and vanquishing the tyrant, he founds the Olympic Games, to bring the people of all Greek cities together. Cleaning the stables and all his other feats are called the *athloi*, the labors, of Hercules.

Connecting both meanings of *athlos*, Stocking says, "You undergo labor for the sake of a prize" when you are an athlete. "The process and the product are named with the same word."

Victory is another main topic of these discussions, because in ancient Olympic contests, second place received no prize. Victory was everything. And because the victor of the first Olympic footrace completed the main ritual offering to the most powerful god at the most sacred site in Greek lands, this victory was an essentially and ecstatically shared experience.

In the words of Gregory Nagy, a Harvard classicist who studies the ritual, sacrificial quality of ancient athletic competition and who is also here at Olympia, "Victory is not victory over the other guy," in the stadion race. "It's victory over death. One person managed to win victory over death, and we're all happy for that guy."

With fine white hair, worn longish, Nagy is a sprightly eminence in his late seventies. Generations of undergraduates have jostled for admission to a class he teaches, "The Ancient Greek Hero." The course is a survey of mythic stories about the life-and-death struggles of heroes such as Hercules, and of lyric poetry such as the odes of Pindar, written to be performed at public celebrations of athletic victories, in which "the ordeals of heroes, as myths, are analogous to the ordeals of athletes, as rituals," as Nagy has written.

"When athletes win or lose in an athletic event, they 'live' or 'die' like heroes," he argues.

Even when athletes' lives were not on the line, in the sense that if they lost today they would die today, still their *lives* were on the line, in that some vital aspect of themselves truly lived or died, manifested or did not, depending on whether and how they engaged in the struggles of competition and of training to compete. The central binary athletic value of life and death shows in the one prize that Olympic victors received at the festival. The prize was a crown or wreath of sacred olive leaves that would soon wither and die.

"It's a struggle *for* life," Stocking says, summing up this view of athletics, based on Greek traditions of sacrifice.

Because the struggle of Greek athletics was expansive, involving not just the contest event but also training for that contest, no individual athlete's struggle was his alone, Gregory Nagy points out. Ancient texts characterize the trainer who prepares the athlete as sharing in the struggle. So, too, the parent who pays for the athlete's training and bankrolls the celebration of victory. So, too, Nagy says, "the ancestors who are benignly looking on when the victorious athlete comes home and has his day in the sun. It's all the friends. It's the celebrants. It's the whole community. It's everything!"

The struggle is everything because of what the classicist and anthropologist Jean-Pierre Vernant called "the dual nature of the Games, both spectacle and religious festival—national spectacle, one might say, which joins and opposes the diverse cities in a great public competition. Each city is engaged in a struggle in which the victor represents his community more than he does himself." Because the competition was also a religious festival, Vernant wrote, "Victory *consecrates* the victor in the full sense of the word. It suffuses his person with sacred prestige," and his triumph "evokes and extends the exploit accomplished by the hero and the gods: it raises man to the level of the divine."

For all the efforts by all the people involved in every experience of training and athletics, Stocking adds, the last word on victory, in ancient writings, always belongs to the gods. "Winning in the ancient world is not a function of your agency. It's a function of the gods' agency," he says. "There's an understanding of chance in competition," a sensibility that makes Stocking think of *Any Given Sunday*, Oliver Stone's movie about a Miami football team where everyone's fate—players, coaches, owners—regardless of work and wealth and power—depends on luck.

"We like to think that it's a function of how hard we work and what we do," Stocking adds. "The Greeks are like, 'No. We understand that *if* you win, you are *blessed* by the goddess Nike.'"

When Charles Stocking started college, he stood six feet tall and weighed 150 pounds. "I was always considered very skinny," he says. He stayed skinny in part because he ran five or six miles a day, often including a steep trail called the Dish, a loop around a giant radio telescope in the Stanford foothills. Stocking ran so much that he developed painful shin splints, and by the time he went home for the summer, the pain prevented him from doing the distance runs he liked to do. His father—who had been a boxer and a serious recreational lifter before becoming a track coach—suggested lifting weights for a while instead of running.

To learn how to lift weights, Stocking read his dad's old copy of *Arnold: The Education of a Bodybuilder*, Arnold Schwarzenegger's first book, published in 1977. By following routines from that book for a few months, Stocking made huge gains of muscle—he put on 30 pounds of muscle—and in the fall, when he returned to college weighing a solid 180, he remembers, "everyone thought I was on steroids." To be clear, he adds, "I did this without drugs." His abstinence was not moralistic, though, it was strategic and ambitious. "The more drugs are involved," he says, "the less sophisticated the training has to be."

Not that his training was very sophisticated at the start. For most exercises, Stocking had been doing three sets of about ten repetitions, lifting weights that felt pretty heavy—the basic protocol described by the U.S. Army doctor Thomas DeLorme, which had since been set in lifting lore not just as a physical therapy prescription, but as a standard basis for novice training. *Arnold: The Education of a Bodybuilder* follows the DeLorme tradition by saying that beginners should do three sets of eight to ten repetitions of most lifts. That protocol reliably builds muscle mass, for most beginning lifters; and it built a lot of mass for Stocking because of what he calls his "genetic predisposition to gain muscle." That protocol reliably builds strength, too. And after Stocking's sophomore year in college, building strength became his main goal, because he had started lifting with a friend at the gym who was a competitive powerlifter.

In a powerlifting competition, each athlete has three attempts to make each of three lifts—squat, bench press, and deadlift—and the weight of the heaviest successful lift is recorded. The tally of those three numbers makes a total, and victory at a powerlifting meet goes to the person with the highest total—the person who lifted the heaviest weights.

Stocking found that powerlifting, a pursuit involving maximal exertions of force, also involved subtle complexities, even in its very name. *Powerlifting*, he maintains, is a misnomer. Where *strength* is conventionally defined with reference to maximal force, *power* is defined as the amount of force transferred per unit of time. Power, in simplest terms, is fast force. But speed counts for nothing in the sport of powerlifting. All that counts is whether the athlete is strong enough to lift the weight. Because of this, Stocking says, "powerlifting should actually be called *strengthlifting*"—and he nudges his joke with a lopsided smile.

Stocking studied his new sport with the same intensity that he studied ancient language. More experienced lifters taught him how to do the squat, bench press, and deadlift with proper form, positioning his body to minimize his risk of injury and maximize his leverage over the weights. For deadlifting and squatting, he learned the hip hinge:

to engage the posterior chain of muscles along his back, from his shoulders all the way down through his legs; to lock his lower back into its neutral position; and to move his body about the axis of his hips, generating most of his force from the muscles of his hips, to reduce the stress of loading on his spine.

He also learned how to make a training plan, to arrange workouts strategically so that his strength would peak on competition days. Highly technical training protocols, written by leading coaches in Russia and Bulgaria, satisfied his craving for complexity almost as well as Greek grammar did. By following Russian training plans he found online during his final year of college, Stocking built up to a squat of 518 pounds—the weight of the engine in a midsize car, with a 200-pound person sitting on top of that engine. The next year, in November 2003, he set his junior state record in the squat, three days before Arnold Schwarzenegger was sworn in as governor of California.

Charles Stocking was born in 1980, the year that Schwarzenegger retired from bodybuilding to concentrate on becoming a movie star, first by filming *Conan the Barbarian*. The movie, released in 1982, sparked a spectacular run of success for its star, a winning streak that lasted all through Stocking's growing-up years.

Arnold Schwarzenegger's commercial, social, and political successes made a lasting change in global popular culture. Schwarzenegger's fame accustomed the whole world to seeing photographs and videos of conspicuously, hugely muscular bodies on an everyday basis, bodies that were obviously built by weight training.

Since the nineteenth-century heyday of vaudeville and circus strength performers, anyone who wanted to see pictures of conspicuously muscular bodies had been able to find them. Yet subcultures centered on weight training had always been marginal, never producing a public figure who had racked up triumphs in as many realms as Schwarzenegger did. The most decorated competitive bodybuilder of all time became a wealthy businessman and a bestselling author, conquered Hollywood, married a member of the Kennedy clan, and was voted leader of the most populous state in what he has called "truly a country without limits." The meaning of Schwarzenegger's life, while Stocking grew from boy to man, seemed to be this: *Lifting can get you anywhere.*

Decades have passed since Schwarzenegger's retirement from competitive bodybuilding and his box-office peak, but he remains immediately, widely recognizable even in silhouette. Dwayne Johnson—The Rock—has multiples of Schwarzenegger's following on social media, but neither he nor anyone who's come along since the 1970s has rivaled the Austrian as the popular exemplar of lifting.

As Schwarzenegger has grown older, though, he has worked to change some aspects of the example he sets for others. When he was a younger man, his example helped to normalize and popularize the use of anabolic drugs. By the eighth decade of life, when asked what he would tell young people about anabolics, he said, "Don't go there." He also said his own reason for lifting had evolved. He said he trains "to stay alive, to be able to do my movies."

Despite Schwarzenegger's evolution, the blazing example of his younger self continues to nurture false impressions of lifting weights as a parochial pursuit. The sports of bodybuilding, powerlifting, Olympic weightlifting, and strongman, along with other types of progressive resistance training—such as weight training for health—are as distinct as the English dialects of New Delhi, Belfast, and Atlanta; but many people still use *bodybuilding* as the catchall name for any regular habit of lifting.

I first heard about Charles Stocking from his colleague Paul Christesen, the Dartmouth classicist, when I asked about ancient Greek semantic equivalents for the English word *strength*. He said Stocking could answer that question better than anybody. "He's a competitive bodybuilder, too," Christesen added.

"I am not a bodybuilder," Stocking clarified, the first time we talked. "I will not wear a Speedo."

To head off such confusion, Stocking was discreet about his lifting when he started graduate school in classics in 2003, at the University of California, Los Angeles. He didn't talk about his training with academic colleagues, and he barely talked about academia with his gym friends or with athletes that he coached. Figuring that the one side of him would never enhance the reputation of the other, he thought it shrewd, most of the time, to keep himself half in the dark.

Classicist by day, coach by night: Compartmentalizing was easy, partly because his research at that time had nothing to do with ancient athletics. He was studying the language of sacrifice in early Greek poetry, but he had yet to start thinking about athletics and sacrifice.

At this point, Stocking still thought of bodies just as bodies, muscle just as muscle, and strength just as strength. He didn't think about how, when people talk and write about the body, their choice of words might have power to work material changes in the body. He had not yet begun to ask the question he now considers primary, when he thinks about the bodies of people in ancient Greece and the bodies of people today: *Are they the same bodies?*

What made him start asking that question?

"Philology," he answers, smirking because he knows it sounds pretentious. "Just reading closely," he adds. "Trying to actually understand meaning in the words, in a very, very close reading."

The belly was a turning point, he says. The ancient Greek word *nēdus* has the same double meaning as the English word *belly*. It can name the stomach or the womb. But since Greek did not also have other words, as English does, for naming the stomach and the womb, the ambiguity of that Greek word is absolute. Whether *nēdus* is a stomach or a womb depends entirely on context.

After noticing this, Stocking watched for other ways that language and art structure people's experience of the body, "and how our bodies are different from the ancients', despite anatomy," he says. "Rather than saying they just didn't understand anatomy, I began to see that they experienced it differently."

The Olympic festivals seem to have started around the same time that people in Greek lands developed a written language. Among the oldest surviving stories from that period are the tale of the Trojan War and the saga of one man's journey home from that war: the epics of Homer, the *Iliad* and the *Odyssey*.

When Stocking says the ancients experienced anatomy differently from modern people, he cites the characters in those epics as examples. At first it may seem beyond belief, but the characters in Homer do not have bodies or minds in the modern plain sense of those words. The only body named in Homer is a dead body.

The singular term for *body*, in Homer's Greek, is *sōma*, but the epics do not use that word in connection with living people. In Homer, *sōma* denotes a corpse that is not yet buried. The word for mind or soul, *psychē*, in Homer, "seems to stand for both the life-breath of a person at the moment of death as well as the 'shadow' of the person which goes to the Underworld" after death, Stocking writes. He points out that in Homer, sōma and psychē are complementary, "not in opposition," except for being "in opposition to life."

The narrator of these epics does have a sense of muscle, however, but just barely. And muscle in these early Greek epics is not at all what modern people think muscle to be.

In passing, the *Iliad* and the *Odyssey* make a few minor references to muscle, using the word *sarx*. According to *The History of Muscle Physiology* by Eyvind Bastholm, *sarx* conveys "the concept of muscular mass or flesh as a whole," meaning in general "the soft parts lying under the skin or enveloping the bones." Remarkably, the *Iliad* also uses a form of the other Greek word for muscle, the term that differentiates individual muscles from overall fleshiness. That more precise word for muscle is *mys*—pronounced "moose."

The narrator uses the word twice, in a climactic battle scene. In the first instance, he says that a warrior charged at a king, but was fended off when the king, with spear in hand, struck first,

> . . . at the top of the leg, where a man's
> muscle is thickest: around the point of his spear
> the tendons were sliced apart, darkness shrouded his eyes.

A few moments later in the same battle, the narrator says, another warrior sought revenge on the man who had killed his brother. But he did not take vengeance fast enough; the one he attacked struck first,

> . . . hit his shoulder: the spear point stripped away
> the base of his arm from its muscles, shattered the bone.
> He fell with a thud, and darkness shrouded his eyes.

These lines of the *Iliad* describe minor events in a major battle, and together they may be a milestone in the history of how human beings have used words to describe the body. They appear to be the oldest surviving literary mentions of individual muscles, as distinct anatomical features.

The narrator of the *Iliad* often alludes to warriors' physiques (rows of broad shoulders on lines of battle) and often cites big bodies as indicating strength, but the poem never explicitly names muscles as an aspect of physical stature. Muscles are specifically mentioned only in those two scenes of dismemberment and death; and such spare mention casts a strange light on muscles, stranger than may first be clear.

At the start of muscle's story, muscles apparently were not a sign of the power to do violence or of any other kind of power. Muscles were not an aspect of a person's appearance. Muscles were not identified with bodily movement. Muscles were known as a distinct substance, located in the vicinity of tendon and bone, but no particular function or significance seems to have been ascribed to them.

When muscle appears on this poem's bloody battlefield, the material connotes little more than vulnerability—in the gore of dying bodies' open wounds.

These two mentions of muscles also dramatize Homer's sense of life and death, while showing how strength can mark the boundary between them.

Homer typically presents dying "as a dissolution" or "a loosening" of limbs, as Charles Stocking's brother, Damian Stocking, has written. Routinely in Homer, Damian Stocking observes, "to suffer death is to have one's limbs and one's knees *dissolved"* or loosened—the Greek verb is *luein*. Such loosening causes "a loss of functioning unity among the body's various members." In Homer's epics, Damian writes, the human self "is nothing other than that functioning," identified with abilities "to act *upon* the world."

The epics depict this loosening and loss of function with a motif that Gregory Nagy explains to me. "There was tension" in the body of a warrior, he says, "and then suddenly loss" of that tension when the warrior dies. Some translators of the *Iliad* indicate the devastating loss of tension with an elegant trope. The warrior who kills another is said to have "unstrung him," or "unstrung his strength," or "unstrung his limbs," or "unstrung the knees."

Tension around the knee joint is especially important for the epic's warriors. *The Origins of European Thought* by R. B. Onians states that in Homer's time "the knee was thought in some way to be the seat of paternity, of life, and generative power, unthinkable though that may seem to us." In the *Iliad*, "the knees are strangely prominent," Onians observes, "as the seat of vitality and strength. 'As long as I live' is expressed by 'As long as I am amongst the living and my knees are active,' and to slay is repeatedly expressed by to 'loose the knees of' the victim."

When a warrior's knees are loosed by his enemy and his strength is unstrung, the wounded man typically falls to the ground. Often, as in the second scene described above, he falls hard—"with a thud"—and he dies.

When characters die this way, in Homer's epics—losing tension in the body and falling—their strength appears to be a binary quality, departing all at once. Yet while they live, these characters experience strength in highly complex ways. The epics tell of strength profusely and make such nice distinctions among various kinds of strength that modern perceptions of it can seem crude by comparison.

In Homer, strength is a gift from the gods. Strength is also a human quality. Strength is given and made, often both at the same time; and people exercise force in modes and degrees as numerous as they express intelligence. There are not as many ancient Greek words for strength as the proverbial Eskimo words for snow, but there are plenty.

The book Charles Stocking wrote about them, *Homer's* Iliad *and the*

Problem of Force, analyzes eight words in Homer that name various aspects of strength. Two kinds of strength are strictly human qualities, according to Stocking, and six kinds of strength are gifts from the gods. In the *Iliad*, god-given strengths are by far the most common forces shaping human events—but humans are not passive recipients of the gifts. Most of the divinely given strengths require human agency to be activated. By contrast, in Homer the two exclusively human strengths, the ones that have nothing to do with the gods, are much rarer, and their effects less decisive.

One of these human strengths is usually named as *iphi*. This is "closest to a 'natural' or 'innate' notion of force," Stocking writes, and it may be related to the Greek word for sinews or tendons. The other human strength is *dunamis*, which refers to personal potential, often described in negative terms. Dunamis evokes what a person *cannot* do. When a god sends a warrior off the battlefield, and the warrior says, *I wish I had the strength to take vengeance on you, but I can't*, dunamis is the strength the warrior says he can't summon, the strength that could let him do what he wants to do—if only . . .

Among the six types of strength gods give to humans, the first, best, most unambiguously divine force is *kratos*. This is the strength of winning, the strength of a contest champion. Kratos makes one person "superior" to others, because it almost always comes from Zeus, who, being superior to all other gods, has the most of it to give away. Several centuries after Homer, Greek democracy—*demos* for "people," combined with kratos—will be built on the winning strength of the people, understood as the most valuable gift from the greatest god.

The second type of strength is *alkē*, "fighting force." Alkē is the strength of battle, suit-of-armor strength. A warrior puts on alkē kind of like putting on a cloak, but a homemade cloak, made by the person wearing it, from material given by the gods. Alkē half rhymes with "shall they"—appropriately, since desire helps decide who gets alkē. To receive this kind of strength, a person has to want it, and not just want it but work for it, by summoning it with thoughts.

The third and fourth types of strength, *menos* and *sthenos*, are a bit like nature and nurture. Menos is life-force, explicitly associated with a person's ancestors. It is an inward, subjective quality that "can be supplemented and increased by the gods" at important times, and it leaves the body at death. Sthenos, on the other hand, is culturally cultivated, based on factors such as family environment or social situation.

A fifth type of strength, the force of subjugation, especially "the paramount form of subjugation, the act of killing," is expressed by the verb *damazō*. Where this verb is used, "co-agency between mortals and immortals is especially pronounced," Stocking writes, citing scenes of Zeus killing one warrior "at the hands of" another. Zeus and the other gods "are the ultimate subjective agents of violent force, while humans nevertheless contribute either as co-agents or as mere instruments."

The most physical, the most violent, and a most personal type of force is *biē* (which rhymes with *plié*, the ballet term for bending at the knees). Biē strength is the defining core of a warrior's identity and reputation, and it is a gift. In the *Iliad*'s description of funeral games, Stocking notes, the athlete "does not win by individual effort alone, but through divine intervention which directly affects an athlete's physical body." The divine intervention of biē strength is such a radical intervention, this type of strength defines and even *becomes*, to the point of standing in for, the person who receives it.

Biē strength illustrates the fundamental paradox of strength in Homer. "The paradox," Stocking says, "is to be defined by force—but force is contingent on the gods."

Contingency becomes undeniable in old age, he adds, when even the strongest characters become keenly aware of how their strength fades. The oldest man in the *Iliad*, Nestor, who was a victorious athlete when he was young, wishes that he could still fight the way he fought then. Nestor laments, "Would that my force were fixed." The adjective conveying constancy here is *empedos*, which literally means "standing firmly on the ground." Yet Nestor can feel his limbs loosening, as the classicist Alex Purves has observed. Nestor can feel old age unstringing his strength. Nestor knows that, as Stocking writes, "Force is not fixed."

Biē strength stands in for the self—but it does not *belong* to the self. The concept is so foreign to my habitual way of thinking, it takes teamwork by both Stocking brothers, Charles and Damian, to help me understand.

Damian says, "You are not Mike! You are the *force* of Mike—*the force that is Mike.*"

Charles jumps in: "There's a problem that I have to mention. Which is that biē marks your personhood. But biē is also given by the gods. That's the paradox."

Damian: "We don't *own* ourselves."

They explain this to me while we are visiting a gallery in Olympia's archaeological museum, where we stand near a series of large relief sculptures showing Hercules performing his labors. The marble panels, each about five feet square, once adorned the Temple of Olympian Zeus, at the heart of the sanctuary. One shows Hercules holding the weight of the world on his shoulders, with Athena standing behind him, easing the burden, lending some of her strength.

Charles Stocking points to the scene and says that yes, Hercules is strong here, but it's not only him who deserves credit for this amazing feat: "It's the god behind, supporting him, as well."

Charles Stocking wrote an appendix to his book, an annotated list, about twenty-five pages long, of every use of every "strong" word in Homer. Analyzing them all, Stocking found a rule that governs the language of force in the *Iliad*.

The epic's narrator never states that one character is stronger than another. All claims of being stronger are made in dialogue, by the characters themselves. Warriors in the *Iliad* say that strength should ratify them. They say that their strength should establish their status and identity. They say this in all kinds of situations: in private conversation, in public assembly while making battle plans, in the middle of combat, in athletic contests. Yet for the purpose of establishing personal advantage, in this poem, Stocking found, "force continuously fails."

On the contrary, the narrator shows force to be always and everywhere at play. Strength is at play in relationships among people, among gods, between people and gods, even among animals and features of the landscape.

This view of strength makes each human character, in Homer, a different kind of person than most modern people know. In Homer, human life is a phenomenon of "interactive interdependence," Stocking writes. The Homeric person is "an intersubjective dividual, one defined by a human-divine symbiosis." With strength always at play, all that happens to the characters in these poems always *depends*.

The *Iliad*'s warriors are what they do, and what they do proclaims their gifts; and the doing, the giving, and the receiving all happen inside this paradox: Enacting force determines identity—but nobody, ever, is completely in charge of their own force.

CHAPTER 2

Break and Build

The first story about athletes told in the first book of history written in ancient Greek, the *Histories* by Herodotus, revolves around a feat of lower-body strength. This is the type of feat, Charles Stocking says, that you might see at a modern strongman competition; but the deed is not accomplished in competition. Instead, it is a glorious act of service.

A long time ago, the story begins, a mother was waiting. She was all dressed up and ready to go out—to go to a sanctuary, to celebrate a ritual for a goddess. Yet when it was time to leave the house, the mother was in a fix. The pair of oxen that were supposed to pull her wagon were still in the fields. Without those powerful animals, she would not make it to the festival on time.

Then her sons appeared—two sons, both champion athletes, each strong as an ox—and they saved the day.

They put the oxen's yoke on their own shoulders, and they pulled the wagon, with their mother as a passenger, the five-mile distance to the sanctuary. When they arrived, a crowd surrounded all three of them, admiring the sons for being so strong and congratulating their mother for giving birth to such men. Buoyed by praise, the proud mother said a prayer to the goddess. The mother asked that her sons be given "the best thing that a human being could have."

As for the sons, what they wanted, after the hard work of pulling the wagon, was rest. They found a place inside the sanctuary where they could lie down for a nap. And they fell asleep, and they never woke up.

They died.

The *Histories* invokes these brothers, named Cleobis and Biton, when the wisest ruler of Athens is asked to talk about some of the best examples of happiness he has seen.

Interpreting the wise man's comments, the classical archaeologist Stephen G. Miller, in his book *Ancient Greek Athletics*, writes that Cleobis and Biton "had died at the peak of glory, loved and admired by all—the perfect gift."

To Charles Stocking, the story of Cleobis and Biton raises a perennial question—*What does athletics have to do with happiness and well-being?*—and answers that question in two ways.

The first answer carries a reassuring air of stability. When you see this pair of prizewinning athletes putting their big, powerful, well-trained bodies to work solving a practical problem for their mother, it seems obvious that athletics can improve your life and the lives of people you love.

The second answer, though, unsettles the first. For Cleobis and Biton, who "immediately died after what is essentially a strongman event," Stocking notes, "extreme physical effort results not in improved bodily condition, but in death—not merely as a tragedy, but in a *good* way." The implicit theme of the story, as Stocking reads it, concerns "transcendence—going beyond the body by means of the body."

Ancient Greek doctors started criticizing athletes for just this type of behavior—striving for peak experiences of excellence, regardless of the risk—in the fifth century BC, the period when the story of the brothers was written down. The most detailed treatment of exercise in the writings attributed to Hippocrates, traditionally called the founder of medicine, says: "Everything in excess is opposed to nature."

Such statements established tensions between athletics and medicine that persist today, in rivalry and conflicts between trainers, who help athletes achieve peak performance in competitions, and doctors, who counsel patients to keep steady regimens in hopes of realizing ideals of stable, balanced health.

Charles Stocking has spent a long time living in the conflict between steadiness and striving, trying to balance its tensions. Through diligent study and experimentation, with a bit of dumb luck, and by dint of some real pain and suffering, he found training regimens that worked for him, and he adapted them to work for others. His preferred regimens, as an athlete and as a coach, have been based on a couple of principles.

The first respects muscles' material quality. Muscle cells grow stronger by carrying tension; and the proper working of whole muscles requires balances of tensions within and among the limbs, and between the limbs and the trunk. Managing those tensions and increasing them—building strength—is a skill.

The second principle concerns building the skill of strength over the long term. The process of growing stronger is not linear, it is cyclical; and no matter how hard people work or how well they plan their training, outcomes of the cyclical process of growing stronger are never more than partly predictable.

That fact shows the foundational myth of weight training—the story of Milo and the calf—to be of limited use, and possibly perilous, if taken literally. If handled creatively, within certain limits, however, the example of Milo can help hone effective strategies for getting stronger. Following those strategies can elicit what may seem to be (as it seems to Stocking) a great gift from the gods: transcendence of limits that surprise even ourselves.

Rewind to the summer of 2002, the summer after Stocking's college graduation, when he was following the workout routines of the Russian national powerlifting coach Boris Sheiko—monthlong training programs involving great volumes of work. "Just crazy amounts," Stocking says. "Like, double squat workouts Mondays and Fridays. Deadlifts on Wednesdays. Double bench sessions Monday, Wednesday, Friday. And I got really strong on it."

But Stocking says he had "no clue about stress and recovery cycles," and he had no idea that "after one of these monthlong periods of really intense training, you're supposed to take a week or two weeks off." He stacked monthlong periods of high-intensity, high-volume, high-frequency training, one on top of another, and he worked himself to the verge of exhaustion.

He had a big powerlifting meet coming up, and his back was hurting. He knew something was wrong. He went to see a doctor, who diagnosed a sprained back and prescribed a muscle relaxant, pills to ease the pain. The doctor, being a general practitioner, may not have seen or treated many other serious lifters, and Stocking, by his own admission, may not have closely followed the doctor's advice. Whatever the reasons, what happened next was, Stocking took a few muscle relaxants immediately before a heavy squat workout. And that was a mistake.

He loaded the barbell, got under it, stepped back, dropped into the squat—and then, on the way up, when he needed tension in his muscles—needed his muscles to contract and stiffen, to exert and transfer force through his body as he lifted the weight—the muscles were not up to the job, because the muscle relaxant had relaxed them.

Stocking felt a *pop* low in his lumbar region. Then a series of pops—whole bunch of them, horrible—popcorn-popping vertebrae, straight up his spine—to the middle of his back.

He did not go back to the doctor.

He rested, and he took it easy, easy, easy in the gym for a while. While he was healing, one day Stocking went to a Barnes & Noble bookstore. Browsing in the Fitness section, he found a tall paperback published by a small press called Dragon Door. The cover showed a studly, shirtless dude in Levi's, standing with his hands on hips, body situated so his left side partly obscured the background illustration, a black-and-white etching of the Cathedral of St. Basil the Blessed, Moscow's iconic polychrome-onion-domed landmark in Red Square.

The title was *Power to the People! Russian Strength Training Secrets for Every American*. The author, Pavel Tsatsouline, was pictured on the cover.

Tsatsouline immigrated to the United States from Latvia in the 1990s. He had earned a degree in sports science from the Institute of Physical Culture in Minsk and had served as a physical training drill instructor in Spetsnaz, the Soviet military special forces. As a new immigrant, he began writing for *MILO: A Journal for Serious Strength Athletes*, published by IronMind, a Nevada manufacturer of strength training equipment. In the pages of *MILO*, he introduced many Western readers to the kettlebell.

His *MILO* articles won a devoted following among martial artists and military and law enforcement personnel, and he was hired as a "subject matter expert" by the Marine Corps, the Secret Service, and the Navy SEALs. *Power to the People!*, published in 1999, is one of more than a dozen books he has written. His books are now published by his own company, StrongFirst, Inc., which Tsatsouline describes as "a global school of strength."

Stocking's education as a lifter, up to the day when he found *Power to the People!*, had been an apprenticeship. "I was taught proper squat form from somebody who was a powerlifter, who was taught by somebody else, in a succession of individuals teaching each other how to squat properly," he tells me. "There was no YouTube." To supplement these firsthand lessons, he wanted good things to read, but had not been able to find much aside from Arnold Schwarzenegger's books, and bodybuilding magazines. Pavel Tsatsouline communicated in a different register. A popularizer of reputable training science, he focused on building strength more than muscle mass (the two are not synonymous—and we'll come back to that later). Tsatsouline filled *Power to the People!* with facts that could have helped Stocking avoid the error of pushing too hard, for too long, without a rest.

Power to the People! defines strength with an equation, set in bold—**"Tension = Force"**—and a note of explanation: "The tenser your muscles are, the more strength you display."

Biceps curls can demonstrate this, Pavel Tsatsouline writes, and he gives instructions to the reader: First, hold a pencil in your hand and bend your elbow, curling it up toward your arm. Then put the pencil down, go and find a vacuum cleaner, and "curl your vacuum cleaner," while noticing how much tighter your biceps muscles become, resisting the greater gravitational force that pulls down on the vacuum. Even after you put down the vacuum, for a moment those muscles will remain more tense, because muscle tone is "residual tension in a relaxed muscle."

Strength training at its most basic is "acquiring the skill to generate more tension," Tsatsouline asserts, attributing the proposition to Thomas D. Fahey, professor emeritus of kinesiology at the University of California at Chico and coauthor of the textbook *Exercise Physiology: Human Bioenergetics and Its Applications*.

Reflexively, Charles Stocking resonated with the concept of strength as tension—and surely he felt this in part because poets, doctors, and physiologists had been expressing that equation for millennia, at least since the *Iliad*'s narrator depicted warriors dying as their strength was unstrung.

The ancient view of strength as tension is finely traced in the writings of Charles Scott Sherrington, a Nobel Prize–winning neurophysiologist who was also an authoritative historian of his field. Sherrington described a tradition of Greek medicine, practiced during the Roman empire, in which "the healthy tautness which our muscles, even when not engaged in executing a movement, still keep, was likened to the tension of a tuned string" of a musical instrument, and called by the name of *tónus*—from *tonikós*, Greek for "pulling."

Delving into the history of the word, Sherrington quoted the Roman physician Galen of Pergamon, whose treatise "On Movement of Muscles," written in the second century AD, described tonic muscular contraction as a paradoxical state in which muscles "are active while appearing not to move in the least," an apparent contradiction, but not a real one. Fifteen centuries later, in Renaissance Italy, Galen's concept of tonic muscle contraction was repeated "with practically no alteration," Sherrington found, in the anatomical writing of Hieronymus Fabricius, a professor of surgery and anatomy in Padua.

Bounding on to the seventeenth century, Sherrington continued tracking the word, finding "the first main departure of tonus from its previous path of meaning" when scientists began to see muscle contraction as "a sort of touchstone for life itself," manifesting "an immanent vital principle."

In the middle of this book, we will come back to that "vital principle." For now, the concept matters mainly for what it shows about progress in the field of muscle physiology. Like most kinds of human progress, knowledge of muscle and its functions has not developed on an unbroken line of enlightenment. Instead, knowledge of muscle has been found and lost and found, as Charles Sherrington showed by analyzing usage of *tonus* in the nineteenth and twentieth centuries, when the word wound its way back to something like its original meaning. In a lecture near the start of World War II, Sherrington echoed the ancient musical simile: "The nerve and muscle are like a tuned string," he said, "always ready to be played on."

Later in the twentieth century, experiments on how to strengthen muscles showed not only that tension is the baseline condition of strength, but that applying tension to muscles is necessary for making muscles stronger. When Richard L. Lieber, a Northwestern University medical school professor of physical medicine and rehabilitation and of neuroscience, reviewed research on what type of stimulus best builds strength, he found unambiguous evidence from experiments as early as the 1980s. "High-tension contractions are necessary for increased muscle strength," Lieber wrote, in *Skeletal Muscle Structure, Function, and Plasticity*, a standard muscle physiology textbook.

The twenty-two-year-old Charles Stocking, browsing the bookshelves at Barnes & Noble, was not consciously aware of all these tributaries of imagination and experiment converging in Pavel Tsatsouline's statement that "Tension = Force." Those old thoughts, nevertheless, made way for this new voicing of insight. As Stocking turned the pages of the book in his hands, one insight led to another.

Some notable research on how athletes develop skills of strength, Pavel Tsatsouline writes, was conducted by Russian sports scientists in the second quarter of the twentieth century.

During that time, under the reign of Joseph Stalin, scientific work in many fields was restricted and distorted by Soviet ideology. Communist leaders who coveted the international prestige of Olympic victories, however, allowed some sports scientists enough freedom of inquiry to make significant progress in understanding athletic training, including strength training.

Those scientists found "that in order to reach top form, an athlete must exhaust his adaptivity, or the ability to improve, greatly," Tsatsouline writes. "That meant the peak had to be followed by an unavoidable drop in performance."

In the late 1950s, the start of the Nikita Khrushchev era, a researcher named Leonid Matveyev "analyzed a great many athletes' training logs and concluded that things worked out a lot better if the athlete voluntarily backed off after a push instead of carrying on at full throttle and waiting for a crash," Tstatsouline adds. "Taking a step back after you have taken two steps forward is the essence of his periodization or cycling approach to strength training."

Periodization, also called cycling—or variation—is a way of working toward a goal, approaching via "gradual buildup of intensity to a personal best, and then starting all over with easy workouts." To keep improving year after year, the best lifters of the 1960s, not just Soviets but also Americans, organized their training in cycles, Tsatsouline writes.

Like the insight that strength is tension, the suggestion in *Power to the People!* that strength is cyclical has force of authority behind it, authority grounded in lines of ancient literature and in experiments of modern science. These diverse modes of observation contribute to a picture of weight training as a rich, complex endeavor that could be summarized like this: *Lifting heavy weights puts lots of stress on a person. Adaptation to that stress takes time. Training programs organize the stress, and the adaptation, in a gradual process that can be structured by the concepts of progression and periodization.*

Of those two concepts, progression is more fundamental, the core principle of weight training. The legend of Milo is a parable of progression. *Every day you lift the calf, until the calf becomes a bull.* But it is a parable and not a model for progressive training, because the legend is physiologically flawed. Human strength does not increase on such a steady upward curve. Even the lifter with every possible natural gift will eventually hit a plateau and get worn out.

"This phenomenon is what prevents the same person from constantly breaking world records," according to the meticulous training manual *Supertraining*, cowritten by one of the deans of Russian sports science, Yuri Verkhoshansky (who moved in 1995 to Italy, to work as scientific adviser to that country's National Olympic Committee), and the South African biomechanical engineer Mel Siff.

Not too long after Charles Stocking read *Power to the People!*, a friend gave him a copy of *Supertraining*, which unpacks the myth of Milo to show the problem of purely progressive training:

> Closer examination of the Milo tale reveals an incomplete ending. Milo, being an enterprising strongman, obviously would have sought further strength increase by lifting progressively heavier bulls. If he had progressed very gradually, the implications are that he should have been lifting well over 500kg after a few years. Similarly, if you began your first bench press with 60kg at the age of 16, then increased the load by only one kilogram per week, you should be lifting 580kg at the age of 26 and 1100kg at the age of 36 years. That this will not happen is obvious. In other words, progressive overloading produces diminishing, and ultimately zero, returns.

Or worse than zero. Continuous progressive overload "produces an increase in injury rate before one's physical limits are reached."

Progression, applied as an absolute principle—as in the Milo myth—is not a strategy but a fantasy.

"The phenomenon of increase in strength and all other fitness factors in response to training is clear evidence of biological adaptation to stress," according to Yuri Verkhoshansky and Mel Siff. Continuing in italics, they write: "*Training may then be described as the process whereby the body is systematically exposed to a given set of stressors to enable it to efficiently manage future exposure to those stressors.*" Whatever form of exercise training a person does, "*the underlying principle to be applied is that of optimal stress and restoration.*" This is true of every mode of training, along the whole spectrum from walking to Olympic weightlifting.

Tipped off by *Power to the People!* and *Supertraining*, Charles Stocking educated himself about stress, as defined by the phenomenon's preeminent investigator, the endocrinologist Hans Selye. The scientist wrote about many aspects of stress from 1936, when he began working in a laboratory at McGill University in Canada, until he died in 1982, as the director of the International Institute of Stress at the University of Montreal. He defined *stress* as an organism's entire response to the whole spectrum of stimuli it encounters. Selye distinguished negative from positive conditions by calling them *distress* and *eustress*—the prefix *eu-* comes from the Greek for "good," as in euphoria.

One of Selye's books, *The Stress of Life*, shows how stress develops in a process called the general adaptation syndrome, consisting of three phases: the alarm phase, the resistance phase, and the exhaustion phase. The alarm phase begins as soon as the body perceives a stressor, a stimulus imposing any kind of demand. The body shifts to fight-or-flight mode by triggering "a generalized call to arms of the defensive forces in the organism." The resistance phase responds to the stressor's demand, as the body draws on reserves of energy and activates various biological systems. Hormones surge, pulse and breathing quicken. But if demands never let up or if demands keep increasing, the body loses its ability to adapt, finally reaching the exhaustion phase, because "no living organism can be maintained continuously in a state of alarm." If a body experiences total exhaustion, Selye ominously concluded, "then death ensues."

In the cyclical process of strength training, the highest points are, paradoxically, made possible during the downtimes. The authors of *Supertraining* write that "tissue repair and growth occur predominantly during the restoration and transition periods between training sessions and not during the heavy loading phases." Moreover, the "adaptive processes" involved in weight training "apparently do not just constrain bodily processes to always produce the same, predetermined maximal level of resistance," but seem to "cause the body to over-adapt, or *supercompensate*, to a somewhat higher level." The levels of achievement produced by supercompensation cannot be predicted with precision, though. Supercompensation might add one foot to an athlete's long jump, or it might add three feet. It might add 5 pounds to a powerlifter's squat, or it might add 30 pounds.

Charles Stocking found vital ideas in *Power to the People!* and *Supertraining* and *The Stress of Life*, and he did his best to put them into practice while starting another training program written by the Russian national powerlifting team coach. But Stocking pushed too hard again, and he hurt his back again—though not as badly as the last time—and he met this injury with a little more knowledge than he had before. "This time I thought, Okay, I need the *opposite* of the muscle relaxant," he recalls. "Which is: I hurt my back from the muscle being overstretched. So I need to tighten it."

Acting on the idea that strength is tension, Stocking began to do a lot of back extensions. The back extension is an exercise that involves a partial bending at the waist, a movement driven by the glutes and hamstrings while the muscles of the trunk stabilize the spine. Stocking did back extensions frequently, in sets of many repetitions, to build endurance in his lower back; and, he says, "doing back extensions until they burned actually made my back feel better."

He came to believe that "pain isn't necessarily a problem with the integrity of the body per se. But it's a function of a balance of tension, essentially," he says. "And that would become my overriding perspective on how to train, especially athletes, but also everyone who even just wants to stand upright: this idea of balancing tension."

The next year, acting on the insight that strength is cyclical, Stocking combined what he had learned from Russian periodized training programs with some central principles of Pavel Tsatsouline's writings and of *Supertraining*, in an effort to harness the power of the supercompensation effect. And it worked. He breezed past his own personal record of 529 pounds in the squat, exceeding his previous best effort by more than 30 pounds and breaking the junior record for his weight class in California.

Stocking's experience of supercompensation, produced by means of periodized training, helped him see the physical performance of ancient athletes and warriors in a new way.

"When you're training, it's hard—and then when this supercompensation effect comes, it's no longer hard," he tells me. "That resistance isn't there any longer, and it's not because you're saying, *I'm going to do this.*

"In the *Iliad*, when it says Zeus gave this warrior *kratos*"—the strength of the contest winner—"it's explaining these anomalies in performance," Stocking believes. "What I think the poet is trying to express are these moments of supercompensation, essentially. These moments of spectacular performance that are above and beyond what the person would be capable of on a daily basis."

While training for his squat record, Stocking says, "from the sports-science perspective, I elicited the supercompensation effect. And we could scientifically try to account for how this was elicited. But the other part about this, the part that's not scientific, is: You can't actually predict the extent of the supercompensation. That's, to me, the sort of transcendental quality of training.

"In other words, the supercompensation argument doesn't deny divine intervention." He lands on that last sentence with a lightness in voice, and the lightness stays as he continues: "When the supercompensation effect can add one foot to your jump, or three feet—five pounds to your squat, or thirty pounds—what's going on there? I think the foundation for the religious experience of athletics is precisely that. It was so intimately tied to religious ritual and experience and worship of the gods precisely because of the factor of the unknown that's in sport, always.

"And the attribution of success and victory to divine favor: That's a way to articulate this participation in the unknown. That feeling of knowing where your limit is, then testing just below it to the point of sheer exhaustion, then recovering, and then feeling like you're unstoppable. *Where did this strength come from?* It's not from my willpower. It's not me willing this: *I'm going to lift this weight.* No. My body just reacted, and I've got this power right now, or this strength, that feels like it's coming from an external source."

For Stocking as a young lifter, these two insights—strength is tension and building strength is cyclical—were the beginnings of wisdom, in an ancient sense of the word.

The archaeological museum at Olympia holds a strigil dating to about 500 BC, not long after Milo of Croton's final Olympic wrestling match. The strigil is inscribed "as a gift to Zeus," offered by an athlete named Dikon, who made the strigil himself, because "He has skill"—in Greek, *sophia*.

Within a hundred years after Dikon made his strigil, Greeks would form the habit of using the word *sophia* in a different way, to name an abstract, philosophical type of wisdom that is now the common meaning of the word. Yet such wisdom, Stocking says, evolved out of "physical craft."

Lifting weights, many have claimed, is a legacy of ancient athletic wisdom.

The relatively few modern physicians who have prescribed heavy resistance weight training to their patients have sometimes justified the regimen with reference to its purportedly antique roots—as when Thomas DeLorme invoked "centuries" of knowledge that "if a person lifts progressively larger loads," muscles will grow bigger and stronger—and there is some substance to this claim. As it happens, the earliest evidence for lifting weights in Greek antiquity comes from the century when Milo lived.

The earliest evidence for ancient Greek weight lifting is a rock. It's a sandstone block about the size of a microwave oven, and it weighs about 315 pounds. When Stocking shows it to me in a museum at Olympia, he points to the pair of deep, smooth grooves worn into the long top side of the stone. The grooves look like handles, and seeing them helps me picture the event described by the phrase cut into the rock:

> Bybon, son of Phorys, lifted me over his head with one hand.

That statement may have inspired a similar inscription on a bigger, heavier rock, found hundreds of miles away, on the island of Santorini in the Aegean Sea. Seven feet long, six feet in circumference, the black volcanic boulder weighs more than 920 pounds, and bears this chiseled announcement:

> Eumastas, the son of Critobulus, lifted me from the ground.

Many classicists have been skeptical of the claims carved on these stones. Few of them have known much about lifting. "Weightlifting in Antiquity: Achievement and Training," the most detailed study of that topic, was published in 1977 by the only scholar of ancient Greek athletics (before Stocking entered the field) who had been a competitive powerlifter. Nigel Crowther, now in his eighties, argued that both stones' claims are plausible because they are comparable to verified modern lifts.

The stone of Bybon at Olympia, Crowther noted, is no heavier than barbells hoisted by a famous nineteenth-century Prussian strongman, Arthur Saxon. The rock of Eumastas on Santorini, he added, could possibly have been raised with a movement kind of like the deadlift, because at the time of Crowther's writing, the deadlift world record was 885.2 pounds, close to the 920 Eumastas lifted. (The record has risen since then. In 2020, in Iceland, Hafþór Júlíus Björnsson deadlifted 1,104.5 pounds.)

In the whole of Greece, archaeologists have unearthed only two other stones that show signs of having been used in some way that possibly resembled modern lifting. Both were found at Olympia. One chunk of limestone, which would about fit in a grocery bag and probably weighs almost 100 pounds, is inscribed, "I am the throwing-stone of Xenareus." There is also "a fragment of a limestone ball, which may have been a similar stone."

Including what little is known about these four stones, Nigel Crowther's definitive review of a thousand years of evidence for weight-lifting-style feats and weight training regimens, from the sixth century BC up until the fourth century AD, fit into nine pages. Those nine pages covered both material and written evidence. All the writing comes from the Common Era, the first century AD and later. Not a single word on these topics seems to have survived from ancient and classical Greece, except for the inscriptions on the lifting stones.

Very little surviving art from ancient Greece shows anything like heavy weight training or weight lifting, either. One painted fragment of a wine cup—also possibly from the sixth century—shows a slender boy, who was probably a slave, struggling to lift a stone.

The same period produced the earliest, best material evidence that some females, too, followed training regimens that built eye-catching amounts of muscle—though the evidence is sparse, about forty small bronze figures of girls and young women who appear to be athletes or dancers; and at least some of these bronzes come from Sparta.

Elite young Spartan women were athletes because physical education was part of their upbringing. "Spartan women's intense involvement in such activities was probably unique in the Greek world," according to the classicist Sarah B. Pomeroy. In her book *Spartan Women*, she writes: "There is more evidence, both textual and archaeological, for athletics than for any other aspect of Spartan women's lives," and "there is more evidence for the athletic activities of Spartan women alone than for the athletics of all the women in the rest of the Greek world combined." Nevertheless, as another classicist, Ellen Millender, observes, scholarship on Spartan women must make do with "a relative dearth of information" in early written sources and "the possible bias and prejudice inhering in such evidence" as does survive.

One of the earliest written histories of Sparta describes athletic training as a matter of gender parity, said to be decreed by a legendary Spartan lawgiver who ruled "that females exercise their bodies no less than males." From boyhood, elite Spartan males were required to follow rigid regimens of athletic training, as part of an educational system designed to produce soldiers with strong bodies that also *looked* strong.

That early written history of Sparta includes this laconic summary of the aesthetic standard for males: "good skin, firm flesh, good health."

Gender parity in Spartan athletic training did not indicate gender equality. Spartan young women received physical education because the men who governed Sparta believed that strong soldiers should marry strong women, and together, they should produce strong children: future soldiers and future mothers of soldiers. "Similar to most other Greek women," Ellen Millender writes, Spartan women "entered marriage under the direction of their fathers or closest male relatives and devoted their lives to procreation and the supervision of their households." Athletics for Spartan girls and young women "ultimately served the interests of the male-dominated community," she adds. According to Paul Christesen, marriage marked the "termination of Spartan girls' participation in athletics."

Spartan girls and women took a lot of criticism for being athletes. The criticism came from people who lived outside their city. Since most Greek writing about Sparta that survived antiquity was the work of outsiders, all of whom were men and most of whom were citizens of Sparta's inveterate enemy, Athens—and few had actually laid eyes on Sparta—the words almost always reveal more about the authors' preconceptions of Spartans than about Spartans themselves. Judgment and stereotypes abounded from as early as the sixth century BC, when Spartan women were called "thigh-flashers." In the fifth century, in a play by Euripides, one character says that Spartan women are inherently out of control: "Not even if she wished to, could a Spartan girl be chaste; in the company of young men they leave their houses with naked thighs and loose tunics," and, he goes on to say, Spartan women run and wrestle, competing with one another just like Spartan men.

The bronze statuettes made in Sparta may offer the only chance to catch a glimmer of how Spartan women experienced Spartan women's bodies. Some of the figures were dedications to the gods or parts of vessels used in rituals. Others were mirror handles. The mirrors belonged to Spartan girls and women and were familiar fixtures of their daily lives, according to the art historian Andrew Stewart. Some of the bronzes are probably older than any of the outsiders' writing about Sparta, too, except for one line by one poet. That dig about the women being "thigh-flashers" was written around the time when a Spartan sculptor shaped this bronze—and it is tempting to imagine what *she* thinks of being called a thigh-flasher.

"Bronze statuette of a girl runner," as she is named in the collection of the National Archaeological Museum of Athens, radiates athletic power. So do her sister bronzes. Before these figures were made, no one else in Greece (so far as we know) had sculpted female bodies with such substantially developed legs; and after this, no one seems to have done it again for centuries. What kind of training did Spartan girls and women do, to build such bodies?

"The earliest and best sources" of written information about Spartan women, Paul Christesen notes, "indicate that the athletic activities of Spartan females consisted largely of running and wrestling." Yet running and wrestling alone would probably not build legs like this for any but a few genetic outliers. "Bronze statuette of a girl running" depicts musculature that, for most young women, could be developed only by progressive resistance exercises, such as weighted lunges or squats, or by the type of vigorous jumping exercises now called plyometrics, which can build as much muscle as lifting weights.

The main character in one Athenian comedy, staged in 411 BC, is a Spartan woman who explicitly mentions a kind of plyometric exercise. The woman looks so strong, another character says that she could probably thrash a bull. "I suppose I could," the woman answers, with a swagger. "It's the exercise. I jump up and kick myself in the butt"—an ancient Greek jumping exercise called *bibasis*, which male pentathletes performed in training.

In addition to their athletic training, elite young Spartans of both sexes participated in choral dancing. Choreographed group dancing competitions and performances were central parts of social and ritual life in Sparta and throughout the Greek lands. The Spartans' dance rehearsals, as demanding and as regimented as military drills, may have involved movements similar to plyometrics, too.

These few details represent the gist of evidence concerning what types of exercise may have produced the awesome lower-body development depicted by the bronzes of Spartan girls and young women—and much of the evidence consists of slights and slanders. Even when it seems possible to discern, in all the harsh words, something of the reality of women's physical culture in Sparta, that reality is ultimately defined by its eugenic purpose.

"Sport and choral dance were ritualized activities that required unmarried Spartan girls to enact social norms that called for them to be graceful and beautiful objects of eroticized desire," Paul Christesen writes. "If athletics did to some degree empower Spartan females," he adds, "it did as much or more to enmesh them in profoundly patriarchal norms that were a fundamental part of the Spartan social order."

Every type of evidence concerning ancient Greek athletics is scarce. Writing from the sixth century BC is especially scarce. Yet we have more artifacts, art, and writing about almost every other aspect of ancient Greek athletics than we have about whatever types of progressive resistance training they did. We even know some rules of competition at Olympia from the glory days of Milo's wrestling career. ("The wrestler should not break a finger"—should not break an opponent's finger—is among the oldest surviving written Olympic contest rules.) Nothing from before or during Milo's lifetime or many centuries that follow, however, provides any real details about how athletes may have trained by lifting heavy weights.

The tale of Milo lifting the calf is probably the earliest glimpse in literature of anything like a modern regimen of progressive resistance exercise; and no one told this tale, as far as can be proven, until about five hundred years after Milo died.

Curious about the myth of Milo's training, Charles Stocking learned that it grew from one line in a Roman textbook, written in the first century AD. The textbook is not about athletics or exercise. It is a textbook of rhetoric, and the relevant passage has nothing to do with athletic training. As an aside, almost a throwaway, the author noted that some people, when giving public speeches, were liable to relate little anecdotes for no reason, with the vaguest semblance of a connection to their main topics. Then he gave, as an example of random things that blowhards liked to drop into their orations, the following remark:

Milo carried a bull, which he had been accustomed to carry as a calf.

Tracking the myth of Milo to its root, it seems, ends with a puzzlement: How can there be so many bulging muscles in ancient Greek art, and so little evidence of ancient Greek athletes lifting weights?

When Charles Stocking thinks about that question, he says he tries always to remember exactly where he stands in time—he tries to remember that "our perspective is historically conditioned. And what I mean by that is, lifting weights is so much an essential part of modern sport and fitness that the two are almost synonymous. But this is a tradition that developed. And it developed tangential to sport. But it was not one and the same with competitive sport."

In early days of ancient Greek athletics, when athletics was not mainly about fitness but about ritual, weight training and other "supplemental exercises wouldn't necessarily take center stage the way they do today," Stocking says. The rocks of Bybon and Eumastas prove that some people did lift heavy weights then—but Stocking thinks a main frame of reference for bodily strength in ancient Greece may have been social, not material.

By Milo's time—the second half of the sixth century BC—the number and type of competitions at Olympia had grown. Wrestling was the first athletic contest at Olympia that did not involve a running race, and it became the most popular Olympic event. By Milo's time, some Olympic events had age classes, too, with separate competitions for boys and for men. (Milo's first Olympic victory was in boys' wrestling.) The boys who competed were probably no younger than twelve and no older than seventeen, but since ancient Greeks did not have birth certificates, Olympic judges decided the difference between boys and men, likely based on physical development. Wrestling later became known as one of the "heavy events," along with other combat sports including boxing, probably because heavier athletes had advantages in these sports, and because those athletes tended to be bigger and heavier than runners.

Olympia's evolution was part of what the historian Donald G. Kyle describes as "an explosion in the number, variety, and geographical scope of contests" that had transformed Greek life by the time Milo came of age. In his book *Sport and Spectacle in the Ancient World*, Kyle writes that "Olympia reigned as the crown jewel of a 'circuit'" of prestigious contests at four sanctuaries. (The others were Delphi, Isthmia, and Nemea.) At all four sanctuaries, the prize for victory was a wreath, like the Olympic wreath of olive leaves, woven of the foliage of sacred or symbolic plants or trees. All four contests drew competitors from all Greek lands, including overseas Greek colonies such as Croton. And all four contests issued a sacred truce, to protect competitors and other visitors traveling to and from the festivals from the ever-present violence of ancient warfare, but the truce was not always respected.

"In the same era, most Greek city-states incorporated, expanded, or refounded earlier cultic games to gods and heroes as local athletic festivals with prizes of material value," Kyle adds. The colonies followed suit, and "like their mother cities earlier, they fostered local contests to build civic consciousness and to proclaim their status and resources."

This was when some Greek city-states became the kind of societies where, as Paul Christesen told me, lots of people in several socioeconomic groups were participating in athletics. In the first half of the sixth century BC, Christesen writes, "participation in athletics became a means by which non-elites could make a regular, public claim to socio-political privilege, and this in turn brought about fundamental changes in the place of athletics in Greek life and consciousness."

Participation in athletics gave more people access to greater privilege in many ways, as shown by the proliferating prizes, rewards, and honors for victory in the new contests. When citizens of some cities won victory on the circuit of the four most prestigious games, they returned home to celebratory receptions and were heaped with gifts such as cash bonuses and front-row seats for public events.

The new rewards for victory were, in Donald Kyle's words, "communal gifts of gratitude" to extraordinary individuals. These reward-gifts served political purposes. As the classicist David Potter writes: "The reason that athletic victory was so useful to a city was quite simply that the virtues of a good athlete were regarded as 'manly': Victory revealed that one had these virtues," making him a worthy "example to others." The communal value of keeping individuals focused on such examples was considerable. Athens eventually passed a law providing victors of the most prestigious games with free meals at public expense every day, for the rest of their lives.

Not long before Milo's Olympic debut, athletic victories started to be celebrated in another new way, too. Statues of some athletic victors began to be put up in the sanctuary. Because the earliest victor statues honored champions in the heavy events, Milo, like all his generation's up-and-coming athletes, would have strived after an honor that may have been all the more desirable for being so new. To be a statue!

Some early Olympic victor statues were made of wood, and then marble or bronze became the standards. Victor statues were decorated with colorful painted details and inlays of lacquer or precious stones; and the sanctuary filled up with a crowd of bright and shining, hyper-realistic sculpted versions of the victorious athletes, possibly at life size. Olympia seems to have had a rule that "victors are not permitted to erect statues greater in size than their own bodies," and one Olympic boxing champion commissioned a statue of himself that was "equal to him in height and thickness." When Milo's own statue was made, it's said that he carried it into the sanctuary with his own two hands, perhaps because he was the only person who could move a body as big as his.

Charles Stocking tells me, "There's a great definition of strength by Aristotle. Which is the ability to move other bodies." Then Stocking quotes his own translation of these lines from Aristotle's *Rhetoric*: "Strength consists in the ability to move another's body as one wishes. It is necessary to move another either by pulling or pushing, lifting or pressing down, or crushing, so that a strong man is strong in all these activities or in some of them."

When modern people think of strength, Stocking adds, "we think of the ability to lift heavy things," but for Greeks, who loved wrestling, it would make sense to believe "you're strong by your ability to move another body. And how do you develop that strength? Well, a lot of it would be from just moving other bodies. As opposed to picking up rocks."

How did Aristotle come to that definition of strength? Maybe he learned it from his former teacher, Plato, who had experience with wrestling and discussed it in some of his dialogues.

Plato and Aristotle, each in turn, opened their schools of philosophy at leading public gymnasia in Athens, the Academy and the Lyceum. Stocking emphasizes that athletic training and philosophy "were actually being practiced at the same time" and in the same place—and the connection was more than proximal. Stocking says the basic character of Greek athletics, as the phenomenon evolved, "also had a very intimate connection to philosophy proper." He says Plato's dialogues, propelled by pushing and pulling between questions and answers, are like the *agōn*—the athletic contest—in being a "struggle or a competition happening between two opponents."

In Athens, gymnasia had begun to flourish around the time democracy developed, at the end of the sixth century BC, in a period of intense conflicts between the popular and the tyrannical ideals of government. The name of this new civic institution where young men trained for athletic competitions and for military service was *gymnasion*, meaning "the place for nudity" or "the naked place"—*gymnos* was the Greek adjective for "naked"—because by this time, it was conventional for Greek athletes to train and compete in the nude. In Sparta, a sign above the door to the gymnasion said, "Strip, or go away."

No one knows for certain what sparked the tradition of athletic nudity, but it may have had something to do with the religious nature of the games. Nudity was a common feature of initiation rites, performed in the same sanctuary as the original Olympic footraces. Athletic nudity had social and civic meanings, too.

Exercising nude at the gymnasium, Charles Stocking writes, "may have led to a greater democratizing effect of sport as a civic good, one in which social cohesion was achieved through communal self-exposure and testing."

If athletic nudity had democratizing effects, it was also exclusionary—because "it made socioeconomic status apparent in bodily appearance," as Paul Christesen writes in his book *Sport and Democracy in the Ancient and Modern Worlds*. He then elaborates:

> Regularly exercising in the nude gave athletes a smooth, all-over tan that was unique, because there were no other acceptable contexts for being fully nude on a regular basis. Men from poorer families typically spent much of their time outside working on farms while wearing a short tunic (very much like the one worn by Spartan girls when they exercised) and as a result had the ancient equivalent of a "farmer's tan." Alternatively they worked indoors, as craftsmen, and hence, were notably pale. These men could come to the *gymnasium* to exercise, but they had to strip down and in doing so immediately made their

socioeconomic status evident to everyone present. That would have acted as a powerful deterrent that kept poorer men from participating in sport.

For most of antiquity, as for most of the modern era, athletics and exercise were practiced mainly by young and wealthy people—although evidence suggests exceptions to the norm, and exceptions became more numerous as time went on. According to Aristotle, the lower classes in Athens may have built wrestling schools of their own. He and Plato wrote of physical training for the elders of Athens, too. Plato mentioned older men working out at the gymnasium, calling them "wrinkled and not pleasant to look at"—though Plato also said that physical training, ideally, should be a lifelong practice. "Exercise for life," he called it.

But neither philosopher mentioned any specific exercises for older people, nor gave any rationale for training in old age. "The first explicit reference to exercises for the old in Greek literature" appears in a fourth-century medical text, according to Nigel Crowther. The text says that older people should run—and they should wear clothes while running, to enhance the warming effect of exercise on the body. The advice aligned with a medical notion that people's bodies tended to cool as they grew older, and that to counterbalance this natural but unhealthy cooling tendency, it was necessary to generate extra heat.

The premise of that notion was a belief, dating back roughly to the time of Milo, and first described by one of Milo's fellow citizens of Croton, that "most human things go in pairs," meaning that human bodies contain a vast number of pairs of opposing qualities or "powers." The opposing powers were believed to include hot and cold, wet and dry, and bitter and sweet. The author of this theory was a physician, scientist, and philosopher—the disciplines were not distinct—and his name was Alcmaeon.

Alcmaeon's writings survived antiquity only in fragments. According to the classicist Robin Lane Fox, a fragment by Alcmaeon provides the earliest "theoretical view of disease and health" accessible to historians. Alcmaeon's fragment says good health consists in balanced equality of the powers, and poor health or disease results from the inequality of one power dominating others.

The exact words Alcmaeon used to state this theory suggest an ancient Greek idea of health—like Aristotle's later idea of strength—that had an essentially social frame of reference, and in this case an evidently political frame of reference.

Alcmaeon's word for equality was *isonomia*. His word for dominance was *monarchia*. These were, respectively, the earliest Greek term for the forms of government that gave rise to what's now called democracy, and the Greek root of the forms of government now called autocracy. Alcmaeon's medical ideal of health "is certainly patterned on a democratic concept of political order," according to the scholar of ancient philosophy Gregory Vlastos.

Appraising the fragment's ingenious phrasing, Robin Lane Fox writes that Alcmaeon "was projecting abstract terms of political language onto the human constitution and explaining its health and sickness as a balance and imbalance of opposites." Alcmaeon "explained health and sickness in the political terms which had governed his own experience."

Fusing the language of politics and health seems to have been Alcmaeon's own original idea, but the concept of health as a kind of balance between opposing qualities such as hot and cold was his more influential theory. Balance of such qualities, later called humors and associated with bodily substances (blood, phlegm, yellow bile, and black bile), became a main principle of ancient, medieval, and early modern medicine.

Still today, many people, including many doctors, say healthy balance is a sensible standard of well-being—even though the hot-and-cold qualities of that standard in Alcmaeon's fragment seem to suggest immense disparity between ancient Greek understandings of the body and ours. What's more, ancient Greek accounts of other aspects of bodily experience, such as basic mechanisms of movement, seem only to deepen the disparity.

After being educated at Plato's Academy, Aristotle opened his own school at the other main public gymnasium in Athens. The ultimate polymath, Aristotle wrote about ethics, politics, economics, music, theater, poetry, rhetoric, physics, meteorology, cosmology, metaphysics, psychology, and biology. His writings include the earliest detailed, mechanical explanations of animal and human movement.

To Aristotle, the heart was the seat of the soul and the center of human experience, including perception, desire, and imagination. The Croatian philosopher Pavel Gregorić says that "it made sense for Aristotle that the central organ is roughly in the center of the body, from where incoming and outgoing stimuli can reach the periphery equally fast."

Aristotle thus believed the heart was the hub of movement. He thought that tiny sinews inside the heart were somehow connected to other sinews radiating outward in the body's limbs and attaching to the bones. He did not differentiate among the stringy structures in the body that would later be identified as nerves and tendons. To him, they were all sinews; and where sinews attached to bones, they could pull on the bones to make bodies move—somewhat like the strings on a marionette, he wrote.

Making sense of Aristotle's ideas has to start with noting basic facts of life, as he saw them. Pavel Gregorić, writing in collaboration with the German philosopher Klaus Corcilius, describes Aristotle's understanding of animal movement in detail. For all animals, including humans, Aristotle thought "there is such a thing as a bodily state in accordance with nature," a bodily state "conducive to the preservation and proper functioning of the animal." Aristotle believed that animals are innately drawn to desire and realize this state of being.

When the body is in good condition, in accord with this natural state, Aristotle believed the soul's perceptions of pleasure and pain cause "thermic alteration" in the body, sometimes heating and sometimes cooling a warm gaseous substance in the heart called *pneuma*. Aristotle's pneuma, Pavel Gregorić says, is "a portion of air trapped inside the body from its very conception," and is "present in all animals and involved in all their movement."

Corcilius and Gregorić write that pneuma "plays the role of a *converter* in the causation of animal motion," converting heat and cold "into pushing and pulling by way of contraction and expansion," in a system of leverage, transferring force to those tiny sinews in the heart and then out through the sinews in the limbs, attached to the bones, pushing and pulling to make people move.

The History of Muscle Physiology by Eyvind Bastholm calls pneuma "the nucleus of antiquity's physiological systems." Belief in pneuma predates Aristotle, going back at least to Alcmaeon of Croton, who thought pneuma was "the motive force" of all movement—and Alcmaeon also believed souls were immortal because they were always in motion. Bastholm reasons that "because pneuma was regarded as a primary source of power and movement, this same theory closed the gate to a deeper understanding of the share of the muscles in the mechanism of movement," with "the question having immediately been elevated to a higher plane."

In the centuries between Homer and Aristotle, Greeks still had made no connection between muscle and movement. Or if they did, no statement of the insight seems yet to have been found. The Greek word *mys*, the one that distinguished muscles from general fleshiness, dropped off the radar after appearing in those two battle skirmish scenes in the *Iliad*. Plato never used the word, nor Aristotle, nor any of the Greek playwrights. Even the authors of Greek medical texts barely mentioned muscles, as the ancient historian Robin Osborne has pointed out while noting that ancient physicians showed a general "lack of interest in what muscles do."

Only two surviving medical texts from the period prior to the first century AD mention muscles by name, as *mys*, according to research by the classicist Tyson Sukava. Neither of those medical texts connects muscles with movement, and both effectively put the Greek word for muscles in air quotes, with phrases designating it as a term that readers were not likely to know. (One refers to "the so-called" muscles of the jaw. The other refers to "the flesh surrounding the limbs which they call 'muscle'"—and in this case, "they" means other medical writers, implying that some people knew of and talked about muscles, but most people did not.)

The near-total silence about muscles in ancient Greek writing seems to contradict the near-ubiquitous visible evidence of muscle in ancient Greek art, as Robin Osborne and other scholars have noted. The disconnect creates more puzzlement and raises another question about muscle, a question parallel to the one about weight training in antiquity raised earlier in this chapter. How can there be so many bulging muscles in ancient Greek art, and so few signs of ancient Greek awareness of muscle?

The odd answer is that when Greeks looked at muscles, they may have seen something else.

Muscle is not self-evident. The cultural historian Shigehisa Kuriyama writes that "attraction to the 'muscular' physique predated the widespread recognition of muscles" among ancient Greeks: "Before they became fascinated with special structures named muscles, the Greeks celebrated bodies that had a particular look—a special clarity of form, a distinct 'jointedness,' which they identified with the vital as opposed to the dying, the mature as opposed to the yet unformed, individuals as opposed to people who all resemble each other, the strong and brave as opposed to the weak and cowardly, Europeans as opposed to Asians, the male as opposed to the female." Robin Osborne, in his book *The History Written on the Classical Greek Body*, adds: "When a body displays what we might describe as soft or weak muscles, Greek texts generally describe it as jointless."

Today when we look at the Spartan "Bronze statuette of a girl running," we admire the strength and power of her leg muscles, but the Spartan sculptor may not have seen her legs the same way. His main concern was probably rendering her knee and ankle joints, especially the lines of articulation around them. Making her knees the main focus would have made sense, because he was shaping the statuette for a world of people who paid close attention to the knees on sculptures.

On fascination with the knees, there is no discrepancy between written and material evidence. As mentioned earlier, Homer often showed warriors dealing death blows by unstringing the knees of their enemies. From about 600 BC, statues of naked young men were a common sight in sanctuaries, cemeteries, and public spaces all over Greece; and these statues, called *kouroi* by art historians, also focused viewers' attention on the knees.

"From the earliest examples," the archaeologist and art historian Brunilde Sismondo Ridgway writes, "sculptors seem to have given

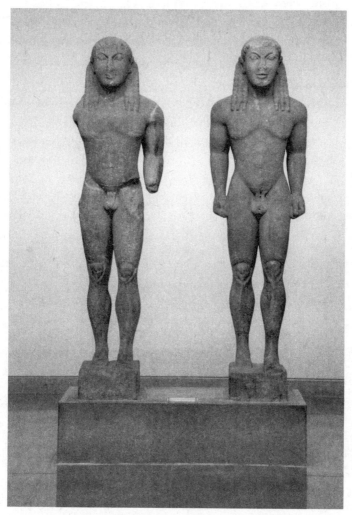

These statues of the kouros type depict Cleobis and Biton.

great attention to the knees." Sculptors rendered other body parts with widely varying levels of care, but across the Greek lands, Ridgway underscores that "powerful knees" were everywhere appreciated.

Two monumental statues of this style depict Cleobis and Biton, the brothers whose shared feat of strength to help their mother brought their lives to an end. When Herodotus told their story in the *Histories*, he ended the tale by mentioning statues of the brothers "to commemorate that they had proven themselves to be the best of men"—and these statues have survived.

Standing seven feet tall, the figures are massively muscled, with kneecaps so deeply chiseled that, if we take a moment to look closely at the joint articulations, while making best efforts to imagine how we might have seen these knees if we were ancient Greeks, the perceptual experience may start to seem almost comparable to a figure-ground illusion, like the drawing in which it's possible to see a vase, and it's also possible to see two faces, but you can't see both at once: Staring at the slabs of flesh on Greek statues like these can make it hard to see the joints—and staring at the joints can make it hard to see the flesh.

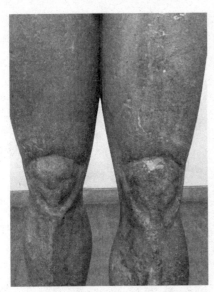 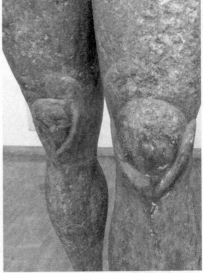

By Aristotle's time, this type of statue had been a fixture of Greek public life for centuries. Continually seeing the fleshy masses surrounding all these magnificent joints did not mean Aristotle perceived the flesh's functional significance, though.

In his descriptions of movement, Aristotle ascribed no active, kinetic purpose to the flexing parts now called muscles. He did not even see the various muscles as distinct. He referred to them all with one word, the blanket term for soft tissues—*sarx*, or flesh; and his writing seems to suggest ambivalence about this flesh. On the one hand, he said, flesh is the organ of touch, and touch is the most basic and universal form of sensation. Therefore, in his words, "all the other parts exist for the sake of the organ of touch (the flesh)." On the other hand, most functions he ascribed to flesh were passive or inert. Flesh that we now name as pectoral and gluteal muscles were, in his view, mainly padding.

Your pecs, to Aristotle, are protective covering for the heart. Your butt, to Aristotle, is a built-in bean-bag chair—"useful for resting the body," he wrote.

He also believed that fleshy padding acts as insulation, to protect bodies from extremes of heat and cold. Yet substantial amounts of flesh, he thought, could diminish intelligence. (Which helps account for the word *thick* becoming slang for "stupid.") Many of these thoughts about the flesh now known as muscle were inherited ideas, mostly from Plato.

The one role in movement that Aristotle seems to have allowed for the flesh was a kind of anchoring, a stabilizing stiffness. He wrote of "inner parts" around the joints "constructed so as to change from solid to supple and from supple to solid." Aristotle did not explicitly identify those inner parts, but Klaus Corcilius and Pavel Gregorić believe them to be muscles. The scholars describe Aristotelian biomechanics this way: "The flesh around the joints must be able to become solid and hard in order to provide joints with the stable parts against which the mobile parts will move."

The center of the whole process of movement, as Corcilius and Gregorić read Aristotle, is the perceiving soul. To clarify, they add, "the activity of the perceptual soul is not to be understood as a process in addition to the physical motions that the animal body undergoes."

Their crucial point here needs to be stressed. Aristotle did not see movement necessarily, or even primarily, as an exercise of will. "Rather, the activity of the soul is that point in this physical process" of movement "at which perceptual awareness kicks in," Corcilius and Gregorić write. "In other words," Gregorić tells me, "animals are constructed in such a way that when they perceive what is good or bad for them, they go for it or shun it."

In *The History of Muscle Physiology*, Bastholm evokes the soul's bivalent, almost surreally kinesthetic role in movement and perception, as Aristotle seems to have understood it: "Indeed, it is probable that he did not distinguish at all between movement and perception in the same manner as we do."

Wrapping up his exposition of voluntary movement, Aristotle took a line from Alcmaeon of Croton's playbook. Aristotle's final word on functional movement was a political analogy. When a body works properly, he suggested, the soul is to the limbs like the leader of a functional community, a community where people abide by the laws not because they're always being watched, but because they know the right thing to do, and they are in the habit of doing it.

He wrote:

> The constitution of an animal must be regarded as resembling that of a well-governed city-state. For when order is once established in a city there is no need of a special ruler with arbitrary powers to be present at every activity, but each individual performs his own task as he is ordered, and one act succeeds another because of custom.

Aristotle pieced together this picture of flesh and movement by dissecting monkeys, pigs, and other animals. Because myriad animals move by such similar mechanisms, he felt confident assuming humans move basically the same way. Aristotle never dissected a human, though, because Greeks of his time considered interment of the dead a sacred duty. They believed that bodies had to be properly laid to rest, or the living who failed to honor them would be polluted.

In the century after Aristotle died, however, in Alexandria, the Greek city in Egypt, a few scholars were allowed to effect what the historian Vivian Nutton calls "a momentous development in medicine, the first clear and systematic attempts to reveal, describe, and investigate the internal organization and working of the human body." The scholars became anatomists. They dissected human cadavers. They also vivisected convicted criminals. Within a few generations, the practice of dissection would again be proscripted, but by then, scalpels had made a thousand cuts to Aristotle's theories and revealed new ways of understanding movement.

Based on their empirical research in Alexandria, anatomists decided that the brain, not the heart, was the seat of what they called the soul's *hēgemonikon*—the "ruling part." They decided that the brain communicated with the rest of the body by means of a certain type of sinew: the nerve. They thought nerves functioned like ducts for transferring a certain kind of pneuma to and from the brain. They identified this special type of pneuma, called *psychic* pneuma, as part of the soul. They said this aspect of the soul, circulating through tiny hollow tubes of nerves, was wholly responsible for two things: sensation and voluntary motion.

Heinrich von Staden, a historian of ancient science and medicine, sums up the view of voluntary muscular movement that emerged from Alexandria: "In the case of voluntary movement of the muscles, psychic pneuma—borne from the brain to the muscles through the motor nerves (*neura kinētika*)—causes the contraction of the muscles by inflating them: when filled with pneuma, the muscles increase in width, but decrease in length, and therefore contract, before relaxing again by a reverse process."

Belief in muscle as vessel of some psychic power found a forceful advocate in the second century AD, in the Roman physician Galen of Pergamon. "Voluntary motion in the various parts of the body is brought about by the contraction of muscles," Galen affirmed, and "nerves, being analogous to conduits, carry power to the muscles from some fount of the brain." Based largely on his own longtime practice of animal dissection, Galen believed that nerves, infused by psychic pneuma, carried psychic power for all kinds of movements including tonic muscle contraction, which he illustrated with a high-flying analogy.

When you look up in the sky and see a bird, wings stretched wide, hovering in place, "Should it be said that it is motionless as if it happened to be hung suspended," Galen asked, "or that it is moved by its own upward motion to the same extent as carried downward by the weight of its body? The latter seems to me to be the truest. Should you deprive it of the control of the brain or the pull of the muscles, you would quickly see it born down to earth."

Today, some scholars read the text that mentions this bird, Galen's treatise "On Movement of Muscles," as a turning point in consciousness of muscle, a turning point in defining human agency, and even as a marker of the cultural borderline between East and West.

In his book *The Expressiveness of the Body and the Divergence of Greek and Chinese Medicine*, Shigehisa Kuriyama shows that "the very notion of muscles—as distinct from flesh, tendons, and sinews—developed uniquely in medical traditions rooted in ancient Greece. Elsewhere, as in China, 'ignorance' of muscle was the rule."

Kuriyama interprets Galen as essentially saying, "Muscles allow us to choose what we do, and when, and how; and this choice marks the divide between voluntary actions and involuntary processes. Muscles, in short, identify us as genuine agents."

Galen's description of muscle tonus seals the argument for muscular contractions as acts of will, in Kuriyama's view. Tonic muscle contraction means that even sitting at a desk, or standing still, "even apparent nonactivity, are genuine acts," Kuriyama writes.

The active nonactivity of tonic muscular contraction, as Galen described it, was ultimately an act of the mind or soul. If muscle prevented a bird—or a human—from falling, Galen still did not give muscle the main credit for the save. "It is clear," he wrote, "that the downward tendency, inherent in objects from their weight, is counterbalanced by the upward force of the power from the brain."

The "power from the brain," Galen believed, was that special kind of pneuma, mentioned above, that fascinated the Alexandrians—psychic pneuma. Galen said psychic pneuma was produced by a sequence of events involving many parts of the body. Air, inhaled by the lungs, began changing into a basic form of pneuma, which flowed into the heart and changed into a second form of pneuma, which then flowed through arteries to the brain, where at last it became psychic pneuma. From the brain, psychic pneuma would pass into the upper part of the spinal cord, and then infuse the nerves with psychic power that would "flow to the soul's instruments" such as the muscles, to make voluntary movements.

What exactly was that psychic power that caused movement? Galen wasn't sure, according to Orly Lewis, who teaches classics at the Hebrew University of Jerusalem. Galen named the power as *dunamis*—one of the Greek words for strength that Homer used, the one that indicated personal potential. But Galen repeatedly admitted that he did not know what dunamis materially amounted to, Lewis tells me: "It is an explanatory concept he often uses" when "one cannot know what exactly it is which makes something work"—as when discussing movement.

Explaining the material nature of dunamis, as we will see, would remain a bedeviling challenge for scientists for a very long time to come.

Galen's views of muscle and voluntary movement raise one final, critical question: What exactly did he mean by "muscle"? Galen's definition of muscle—his understanding of the anatomy of muscles—may be the most concise proof that muscle is not self-evident.

A muscle, to Galen, was a whole complex of materials including nerves, ligaments, blood vessels, and the flesh that surrounds them. As nerves entered one end of a muscle, Galen believed, they split into many small fibers, like threads, tunneling *through* the fleshy middle to the muscle's other end, where they rejoined each other and fused with ligaments, emerging transformed. This fusion of nerves and ligaments created tendons, the final activators of the psychic power, infused by psychic pneuma, that pulled and released people's bones, producing movement.

Galen's magnum opus, *On the Usefulness of the Parts of the Body*, written to be understood not only by medical colleagues but by all the educated elite, put these points about muscle and movement in plain terms. Galen wrote, "a tendon is the principal instrument of motion. The muscle is formed for the sake of producing it."

Galen articulated more detailed views of muscles and their function, and by today's standards more accurate views, than his predecessors had done. Yet he still followed Aristotle, who had followed Plato, in saying the fleshy part of muscle played only a passive role in movement; and was a kind of insulation from extremes of heat and cold; and was padding and protection against the impact of falls. The fleshy part of muscle was, Galen wrote, "like a cushion" for the stringy parts of muscle—the nerves, ligaments, and tendons that, he believed, did the real work of movement.

For many who came after him, for more than a millennium, Galen's statements about movement and muscle were considered to have absolute authority; and his perspective on voluntary movement remains highly influential today. A lot of us are so used to thinking that our muscles can or should be able to just do what we will them to do, it would not be hard to argue that psychic pneuma, as Galen described it, is the reason Nike—the corporation, not the goddess—may never stop running ads that say *Just Do It*.

Claims of absolute authority always fade with time. Charles Stocking knew this from his study of history, and it helped him take a long view of some authority figures he encountered in the next phase of his education: head coaches of some of the Bruins athletics teams at the University of California, Los Angeles.

At UCLA, where Stocking began work on his doctorate in 2003, he spent a lot of time in the gyms. Sometimes he went there to lift weights on his own, and sometimes he went there to make money to pay for school, doing his job as an assistant strength and conditioning coach for Bruins teams, including women's soccer and volleyball and men's football and water polo. As a strength coach, he got to know individual student athletes while also learning the ropes of the giant university sports system, and he came to believe there were some tensions between the best interests of those individuals and of the institution. "Health and well-being aren't always the first priority of university athletic training," he observes.

Many coaches, as far as Stocking could tell, had little empathy for athletes. "They knew the results they wanted, and they weren't able to hear the complaints of suffering."

Once he overheard an athlete complain to a coach of being sore from the previous workout. As Stocking recounted their conversation in an essay published later, the coach "responded that the athlete should learn not to complain, and if he learned not to complain, the training would be easier for everyone."

The athlete answered, "I can't help it, Coach, it's my body."

The coach shot back, "Your body? No, it's not *your* body. For the next four years, your body belongs to the university."

But then instead of resisting or taking a stand against the coach, the athlete "had a sort of *aha!* moment and simply accepted the yoke of necessity," according to Stocking, and "even thanked the coach," professing to feel better about the training.

After this, T-shirts emblazoned with the phrase PROPERTY OF UCLA never looked quite the same to Stocking.

The conversation Stocking overheard made him ask, "How much autonomy and freedom does the student athlete actually have?" And it made him wonder how much autonomy and freedom the student athlete truly wants. Talented athletes intent on maximizing their gifts tend not to claim absolute autonomy, at least in training, he thinks, because "Part of the joy of athletics is the struggle to work against that voice you hear in your head that keeps telling you to stop."

If you ratchet up the absolute epistemological power dynamic of Coach Knows Best, though, and meld that with a central tenet of Galen's muscle physiology—the belief that movement is most substantially an act of will—athletic training can be dangerous.

During graduate school, while working as a strength coach, Stocking had a superpower: his ample experience with injury. When he says, "I became a specialist in injury prevention as a function of my own experience and stupidity," it's only partly a joke. Having repeatedly hurt himself in the gym, and having methodically treated each of his injuries with pain-free movement, he devised a simple explanation of why injuries sometimes happen in training and in competition: his theory of the sport-specific paradox.

The sport-specific paradox encapsulates a problem that strength coaches are supposed to solve. The problem is, *Skill seeds its own ruin.*

"How do you get better at a sport? By doing it. But the more you do it, the more you overuse certain muscles in a certain range of movement. And eventually you reach a point where you can no longer do your sport." To solve that problem, "one of the first jobs of a strength coach is to rebalance the body so that it can continue to, say, throw a baseball or kick a soccer ball," he says.

Stocking reverts to that example of an athlete continually kicking a ball in practices and games so that the hip flexor muscles grow bigger and stronger; they "tighten and shorten, being moved in only a single direction over and over." Some people, he says, think that if an athlete wants to kick more, and kick harder, "you need to work your hip flexors a lot because you use your hip flexors a lot." But, Stocking says, "No, it's the opposite." Among soccer players, the muscles on the back of the lower body, "lower back, glutes, and hamstrings are very weak, compared to the front side." That strength imbalance is a risk for injury. So athletes who want to be great at kicking "have to do more of the opposite movements" at levels of intensity and in volumes of work comparable to the intensity and volume of their work on the field.

One of Stocking's main goals as a strength coach for the women's soccer team was to reduce the risk of that sport's most disruptive injury: injury to the anterior cruciate ligament of the knee. ACL tears are not the sport's most common injury—fewer than 5 percent of players suffer them each year—but of all injuries, they are associated with the longest periods of time loss from games and practices. ACL tears can also create ongoing and life-changing health problems. Younger female athletes who suffer ACL tears increase their odds of recurrent injury in the following year by 700 percent, compared to their peers who have not been injured. Women in their twenties who suffer ACL tears increase their odds of later developing osteoarthritis of the knee by at least 400 percent. And female athletes who suffer knee injuries such as ACL tears more than double their risk of requiring knee replacement surgery within fifteen years.

ACL tears illustrate the sport-specific paradox in full, malignant bloom. The muscles on the front of a soccer player's legs are highly developed from constant kicking and zigzag spurts of sprinting, so her forward momentum is supercharged. Racing toward a goal, suddenly she has to stop and change directions, and her posterior chain of muscles can be overwhelmed by the force of her quads and hip flexors, force augmented by the speed of her stride. When all this force in front exceeds the force in back, to the point where those muscles in the rear can't pull hard enough to keep the joint protected, the shinbone in the lower leg goes out of alignment with the thighbone in the upper leg, and the ACL, which stabilizes the knee joint by connecting those two bones, can tear.

When Stocking studied the research on soccer biomechanics, he found that "lower back endurance and hamstring strength are directly related to decreasing the incidence of ACL injuries." So his training for the soccer team emphasized these qualities—and to the best of his knowledge, no one tore an ACL after implementing his muscular balance protocol.

Stocking's research on biomechanics for all the sports he coached was heavily influenced by the writings of Stuart McGill, an expert in spine biomechanics at the University of Waterloo in Canada. After reading McGill's books and papers, Stocking decided that, as a strength coach, traditional strength training was not necessarily always his main job. McGill showed that athletes need to do more than build balanced strength in their limbs. His research convinced Stocking that normal prerequisites for high-level athletic strength training should include balanced *endurance* and coordination in the trunk muscles, the muscles surrounding and attached to the spine.

These are almost certainly the muscles in the "inner parts" that Aristotle wrote about, the flesh that is "constructed so as to change from solid to supple and from supple to solid" so that, as the philosophers Klaus Corcilius and Pavel Gregorić put it, "a stable supporting point is either formed or dissolved."

Core endurance is the precondition for building limb strength, according to McGill, and the reasons are biomechanical. He presents evidence for this view in three books, synthesizing more than thirty years of research. *Low Back Disorders: Evidence-Based Prevention and Rehabilitation* is a textbook for health professionals; *Ultimate Back Fitness and Performance* describes back exercises for athletes that can prevent or rehabilitate injuries or enhance performance; and *Back Mechanic* is a step-by-step guide for readers (including nonathletes) who want to understand and fix everyday problems of back pain.

In all types of exertion—lifting, throwing, kicking, jumping, running—force is transferred through the body. McGill's books state that force transfer is most efficient in a body with a stiff and stable trunk. "Consider the sprint start," he writes. "The core is stiffened so that when the pulse of hip muscle activation begins," with the sprinter's first stride, the force of that stride "propels a core of stone rather than pushing a soft rope. Pushing rope is a massive energy loss."

Spinal biomechanics are shaped by muscular anatomy. "The spine is a flexible rod," McGill tells me. "But no engineer would design a flexible rod, stand it upright, and balance a load on the top. The rod would buckle. To enable people to support or lift heavy weight, and perhaps carry it, the flexible rod is stiffened." McGill sometimes compares people's spines to radio towers with guy-wires attached to the ground. He writes: "The function of these guy-wires is similar to that of the network of muscles and ligaments that surround our spinal columns: they provide strength and support. In the case of our backs, these 'anchoring' muscles also facilitate mobility."

The muscles that anchor us work in a different way from the muscles that move us. In McGill's words, "limb muscles create motion while torso muscles primarily stop motion." Muscles in the limbs work by contracting: When the biceps flexes the elbow, or the quadriceps flexes the knee, movement happens because those muscles get shorter. Muscles in the trunk work by stiffening: When abdominals brace for a punch or obliques tighten to help maintain balance and stop a fall, those muscles do not get appreciably shorter, they mainly firm up.

Because "the core is different," McGill writes, "it needs to be trained differently." Muscles in the limbs need to be trained through a range of motion, to practice doing what they are designed to do: exert force about the fulcrums of our joints. Muscles in the core need to be trained to be stiff and hold still, to practice what they are designed to do: hold stiffness over time.

Core stiffness may be the ultimate active nonactivity. In spinal biomechanics, as described by Stuart McGill, Galen's concept of tonic muscular contraction is foundational to healthy mobility.

The entire trunk musculature—front, back, and sides—needs to be trained mainly for endurance, according to McGill, to build a foundation for whole-body strength. Ignoring the trunk muscles or training only some of them, such as the rectus abdominis, commonly known as the six-pack, and not others, such as the quadratus lumborum—in the deepest part of the trunk, it links the lumbar vertebrae to the rib cage and the pelvis—leads to imbalances of tension in that guy-wire system surrounding the radio tower of the spine. Over long periods, those imbalances incrementally compromise the integrity of the back. Episodically, when a person bends the wrong way or lifts a barbell, a bucket, or a baby or any object that weighs more than the tensions of those interconnected elements of muscle, tendon, ligament, cartilage, and bone can bear, some part of that system is degraded, and the result is pain. Eventually, those systems can even give way, causing instant extreme pain, and sometimes life-changing losses of function.

Stuart McGill's books for lay readers so nimbly shuttle back and forth across the boundary between athletic training and the rest of life, the boundary can disappear. "All daily activities are opportunities to spare the spine or make it sensitive to pain—the choice is yours," he writes in one book, and in another, he describes the example of a record-setting powerlifter who lived in constant pain. The man's back hurt even when he bent over the sink while brushing his teeth. "Incredibly, he had forgotten how to bend," McGill writes, and then says he worked with the lifter to develop new habits of "spine hygiene throughout the day," in all the tasks of daily living. "For brushing his teeth we reviewed powerlifting 101," the basic movement pattern of "the hip hinge coupled with spine stiffness and control." When the athlete understood that he had to bend "with the same discipline, regardless of whether he was brushing his teeth or lifting 200 kilos, he was able to restore his training."

McGill's scientific publications, too, reflect his practical mindset. One of his most influential articles established "normal ratios of endurance times" for the muscles of the trunk. If there is a key to lower back health, the research suggests, it may be the endurance of extensor muscles in the lumbar region—that group of muscles Charles Stocking diligently trained after his first serious back injury, by doing back extensions until they burned, because high-volume repetitions of that movement made his back feel better.

"I had intuitively been doing these kinds of things," inspired by Pavel Tsatsouline's equation of strength with tension, Stocking remembers. "And then as I started to read the studies, I started to see this idea of the balance of tension being the issue, the importance of balancing it out in the body." Stocking administered McGill's back endurance tests to Bruins athletes, and the tests became part of his training programs for several teams.

As a coach, Stocking applied his hard-won training wisdom to protect athletes from injury in simpler ways, too—sometimes just by knowing when to say no.

The men's water polo players, he says, had tremendous stamina both in the pool and out of it. In 2007, after an all-night celebration of a national championship victory by their sister team—the 100th NCAA national championship won by the UCLA women's water polo team—the men showed up to practice at seven A.M. reeking of booze and raring to train. But as soon as Stocking smelled them, he said, *There's no way I'm going to have you guys lifting heavy, when most of you are still drunk*, and sent them to the track to jog it off.

That judgment call was typical of him, according to Chris Joseph, a Bruins football player he trained. If Stocking could tell an athlete wasn't up to lifting, he did not push it, Joseph says. If that meant stopping in the middle of a workout, the workout stopped.

Stocking explains his rationale to me: "I had very detailed planned and periodized workouts for a whole season. And on any given day, I was always ready and prepared to scrap the whole thing. For me, there was no objective measurement to determine whether a workout needed to be changed. It was strictly based on observation." He says, "Any plan or program, no matter how scientific, always had to be subject to modification at the last minute."

Chris Joseph played center on the football team's offensive line. Joseph was objectively huge—six-four, 290 pounds—but, he says, "I was still underpowered and undersized for my position," where he had to protect the quarterback from the other team's defensive line, an onrushing wall of men, each weighing 300 pounds plus.

Wanting to be stronger, Joseph says he sometimes tried to "overcompensate with effort" in the weight room, but Stocking reined him in. He remembers Stocking saying, "*You don't need to do so much. Doing more is not better. Just do it smarter. And do it more effectively.*"

One of Joseph's main intentions was to get bigger, which was necessary because he wanted to play in the NFL after graduation. He asked Stocking to help him put on muscle while building functional strength for his position.

Joseph also had a history of injury, with ACL tears in both knees that had been surgically repaired, which increased for him all the associated risks, from reinjury in the short term to knee arthritis or knee replacement in the long term.

While wrestling with those risks, one day in the weight room he said to Stocking, "Do I want to play in the NFL, or do I want to be able to walk when I'm forty?"

"And so instead of training him for the NFL tryout, I helped him with his essay for the Rhodes Scholarship," Stocking says. "Which he got."

Their training together, Joseph tells me, changed how he experiences muscle and strength. Before their training, he says, muscle "wasn't a fun thing for me to think about, because I always felt a little deficient." After their training, "muscle became kind of like a realm to explore. Like a place to discover new things, where there's always opportunities to achieve. And even when there's some backsliding, there's no guilt, or anything like, *You've lost something.* It's kind of like, you just reset the mark, and there's more to achieve."

A focus of their workouts, Joseph says, was building stiffness in the muscles of his trunk and building strength in some of the smaller muscles in his lower back and hips, which, he learned from Stocking, "were all really lagging behind" the bigger muscles. "So I had no linkages to transition the power that I had in my lower body *up*, into moving a nose guard or walking that guy off the ball, when we were running an inside zone play." To make those links, Stocking worked with Joseph on "building that whole posterior chain, so that I finally had the transmission in place, to hook all the strength together."

When he was twenty years old, Chris Joseph built muscle and strength so he could throw other football players around, and fifteen years later, at age thirty-five, he says his early experience of training still protects him from some constant little strains of daily life—like washing his hands or doing dishes. Being tall, he has to lean farther forward than most people do, in order to situate his hands at the standard height of countertops and sinks and hand dryers. "So every time I do something like that, I focus on locking my posterior chain, hollowing out my lower back, pushing my hips backward, and leaving my knees over my feet," he says. "It's something I consciously think about every time I wash my hands. Just to preserve one of those weaker points that a lot of people start to have trouble with, especially at my age, with lower back pain. That's something I do every single day."

When I hear him say this, I think it sounds like a story from one of Stuart McGill's books; and I can see how McGill's research, as interpreted by Stocking, became a gift of wisdom that changed Chris Joseph's life.

This gift of wisdom also has a larger meaning. It points to another paradox, defying old conflicts between medicine and athletics that shape ongoing conflicts between doctors and trainers and their respective values of steadiness and striving. This is the paradox of training for the long term: Excellence requires stability, which is a kind of excellence, too.

CHAPTER 3

Live and Die

The site of the *stadion* sprint race, the original and central Olympic contest, was relocated several times during the athletic festival tradition's first few hundred years. By the fourth century BC, the stadion took place completely outside the main sanctuary, away from the old altars to the hero and the god. The stadium track was resituated to its present location, in a shallow basin with grassy green embankments on three sides.

Greeks sat here on the grass to watch the contests. This is where Aristotle sat when he watched the games. Aristotle, in a crowd of perhaps 45,000 spectators, most or all of whom were men. Only one adult married woman was officially allowed to witness the competitions, a priestess of Demeter, the goddess of the harvest. The priestess sat in a special seat—upon a marble altar to the goddess—to one side of the stadium, midway down the length of the track. It is possible that younger women, especially if they were attendants to the priestess, were also spectators at the games, but the evidence is unclear.

About four hundred years after Aristotle died, a traveler from a Roman colony in Asia Minor visited Olympia, a writer who made incomparably detailed descriptions of the place, its statues and monuments, history and myths. Much of what we know about the ancient Olympics comes from the writing of this one man, Pausanias.

When Pausanias came to Olympia in the late second century AD, the quadrennial athletic competitions had been held for a thousand years, without interruption; but Greece had changed. Conquered by

Rome, Greece was a province of the Roman empire—the most prestigious province imaginable.

In upper echelons of Roman imperial society, people competed to see who could seem to be most Greek. To be like Greeks was to be legitimate, established, connected to history.

In the best Roman schools, still called gymnasia, students learned to speak the Greek dialect that Athenians spoke seven hundred years earlier. (Pausanias wrote his travel book in this dialect.) They also learned techniques of exercise and care of the body from two kinds of experts, trainers and doctors, whose rivalry maintained and intensified ancient tensions between athletics and medicine.

Charles Stocking has translated a lot of what Pausanias wrote about Olympia. Some of it is uncanny, the text like a place where far-flung times converge. One day at Olympia, Stocking and I talked about Thomas DeLorme, about how DeLorme's groundbreaking medical experiments with weight training in the Army hospital had grown from his personal experiment as a young man, lifting weights to rehabilitate himself after suffering from rheumatic fever. DeLorme's story reminded Stocking of something in Pausanias, and by the next morning Stocking had translated a few lines about a man named Hysmon who became an Olympic victor in the pentathlon.

"It is said that when Hysmon was still a boy he was attacked by a flux in his nerves," Pausanias wrote—the Greek word for "flux" is *rheuma*—"and he practiced the pentathlon in order that he be a man healthy and free from disease because of his labors. So his training was also to make him win famous victories in the games."

In the mornings at Olympia, or sometimes in the evenings, when the archaeological site is empty of tourists and when Stocking can break away from his duties leading the conference, he walks down to the ancient stadium, where he runs sprints on the ancient track. Sometimes he goes there alone, and sometimes he goes with his wife and their young daughter Stella. This is a rule he keeps when he's in Greece: If he visits an ancient sanctuary and the sanctuary has a track, he has to run on it—he has to run sprints.

When I ask him to explain why he likes sprinting on the ancient track, Stocking laughs, a little embarrassed. "Give me a second here," he says, and he pauses for about ten seconds.

Then he reiterates something he said while leading the big group tour of Olympia's sanctuary ruins. "The point of connection between us and the ancients is that we all have bodies. It's a joke I make, you know: *I've never met somebody without a body.*

"Now, granted, it's true that an ancient Greek's embodied experience would be very different from ours—how they thought about it, their language: Lots of things are different. But there is still a physical body that is more or less the same. And running on the track is the closest that you can get to the lived religious experience of an ancient Greek."

He says, "Almost every sanctuary has a track. A lot of folks would downplay that. They say everything was religious, so sport was not particularly religious, for the ancient Greeks. And I disagree with that entirely. There's something unique about it. There's a reason why tracks are at these sanctuaries and not often found elsewhere. Through physical practice, athletes would get access to a sense of potential, and that was projected to a sense of limitless potential for the gods. So you, the athlete, get to participate in that temporarily, you get to feel what it's like to run fast. Now imagine if you're a god, without a limit to your speed. For an ancient Greek, to run or to participate in athletics is to participate in that divine potentiality.

"So for me to run on an ancient Greek track, where they were participating in that, allows me to participate in that experience—at a third level removed.

"The first level is the god and his speed or his strength, the god and his physical ability. The second level removed are athletes who are compared to gods, in their speed and their abilities: This is them participating in divine potential. And that's why these sporting events are taking place on tracks at sanctuaries, because it's a form of divine participation in this ability."

The third level, Stocking says, is our lives—his and yours and mine—right now.

"To be at a sanctuary as a tourist—First off, to think that there were ancient athletes here is incredible, right? There's something about the lived experience that you can start to feel. Even if it's just romantic reconstruction, there's something there. To just be here is one thing. But to put your body through those movements and those mechanics that ancient athletes experienced is the closest you can get to that feeling they had. Because the biological process that one experiences when running or wrestling hasn't changed: the heart rate, ATP, glycolysis. And lactic acid thresholds, for the longer runs—which would have been painful. You can feel that.

"And you can feel it from an ancient perspective, and not just your own. And that provides a context for sport that's really different from the modern context. Because, you know, you presume a god is not going to feel the lactic acid buildup. So you can participate in the divine at the same time that, in that participation, you're experiencing your own limits compared to the divine."

Later, thinking about what he said, I realize it was a refrain. When Charles Stocking sprints on the ancient track—The gods are *there*.

Late one morning at Olympia, a few hours after his morning sprint workout, Stocking returns to the archaeological site, this time with dozens of students in tow. The students want to take their turns on the ancient track, too. They want to run where ancient athletes ran.

For the students who have not done much running lately, Stocking gives a warning and advice. "There are probably more torn hamstrings on this track than on any other track in the world," he says. "Everybody gets inspired and just starts running out of nowhere. With horrible form. So I would recommend warming up."

After he says this, the students start to come alive in their bodies. Before, they might have seemed to be all eyes and ears. Their bodies could have been roving brain-pillars, gliding attentively behind their teacher. After his cue to warm up, most of them start to shift their bearings, as if morphing into wiggly, wheeling hubs of levering limbs, doing high kicks and hip circles and jogging in place and jumping—while Stocking tells them, "Just think, people were running on this track more than two thousand years ago. And now you're running on it. And whatever is happening to your legs and lungs was happening to their legs and lungs. But they may not have thought about it the same way that you do."

He herds the students to the far end of the track. Then he loops around the group so they all turn to face him and, more to his point, so they face the sanctuary ruins that stand scattered behind him. As in early athletic rituals of sacrifice, he says, they're going to be racing "*toward* the sacred, not *away* from it."

Stocking counts down to start the race . . . and when he yells GO! a scrum of jocks bolts forward, pounding hell for leather to be first—one tough guy runs barefoot—but most of the runners take it easier, and the finish line's a splash pool of laughter. Laughing in lots of tones—loose and forced, embarrassed and free—the students smile, and afterwards seem slightly more comfortable in their skin and with one another.

The shift in demeanor was subtle, but it seems real. As if all it might take to make us a little more present to one another is the force and breath and sweat of efforts to move ourselves, together, fast.

Stocking wishes the race could have played out somewhat differently.

"What I would do first, with my students, if I had my way, is I'd require them all to learn proper sprinting mechanics before they hop on the track and start running," he says. "Because there's a big difference between somebody who's just trying to run fast, and feeling yourself being propelled forward with proper mechanics."

Efficient biomechanics, he says, can bring the experience of running on this track a little closer to the divine, owing to "the absence of effort. You hear a lot of great athletes, after amazing performances, say they feel like they weren't the ones in control, or they were just being pulled forward. Part of that's the technique. And part of it is an experience that is almost out of body, not a function of one's will. And you can completely feel that in a religious sanctuary, where that's the point."

Then he walks back his point a step. "Perhaps," he adds, detaching cheerfully. "I mean, this can't be argued in an academic paper."

But he can argue on evidence that ancient athletes knew the same sprint technique he practices and teaches; and another day, during another museum visit, he does.

"Have I talked to you about the biomechanic representation of sprinting on Panathenaic amphoras? It's dead on," he says—it's like a freeze-frame shot from video of good running form today. An amphora is a type of Greek vase, a big storage jar, about two and a half feet tall, with handles on either side of a narrow neck. Panathenaic amphoras were prizes for athletic victors at the Panathenaic religious festival, where Athens and its allies offered sacrifices to Athena, the goddess that protected them.

The value of the prize was not the jar, but what was in it: almost forty liters of olive oil, more than ten gallons. The man who won the stadion probably received eighty of these amphoras. Following a formula laid out by David Young in his book *The Olympic Myth of Greek Amateur Athletics*, the total value of the men's stadion victory prize may have been equal to something like $185,000 in today's currency.

Victory in the stadion, or in any athletic event at the Panathenaic games, Young writes, would have "paid noticeably more money than a full year's work."

Stocking's favorite Panathenaic amphora is decorated with a painting of five men sprinting, naked and barefoot. Describing the painting, he ticks off the strong points of the sprinters' form—"Foot strike is on the ball of the foot. Arm action. Knee action"—and he faintly mimes the moves. "Nobody's heel-toe running," he points out. The depiction is "stylized, but at the same time it's clear they understood."

Ancient Greek sprinters understood what in 2010 was reported by Daniel E. Lieberman, the evolutionary biologist: Barefoot running can be much easier on the body when you land each stride on the front of the foot, not the heel. Data collected by Lieberman and his colleagues showed that each stride of heel-toe running can send up to 300 percent more force up through the joints compared to forefoot-strike running. Because landing on the forefoot can reduce peak vertical force impact, the researchers said this technique may help runners reduce the risk of repetitive stress injuries such as plantar fasciitis.

Consistently, "you'll never see heel-toe running on an ancient Greek vase," Stocking says, with some of the comfortable awe that you feel when you find art that shows the world the way you experience it.

Stocking's insight about sprint biomechanics depicted on Greek pottery, like his interpretation of divinely given strength in Greek epic poems, is grounded in his experience and knowledge of athletic training—which are useful when studying ancient athletics, because the literature is full of gaps. "The biggest difficulty in studying ancient athletics is the lack of sources," Stocking says. Much contemporary knowledge of ancient Greek athletics consists of inferences from asides made in works on other subjects, such as philosophy, politics, and medicine.

The few ancient writings that take athletics as their main subjects tend to be brief, celebratory, and focused on best possible outcomes: inscriptions from the bases of statues of athletic victors, for instance, and praise poems written to celebrate athletic victories. (Lyric poetry flourished as an art form when poets wrote about athletes—when poets tried to make athletes' fleeting victories last forever, in the form of words.) The only precise description of physical training from antiquity that so far has been found is a crib sheet of wrestling moves, scratched on a fragment of papyrus.

Training manuals are mentioned, in passing, in writings on other subjects, so we know that training manuals existed; and there were lots of them, on specialized topics. In his book *Athletics and Literature in the Roman Empire*, the classicist Jason König writes that many training manuals "seem to have been relatively complex and erudite, and some of them may have been similar at least in form" to medical texts. Also according to König, "we have evidence for treatises on different kinds of sweat and on different kinds of tiredness," but all these books are lost.

Only one long piece of writing fully focused on athletics has survived from antiquity. It is an incomplete book about training from the third century AD called the *Gymnasticus*. For most of the time since it was written, the *Gymnasticus* has been inaccessible and obscure. The first full translation of the *Gymnasticus* into a modern language was published in 1858, in French. A German version followed in 1909. The first full translation of the *Gymnasticus* into English was not published until 1936, and then it appeared in a journal for physical educators. Ensuing decades brought translations into other languages—modern Greek, Italian, Spanish, Swedish—but there was no easy way to read the whole of the *Gymnasticus* in English until 2014, when Jason König's translation was published in the Loeb Classical Library's bilingual format, with Greek and English on facing pages in a sturdy green hardback, small enough to hold in the palm of your hand.

Around the same time, independently, Charles Stocking was completing his own full English translation of the *Gymnasticus* as part of another project. With a senior colleague, Susan Stephens, a Stanford professor of classics, he collected many fragmentary writings about ancient athletics into a one-volume sourcebook, *Ancient Greek Athletics: Primary Sources in Translation*.

Most similar anthologies have been organized by topic, with bits of writing about running or wrestling, for instance, batched together, no matter when they were written, in 600 BC or 400 AD. Those anthologies make it easy to find details about specific subjects, but they can also make it easy to ignore the complex evolution of ancient athletics and to imagine it all happened in a hazy blob of oldness.

Stocking and Stephens assembled their sourcebook in a different way. They organized it chronologically, and Stocking says, "what's really fun about that is you can start to see themes and phraseology being picked up and repeated. Like mind-body dualism, which pervades the discussion" of athletics after Plato and Aristotle.

While editing that book, Stocking studied the concept of a mind-body split with increasing curiosity. "Most people today take dualism for granted and assume it is a universal," but that's a false assumption, he says. "Mind-body dualism has a history. The fact that it has a history means that these are not essential categories." Far from being discrete, objective realities, *mind* and *body* are products of a long series of decisions that many people have made through history.

The students at Olympia, a few hours after running on the ancient track, assemble in a meeting room to hear Charles Stocking talk about that series of decisions. Ancient writings about athletics helped forge the notion of mind-body dualism, and even now the language of those ancient texts still shapes "the actual matter of our bodies," he says.

At a lectern, Stocking presses a key on his laptop, a projector casts a news-flash headline from *The Guardian* in foot-high letters on the wall behind him, and he reads it aloud: "HOW PHYSICAL EXERCISE MAKES YOUR BRAIN WORK BETTER." Then he raises a question: "Why is this considered breaking news, when the connection between mental and physical training, we know, goes back to ancient Greece?"

His short answer is that "a very long history of language sets the mind in opposition to the body," language that also goes back to ancient Greece—when, as discussed earlier, mind and body were called *psychē* and *sōma*.

His long answer deploys the gamut of Stocking's experience and education, academic and athletic, from his high school phone calls with his older brother Damian about Plato, to his powerlifting injuries after college, to his translation of the *Gymnasticus*—and it all starts with Homer, with a reminder that in Homer, psychē and sōma were singular and separable only in the context of funerals, only in death.

Then Stocking leaps ahead a few centuries, to the time of Plato. Stocking says Plato's worldview was defined by a hard line between psychē and sōma. Though not everyone shared that worldview, the notion of this split eventually came to organize Greek education. But even Plato, who did not regard the athletic body (or any body) as good and valuable for its own sake, considered athletic training as an essential form of bodily knowledge and thus a pillar of education.

Gymnasium students kept psychē and sōma in excellent condition, for the good of the city and the might of its military—while in the same era, another discipline came into its own: medicine, devoted to preventing and curing sickness of the sōma. Plato considered medicine as the other main aspect of bodily knowledge, coequal with athletics.

During his lecture at Olympia, Stocking reminds the students that athletics was not a major focus of early medical writings, and when athletics was discussed, it was sometimes denounced. A treatise on nutrition says, "The athletic disposition is not natural, better the healthy condition." A collection of aphorisms says, "For those engaged in gymnastic training, peak condition is dangerous, if it has reached its limit." To a trainer or an athlete, these must have been fighting words.

"And here's where we start to see a transition," Stocking says. "A competition started to exist between medicine and athletics"—the competition between ideals of bodily condition. Medicine aimed for balanced, stable health, while athletics prized the peak fitness that helped win victory.

Stocking continues: "What comes after a pinnacle? A decline." For that reason, in the early writings on medicine, peak condition "cannot be a viable definition of health."

In the lecture, Stocking leaps through time again—five hundred years—to land in the Roman empire. During the first century AD, Stoic philosophers in Rome still named body and mind as sōma and psychē, yet the meaning of those words continued to evolve.

Where Plato believed the soul is purely spiritual, Stoics said souls have a material quality, too. What bodies do, according to some Stoics, helps determine what souls become. Consequently, these philosophers suggested, athletics could be a practice of the soul. Athletes could participate in the most valued philosophical practice *as athletes*—not by analogy, but in fact. Training could be soul-craft.

As long as athletes train with care.

Because the soul has a material quality, some Stoics wrote, mind and body have limited resources. The philosopher Seneca, for instance, cautioned that anyone who spends too much time at the gym and ignores the study of philosophy "is well only in the way that a madman or a lunatic is well." Seneca drew an extreme contrast between the sane philosopher and the person who trained only his body "as literally somebody who is mentally unhinged," Stocking notes.

The worst kind of training—the most spiritually hazardous—involves building body mass, Seneca wrote, in a letter to a friend:

> It is stupid, my dear Lucilius, and it is not a fitting pastime for an educated man to practice developing the shoulders, broadening the neck, and strengthening one's sides. For although your heavy feeding

produces good results and your sinews grow solid, you can never be a match, either in strength or in weight, for a first-class bull. Besides, by overloading the body with food you strangle the soul and render it less active.

Stocking rephrases Seneca's main point, to drive it home: "The physical adding of mass, material mass and muscle, is somehow suffocating the actual mind and spirit, it's sort of being weighed down." As an aside, Stocking tells the students that when he was in college, his grandpa said that lifting weights could make a young man's pecs so big, it would damage the ability to breathe—a fallacy that sounds so much like Seneca, it makes the students laugh.

In Stoicism, mind-body dualism was no longer metaphorical, as it had been for Plato. Writings by Stoics such as Seneca mark "the beginnings of hostility between mind and body," Stocking says, "while at the same time they want to promote some type of training of the body."

About a hundred years later, Galen became the best-known doctor since Hippocrates, the most ambitious biological researcher since Aristotle, and one of the most fervent public speakers of antiquity. Galen was a dominant figure in the Roman imperial culture of argument, in which crowds gathered for public debates involving long, fiery, complex speeches on many subjects, such as medicine, science, philosophy, and art. Galen's arguments survive in one hundred and fifty books (though he wrote double that number), a body of work shaped by his cardinal belief that health consists in a balance of opposing qualities like hot and cold, moist and dry.

Following Plato, Galen said medicine and athletics are two forms of one type of expertise that keep bodies in good condition. Galen called that inclusive expertise by the name of "health," and he said medicine, not athletics, owns the field of health.

This was more than an intellectual and philosophical assertion. For Galen, as for other doctors at that time, it was a point of professional self-promotion. While "doctors and trainers often worked together within the ever-expanding culture of the gymnasium" education in imperial Rome, as Stocking has written, "doctors and trainers also seemed to have viewed themselves in direct competition with each other over bodily knowledge."

Athletics is the opposite of health, Galen said, because athletic training creates imbalance, involving "much thick flesh and a great amount of blood." Like Seneca, Galen thought mass-building was the most deplorable aspect of athletic training, ruinous to the whole person, body and soul.

When Galen talked about building mass, he spoke less about exercise than about how much food athletes ate. In one book, Galen said that eating great quantities of food pushed extra blood into the veins, heating the body into a state of dysfunctional imbalance, making for ghastly scenes:

> . . . some have lost their voice, others have lost their sight, or have become paralyzed and completely crippled from their unnatural size and mass, which blocks the naturally hot air and exhalation of air. The softest harm some of them suffer is breaking a blood vessel and then vomiting or spitting blood.

In another book, Galen pulled out the stops. "Now it is clear to everyone," he declared, "that athletes have never taken part in the goods of the soul, not even in a dream. In the first place, they do not know whether they even have a soul, so much are they lacking in knowledge of its rational quality." Again he sounded like Seneca when he added: "Always engaged in increasing the amount of flesh and blood, they have a soul that has been extinguished entirely, as though by much filth. Such a soul is unable to think clearly but is mindless in a way similar to the irrational animals."

Galen's writings about the soul are, in Stocking's words, "famously ambiguous. In his works, Galen is often ambivalent about what precisely the *psychē* is, and yet this does not prevent him from talking about the *psychē* and its relation to the *sōma*" recurrently in his writings.

Galen received philosophical and medical traditions that, as discussed, attributed two main purposes to the soul: cognition and voluntary motion. The philosopher R. J. Hankinson notes that "one of the earliest surviving philosophical accounts of the soul comes from a doctor: Alcmaeon of Croton," who believed "the soul was immortal because it always moved." Many spins were later put on Alcmaeon's idea, but the basic notion stood: Movement had a sanctity. This conviction animated Galen's writings about athletics, just as the sanctity of the knees animated the battlefield scenes of warriors in the *Iliad*.

Stocking says, "Galen is not opposed to physical activity in general but only to the institutionalized forms of sport training in his era." Galen sounded downright enthusiastic about exercise in a short work called "Exercise with a Small Ball." Exercising with a ball is better than any other form of exercise, Galen wrote, because this activity "is able to benefit the soul in every way and trains all parts of the body equally."

Exercising with a small ball has great social value, too, Galen said, because it cultivates the qualities that make for competent military command. "To speak plainly, it is necessary that a general be good as both guard and thief. This is the most important part of his expertise," Galen wrote. "Is there any other exercise that accustoms one sufficiently to guard what one has obtained, to recover what one has previously given up, or to anticipate the thoughts of the enemy?" Many other exercises "have the opposite effect, making people slow, sleepy, and heavy in their thoughts," he contended, pointing to elite athletes, such as Olympic competitors, who "promote more flesh rather than the training of excellence." Then Galen fell back on his old familiar tune: "Many have been so beefed up that they cannot breathe without difficulty." Big, beefy athletes could never become good generals, he added, with a flourish: "One would sooner look to pigs than such men."

But in praising those who exercise with a small ball, as Stocking observes, "Galen insists that *psychē* and *sōma* are indeed interdependent."

Exercise with a small ball, as Galen describes it, is an ultimate key to health: a boon and balm for mind and body, for individuals and for society. The contrast with Galen's tirades against athletic training that involves building body mass could not be sharper; and to most of us today, the contrast probably sounds familiar.

For a long time, a lot of us have been inclined to view exercise much as Galen did, tending to view physical activity on a spectrum between recreation such as exercise with a small ball on one side, and sports such as wrestling on the other. Many of us, also like Galen, have been at least somewhat or occasionally inclined to assume (even if, on reflection, we reject these assumptions) that people on the small-ball end of the spectrum—smaller and thinner people, the agile and the spry—tend to be smarter or healthier than people who are bigger and heavier, the bulky and the strong.

The dichotomies structuring such ancient and modern views of physical exercise and activity lock us into cultural preconceptions, and this can happen even when as individuals we make concerted efforts to resist them. During Charles Stocking's talk at Olympia, when I heard him read from Galen's tirades against athletic training that involved building body mass, I believed they were tirades against training to build muscle. I did not consciously agree with what Galen was saying, but I assumed I basically understood *what* he was saying, because it sounded so much like the common judgment about big muscles that I've heard all my life. After I came home from Olympia and read some of Galen's writing about muscular anatomy and function—especially his accounts of muscle's role in voluntary movement—I grew cautious about connecting those dots.

I grew cautious because of a question raised earlier, the question of what exactly Galen meant by *muscle*. To recap that question's answer: While Galen's scientific writings about muscle are more accurate, by today's standards, than his predecessors' surviving works, Galen also made what modern physiologists would consider a categorical mistake. Like Aristotle, and like Plato before him, Galen dismissed the fleshy middle parts of muscles—the muscle bellies—as playing only passive or secondary roles in the work of movement. When Galen looked at the big fleshy muscles of a wrestler, he saw padding and insulation, where later scientists would see contractile tissue, indispensable for effecting movement. Galen believed that a muscle did its work to move us in spite of the part of itself that scientists now believe does most of that work.

Given all this, the answer to Charles Stocking's question about ancient and modern bodies—*Are they the same bodies?*—as far as muscle is concerned, where Galen is concerned, seems to be no.

Based on *his* understanding of muscular anatomy and function, Galen had reason to denounce athletic training to build mass. This may have been his real and serious beef with heavy athletes and their trainers: Having too much of the less important part of muscle blocks the soul from fulfilling its purpose.

Some elements of Galen's own personal history and experience provide more context for his fervid denunciations of the big, meaty bodies of heavy athletes and raise more questions about those judgments. In Galen's hometown of Pergamon, where the city of Bergama in western Turkey now stands, his first job as a doctor, when he was in his late twenties, was in a school for gladiators.

"They were bulky men, fattened on a special high-carbohydrate, vegetarian diet," the ancient historian Susan P. Mattern writes. The gladiators' diet, and the large bodies that diet helped build, inspired a derogatory nickname. Romans called gladiators *hordearii*—Latin for "barley eaters." Some translations play up the sneering tone of that nickname, rendering it as "barley boys."

The gladiators' thick flesh, heavy with fat, gave "insulation from serious injuries, while allowing for showy surface wounds: bleeding from a gaping gash that affected no major nerves or muscles, the gladiator fought on undaunted while the crowd went wild," Mattern adds.

It was part of Galen's job to heal those wounds, and he bragged that he could heal them better than any other doctor had ever done. This work would have given him an intimate sense of the burdens that people experience when they carry so much extra weight that they could take a few slashes of a sword and keep on going.

Galen's view of bulky bodies was not just a view from the sidelines, though. A few years after Galen left the school for gladiators, when he was in his early thirties, he lived in Rome; and, his writings suggest, he liked to wrestle.

During one of Galen's wrestling workouts, when he was thirty-four years old, he very badly hurt one of his shoulders. His collarbone was dislocated. Susan Mattern writes: "Galen's injury was severe and probably involved the muscle attachments, as well as the ligaments, which were totally separated. Three fingers' worth of space could be felt between the clavicle and the tip of the shoulder." Trainers and bystanders at the gym tried to fix the problem, but they may have made it worse. For the next forty days, Galen endured an excruciating treatment—and finally he did recover, with a shoulder good as new, he later wrote.

In the same way that Galen's arguments about incompetent athletic trainers were never purely intellectual and philosophical, but always also part of his indefatigable efforts to promote himself, it seems plausible that Galen's firsthand experiences of sports injuries—treating the gladiators' wounds and recovering from the trauma to his shoulder while wrestling—might help account for his philosophical distaste for hefty athletes.

But Galen's warnings about body mass may in the long run have created at least as many problems as they originally aimed to avert. His powerful rhetoric marked big slabs of flesh on athletes' bodies as being inherently questionable and dangerous. The organ of voluntary motion was on course to become a locus of anxiety.

In the traditional account of ancient Greek athletics, Christianity spoils everything. Near the end of the fourth century AD, a Christian emperor of Rome outlaws the Olympic Games; and the whole culture of athletics is blown away by huffing, puffing, body-hating priests, monks, and popes.

But in fact, Greek athletics faded gradually. Olympia weathered many storms, as the ancient historian Sofie Remijsen has shown. When Germanic and Goth barbarians invaded Greece in the third century AD, Greeks built a wall to protect Olympia's sanctuary. In the process, they permanently defaced their own most sacred space. By the start of the fifth century, all the most important athletic contests in Greece, including Olympia, had ceased. But one athletic competition in all the Roman empire, at Antioch in Syria, continued until sometime in the sixth century, "as a remnant of this antiquated tradition." (To imagine the atmosphere of the final ancient Greek athletic competitions, picture a Renaissance fair in the desert near Phoenix, Arizona—or in the suburbs of Brisbane, Australia—and you may not be too far off the mark.)

Even after the games are over, Stocking says, "the language of athletics continues" to be repeated, with Galen's take on the matter reproduced more than any other ancient writer's. Things might not have turned out that way. After the fall of Rome, Galen's books were all but lost to the Christian West, even while Islamic and Byzantine scholars preserved, studied, and translated the manuscripts through late antiquity and medieval times. That is why, in the Renaissance, when Europeans started reading Galen again, many respected his authority as absolute; and Galen's borderline-infallible reputation lasted well into the nineteenth century.

No one has described Galen's achievement more succinctly than Heather Reid, a philosophy professor at Morningside University in Sioux City, Iowa, who is with us at Olympia.

Reid says, "Galen won."

Because Galen won, some today still take his scathing attacks on elite athletes and heavy training at face value, and as statements of the spirit of his time. Charles Stocking says that's a mistake. "It's not as though the ancients thought of performance and health on opposite ends of the spectrum all the time. Galen wants to present them as opposites. But that's because he's working against the assumption that they actually go together," an assumption that went back to Plato and was still shaping elite Roman gymnasium education. When Galen wrote things like "Athletics is the cultivation, not of health, but of disease," it was provocative because so many other people believed, to the contrary, that athletics was the cultivation of health.

Yet the ancient voices that valued and defended athletics are mostly lost to history. Near the end of his lecture to the students at Olympia, Stocking says he believes ancient proponents of athletics were "intentionally silenced by philosophers and doctors," and silenced so authoritatively that today they still struggle to be recognized as intellectually legitimate.

Because Stoic philosophy was adapted, in Roman medicine, to create "mind-body dualism that is specifically anti-athletic," and "because this language has been reproduced throughout the history of the modern academy," the notion of a positive and mutually reinforcing relationship between mental capability and athletic practice "has become breaking news," he says.

As long as news organizations like *The Guardian* provoke surprise or skepticism with headlines like the one Stocking quoted at the start of his talk—HOW PHYSICAL EXERCISE MAKES YOUR BRAIN WORK BETTER—and as long as that claim's appeal remains even faintly quirky or offbeat—Galen still wins.

Charles Stocking's concept of the sport-specific paradox, which says that training for sport builds skills, while forging physical imbalance that deteriorates those skills, is an aspect of a larger paradox, he says: "The paradox of athletics," for people in all times and places, is that "one trains and competes for improved life-force, but always at risk of death."

Stocking's observation is clearly true of violent sports and at extreme limits of human performance, where building strength can break people, as in the story of Cleobis and Biton. This existential athletic paradox infuses all training, however, as shown by the research of Hans Selye and others who have studied the physiology of stress. That research shows, as Stocking says, that any exercise regimen for improving health or performance "is a delicate balancing act between the application of stress and the ability to recover from it. Too much stress, and the athlete's condition will deteriorate rather than improve."

When Galen said athletics cultivates disease, not health, he warped the athletic paradox by resolving it. Discernment of risk involved in training became, for Galen, a pessimistic dictum, a confident prediction of worst possible outcomes. Many modern athletes who take performance-enhancing drugs warp the paradox, too, in a different way. Obsessed with best possible outcomes, they will run any risk in pursuit of victory.

Stocking has spent a lot of years navigating the athletic paradox. In his twenties, when he sustained all those serial weight-room injuries, his powerlifting proved Galen's point. Training hurt Stocking at least as much as it made him healthy. Then through his time working as a coach, earning his doctorate, and becoming a professor, and especially after his marriage in 2011 and his daughter's birth in 2017, Stocking has restructured his experience of the athletic paradox.

Through most of this time, he has been studying, translating, and writing about the sole surviving ancient book about athletic training, the *Gymnasticus*. Reading the *Gymnasticus* by the lights of his own athletic experience, Stocking found this ancient text to have a finer grasp of physiology than other scholars had noticed. The *Gymnasticus*, he discovered, anticipated some principles of his own training, such as the sport-specific paradox. The *Gymnasticus* also anticipated concepts of twentieth-century exercise science, such as Selye's general adaptation syndrome theory of stress response, the basis of periodized lifting programs. The author of the *Gymnasticus* was at the same time highly critical of any trainer's "attempt to create an overly objective form of bodily knowledge," Stocking writes.

As Stocking read the *Gymnasticus*, the text seemed to be saying, "Yes, athletics has a basis in empirical knowledge," but a "level of uncertainty and contingency must always be factored into training and competition." Finding these sophisticated ideas at play in the ancient book, Stocking came to believe, as he wrote: "the athletic paradox cannot be solved. It can only be negotiated and balanced in successive individual moments"—moments that, together, make up our lives.

Not an instruction manual, but "a defense of athletic training," in Stocking's phrase, the *Gymnasticus* presents athletics as a "socially and culturally legitimate form of bodily knowledge."

In its first paragraph, the book states its main claim—"concerning athletic training, let us consider it a form of wisdom inferior to no other expertise"—while also granting a valid objection to that claim: Most readers would not see training as wisdom, because athletics was in decline. In ages past—the time of virtuous heroes, the time of Greek athletic champions like Milo, and more recently in "the time of our fathers"—athletes did impressive things and won victories worth remembering, the book says, where by contrast contemporary athletes had fallen far and fast.

The decline of athletics had so damaged the reputation of training, most people had come to find it slightly annoying even to think about "exercise enthusiasts," according to the *Gymnasticus*. The book's purpose, therefore, was to restore the reputation of athletics through a wide-ranging exploration of the topic, going back to the origins of Olympic events—including the stadion's origin as a race from one altar to another, to light the flame of sacrifice to Zeus—and with special emphasis on appearance and regimen: the build of athletes' bodies and the structure of their workouts.

The book's author, Flavius Philostratus, was a writer of fiction and nonfiction whose books explored many subjects, including philosophy, religion, art, and travel. These books offered readers new perspectives on history, creative involvements with history as ways to shape contemporary life. Philostratus was born to an Athenian family, probably around 170 AD, in the prime of Galen's fame and influence. (Galen was then personal physician to the emperor Marcus Aurelius.) Philostratus did not write the *Gymnasticus* until after Galen died. But still, Stocking believes the *Gymnasticus* "may be read in direct response to Galen"—as if pushing off prior generations could launch this book toward its audacious goal.

The *Gymnasticus* meant to change how people see and experience athletic training: its purpose, and its very nature.

Where Galen had said athletic training inherently tends to throw bodies and souls out of balance, Philostratus now said the whole point of training is to create balance—which, he believed, could be seen in an athlete's symmetrically developed physique.

Most of the *Gymnasticus* "is devoted to a description of the athlete's body," Stocking writes. The book asserts that a trainer should be "a type of art-critic of the body." (The Greek phrase is *phuseôs kritês*, "judge of nature.") Ideally, Philostratus said, an athlete should look like a sculpture of an athlete. His ideal was related to a traditional sculptural ideal involving a certain set of proportions, based on *symmetria*: that is, symmetries among the sizes of body parts. For instance, he thought that ankles and wrists should be the same size; and so should forearms and shins, arms and thighs, and shoulders and glutes. Yet Philostratus may also have been ambivalent about symmetrical standards of beauty. His writing about the links between performance and appearance is marked by inconsistency and self-contradiction.

While propounding one ideal physique, Philostratus also approved a diversity of athletic physiques, because, as he asserted, different kinds of bodies suit the different movements involved in various competitive events. For example, the *Gymnasticus* says a sprinter's "muscularity should be proportionate. Excessive muscles are the bonds against speed." The *Gymnasticus* also says a boxer should not have "bulky calves" (the text does not say why), but a substantial belly on a boxer was fine with Philostratus, since the belly could "ward off blows against the face by sticking out in a way that impedes the forward motion of the punching opponent," he wrote.

Typically, however, "Philostratus insists on aesthetic form over athletic function," as when describing wrestlers' bodies, Stocking observes. Philostratus described wrestlers in more detail than he described any other athletes, probably because of wrestling's supreme popularity. The *Gymnasticus* says wrestlers should have straight backs because straight backs are "pleasing," despite the fact that "a slightly curved one is more suited for training, since it is better suited to the position of wrestling which is curved and bent forward."

And that right there, Stocking says, is the sport-specific paradox. Wrestlers spend so much time grappling, bent forward toward their opponents, that their latissimus dorsi muscles—lats, for short—become highly developed, fanning out and tapering in a V-shape down the back. As a result, many excellent wrestlers develop a hunched posture.

These wrestlers have the curved back that Philostratus disapproves of.

Despite his inconsistencies, Philostratus generally emphasized form over function because he wanted "to cash in on the cultural capital of the Greek tradition," Stocking says, and he sketches the ancient rhetorical strategy. Having named the likes of Milo as "the pinnacle of physical perfection," Philostratus logically recommended "the emulation of sculpture that is based on the Greek tradition" as the way for Roman imperial athletes to get in top shape, to excel in their sports—and thereby to experience ancient Greek virtue.

Many centuries later, these ancient beliefs that athletes should look like sculptures took on very different meanings. As Charles Stocking says, "Greek sculpture has been appropriated for body fascism in the modern era: conforming to a certain standard of beauty." For many people, these appropriations have spoiled ancient Greek sculpture, if not all of ancient Greek culture. Stocking recognizes their dilemma and the questions it raises. "Is it possible to appreciate and even praise the beauty of sport in a way that is free from the force of earlier ideological discourse?" he writes. "Can we appreciate the allure of sport without concern about the ethics of doing so?" He thinks we can, if we take care to avoid false equations of ancient and modern standards of beauty.

"The ancients saw the beauty of the body as a signifier of *movement*," he says. "They understood the image" seen in sculpture "in reference to exercising bodies, not as an image unto itself," the way many modern people have tended to do since photographic images became ubiquitous in everyday life. "So the original understanding of these things works against modern concepts of body fascism," he says. "Because beauty is a function of movement, not appearance, as the priority."

No blanket statement about standards of beauty can strictly encapsulate all the attitudes of any era, but evidence does support Stocking's view of beauty and movement in antiquity. Another passage he translated from Aristotle's *Rhetoric*, for instance, reflects on "Bodily Excellence" by specifying that "Beauty is different for each age." For each stage of life, Aristotle named the predominant purposes of beauty in his society. In youth, Aristotle wrote, beauty was "for labor, running, and acts of violence." In the prime of adulthood, beauty was for "acts of war," to protect the city-state. "For an old man," Aristotle added, "beauty means being sufficient for the labors of necessity and to be free from pain by not having the typical ailments of old age."

There is only one type of Greek sculpture, Stocking notes, that Philostratus deemed "to be a *bad* role model for ancient athletes." The *Gymnasticus* warns that wrestlers should not have "a neck yoked to the shoulders" like "the statues of Hercules." The "yoke" here probably refers to highly developed upper trapezius muscles that slope from the neck to the shoulders. While Philostratus admitted that such a neck is "naturally suitable" for wrestling—another nod to the sport-specific paradox—he thought the look unsuitable for wrestlers, for making them appear "more similar to one punished than trained."

The "punished" Hercules statue—showing the neck that, Philostratus said, athletes should not have—probably refers to a particular style of sculpture, Roman copies of a Greek original, that dominated a large space in the main public bath in Rome around when the *Gymnasticus* was written. Two versions of this sculpture of Hercules, each more than ten feet tall (almost double the height of the average ancient Roman man), flanked the building's central hall.

Baths were the most popular, lavish, expensive public buildings in Rome, and they were not just places for training and bathing. Some also had libraries, lecture and music halls, and miniature parks. No other indoor space in the imperial capital, and probably in the whole empire, had sculptures so big and so dazzling. This colossus of Hercules, to ancient Romans, was ten thousand percent *wow*.

Even readers of the *Gymnasticus* who lived outside Rome and never visited that bath would have known this style of sculpture. The hero had been sculpted in similar poses for five hundred years by the time Philostratus wrote. Describing this Hercules as "punished" would have instantly made sense to ancient people, who knew his labors were punishments imposed by the goddess Hera.

Often called the Weary Hercules, this depiction of the hero shows him worn out, looking reflective after performing the last of his labors. In effect, the Weary Hercules "is resting after exercise," Stocking says. "He's not just posing. What the ancients saw was somebody who's exhausted after training."

I ask, "This is showing what movement does to you?"

"Right," he answers. "You need to recover." The point was underlined by the statue's placement in the baths, in a room called the *frigidarium*, a space for cooling down, where every man went after exercising. To any Roman seeing this statue, Stocking emphasizes, "movement is always there. And it's always already there. He's not just posing on a column."

Lost when the Roman baths fell to ruins in the sixth century, this statue of Hercules was found again in the sixteenth century. "From the moment of its discovery it achieved an instant fame," and was known simply as *Lo bello*—Italian for "the beautiful one"—according to the archaeologist and art historian Miranda Marvin. In the late eighteenth century, when Napoleon ransacked Italy to steal works of art for display in Paris at what would become the Louvre Museum, this statue of Hercules was the item he most wanted but did not manage to acquire.

In modern physical culture, there has been no more influential figure than the Weary Hercules. The Prussian strongman Eugen Sandow, who "invented the business of bodybuilding" in the late nineteenth century, "seemed to be a Greek statue come to life," as his best biographer, David L. Chapman, has written. To many of Sandow's admirers, he seemed especially to be *this* particular statue come to life. Popular photographs of Sandow showed him posing as the Weary Hercules.

SANDOW.
Copyright 1893 by Napoleon Sarony
37 Union Sqr., N.Y.

Almost one hundred years later, Joe Weider, the bodybuilding impresario who sponsored the competitions and published the magazines that made Arnold Schwarzenegger's reputation, said the Weary Hercules statue "personifies power," and exemplifies "what a bodybuilder should look like." Weider described the figure's magnetism with a phrase of almost animal desire: "What he has is what we want."

Philostratus wanted something else. The *Gymnasticus* tells readers they do *not* need to try to look like Hercules. They do not need to try to look like the most monumentally impressive body they have ever seen. "Rather," Stocking writes, "Philostratus offers a type of sociology of muscle—too much muscle signifies punishment," and symmetrical muscle symbolizes freedom.

The *Gymnasticus* says training is not punishment, but freedom.

Among the many types of expertise that humans possess, Philostratus believed there was only one legitimate distinction to be made—between material craftsmanship, such as carpentry and cookery; and the abstract arts, which lay hold of "that which is not visible." Music, geometry, philosophy, poetry: All these are forms of abstract expertise—considered as types of wisdom—of *sophia*, Philostratus wrote—and so is athletic training, "inferior to no other expertise."

Philostratus knew that a lot of people did not see training this way. He knew they thought exercise was about following directions, like a formula or recipe. Those people were in thrall to sets of rules for training called the Tetrads, he wrote, "because of which all aspects in athletics have been destroyed."

Tetrads were a type of training program, structured by four-day cycles of workouts. Each Tetrad, as Stocking summarizes the scheme, progressed through "preparation day, intense day, recovery day, mediating day." Studying the *Gymnasticus*, Stocking was struck by similarities between this ancient Roman regimen and the periodized training cycles written by coaches such as Boris Sheiko that he had followed as a young man. He recognized that by modern, scientific standards, the Tetrad is a sensible training program.

So why did the *Gymnasticus* condemn a workout program that works?

Philostratus was not dismissing the rules themselves. It was common, Stocking says, for ancient bodily understanding to be organized in four-day cycles such as the Tetrad. (For instance, in medicine, from Hippocrates to Galen, physicians understood disease to operate on four-day cycles.) Philostratus dismissed a hidebound, rules-based mentality that fails to see training as subject to uncertainties of circumstance.

To show why the Tetrads had "destroyed" athletic training, Philostratus told the following story:

A wrestler named Gerenos won victory at Olympia. To celebrate, he got drunk and stayed drunk for a couple of days. Then, with a massive hangover, Gerenos showed up to the gym for training.

Seeing the athlete's condition, his trainer became irate—did this punk have no respect for authority?—and, in anger, he forced the wrestler through his workout. Because the schedule was the schedule. They were sticking to it, period. But the stress of that training session, at a moment when the wrestler was not ready to perform, was too much for the athlete. During that workout, Gerenos died.

The death was tragic, Philostratus wrote. This wrestler was a bona fide champion. There was no reason to force him to stay on schedule for schedule's sake. No reason to put him in crippling distress. But despite the catastrophic outcome of the trainer's wrong decision, Philostratus did not call the trainer a bad person. He said the trainer killed the athlete "through ignorance of training"—and he said the best trainers, preparing athletes for the best competitions, never worked the way that trainer did. When athletes arrived at Olympia, Philostratus wrote, they would train for thirty days under the sanctuary's own staff of trainers, who "do not train by prescription, but provide exercises 'all improvised for the right time.'"

The key word there, translated as "time," is *kairos*—and if you remember just one Greek word from the book you now hold in your hands, let *kairos* be it.

Kairos has no precise equivalent in English, but Stocking interprets it as meaning "the window of time or opportunity to act." Another scholar, Catherine Eskin, describes kairos as "the ability to recognize the 'right' moment, and, knowing that right moment, to take decisive action."

Kairos was an important concept in ancient athletics. Pausanias, the

Roman traveler who wrote about Olympia, said there was an altar to kairos at the entrance to the stadium there. Kairos was also important in rhetoric and medicine. The first readers of the *Gymnasticus* may have heard some grace for Galen in the emphasis Philostratus put on kairos, since Galen, despite his rhetorical excesses, always argued that doctors should treat patients as individuals and should not assume disease worked on a fixed schedule. Galen argued that doctors should always be flexible in caring for patients, using logic and observation to guide treatment, never making cookie-cutter prescriptions.

Considering the cultural prominence of kairos helps Stocking appreciate why Philostratus could not abide blind faith in training formulas such as the Tetrad or in the authority of trainers like the one who killed Gerenos. Stocking says respect for kairos "is necessary to be able to react appropriately to the contingencies of time. The Tetrad, by contrast, ignores those contingencies," because the Tetrad "seeks to control the body through the control of time." Philostratus praised training that was based on "skills of analysis and improvisation," Stocking adds, which "render sport a valid form of knowledge and cultural education."

To Philostratus, the one essential element of wise training was doing the right thing at the right time. Liberating and demanding, this transformative vision of training described by the *Gymnasticus* promised to change even the places athletes inhabit. "So much for the Tetrads," Philostratus wrote, "and if we follow the advice I have given, we will demonstrate that athletics is a form of knowledge, and we will strengthen athletes and the stadiums will grow young from training well."

Philostratus, even when his prose grew lush with swelling chords, never lapsed into the feel-good sentimentality of "you do you."

When stakes are high and pressure is real, Philostratus said, athletes should be expected to show up and give everything they've got, to strive for victory—and athletes who do not give everything are liable to lose everything. Philostratus made that point with another vivid anecdote. "Some say that a coach at Olympia once killed his athlete with a sharp strigil, because he did not strive for victory. So then, let the strigil serve as a sword against bad athletes," he wrote.

Stabbed with a strigil!

Savage as this may sound, the assault was anything but gratuitous. The severity makes some sense when you remember that athletics, for Philostratus, was by definition striving for victory in competition for a prize.

But what's the prize?

To early readers of Philostratus, it was victory in the games.

To Charles Stocking as a young man, it was victory at powerlifting meets.

And for Stocking today, the prize is both grander and simpler than that. The prize is life.

The prize of training is a high level of function that keeps life running as smoothly as possible, in a world that is inherently unpredictable. A pain-free day of working at his desk helps him maintain the equilibrium it takes to teach, to do scholarship, and to care for family, friends, students, and colleagues—an equilibrium that, following Galen, people have liked to call by the name of health.

Unlike Galen, however, Stocking does not see performance and health as necessarily opposed. He squats and bench-presses more weight now, in middle age—with maximal care to minimize risk of injury—than he lifted when he was a young man. Informed by the already substantial and growing body of mainstream medical research on how progressive resistance training can reduce the risk of many chronic diseases and preserve or improve a person's physical and mental functioning throughout the lifespan, into oldest age, Stocking expects to follow lifting regimens for as long as he lives.

Athletic training, Stocking says, "is a life-and-death issue. This isn't just about looking good on Instagram or feeling good about yourself. No, this is life-and-death. It really is. And that isn't rhetoric."

Stocking's own regimens of weight training and sprinting also speak to the question he raised on our first day at Olympia, that question about ancient Greek bodies and our bodies today: Are they the same bodies?

In his training, Stocking strives in earnest to experience a religious aspect of ancient athletics, an aspect he names as "moving beyond your current state." Explaining this to me, he goes back to the fierce heart of all he knows about ancient religion and athletics: the physical reality of divinity, the sense that the gods are *there*.

"What are the gods?" he asks out loud. "The gods are a representation of human possibility. The Greeks are thinking, *What is possible?*" and their thoughts on that question involve belief in gods with physical bodies capable of doing more than any mortal can do. "But that potential is seen within the body," he says; and in a similar way, "this idea of going beyond is what's also driving athletics. This is why you run to the altar. That's going to make you run faster than just going for a run."

He continues: "To think about what's possible allows for us to do more. To exceed your current limits, which will actually simply allow you to live in a better way." Any physiological adaptation to any stress means that "your body wasn't able to do something and is now able to do it. To accurately respond to the demands of stress. And that's victory, right? Victory is a symbol of that.

"Thinking about life in those terms—in terms of movement, action, and accomplishment—victory is *accomplishment beyond what you were capable of*. So victory might be winning the Olympics. Or victory literally might just be taking your groceries to your car and getting them back to your house."

The purpose of athletic training comes down to this, for Charles Stocking: "You can see what's actually physically possible for you. That is inherent in the human organism: possibility."

This vision of training is not about bending the world to your will. It is not about realizing some kind of heroic divinity that you possess. It is about doing what you can do to make yourself able to act upon the world, while recognizing and respecting those abilities' mysterious emergence and transience, which many ancient Greeks attributed to the gods.

Athletic training, as Charles Stocking experiences it, is a practice of moving beyond limits of past and present circumstances. A practice of creating futures in which possibilities can always be recognized and in some measure can be fulfilled. In this kind of athletic training, preparatory regimens are inseparable from outcomes of victory. Aspects of personhood that some like to name as mind and body are inseparable, too; and the muscles and strengths we build, allowing us to act upon the world, are more than instrumental: They are us, as we are the gifts of others.

Charles Stocking learned to see athletics as a practice of possibility, and as matter of life and death, from the writings of Philostratus—with help from his brother and teacher, Damian Stocking; the classicist Greg Nagy; the trainer Pavel Tsatsouline; the biomechanical engineer Stuart McGill, and a host of others—including many people he's never met. Even Arnold Schwarzenegger unwittingly gave Stocking a hand, by showing how lifting weights can create new possibilities in life.

To become the athlete and the scholar that he is, Stocking needed many kinds of help and inspiration. He needed examples of lifters who reached the highest levels of excellence without using performance-enhancing drugs, and of scholars who study physical culture with the same respect that academics more typically pay to forms of culture such as art, music, and literature. Few people in the world have done both of those things. One of those few—whose example has guided Stocking along his way—is a woman named Jan Todd, who started lifting weights by lucky accident.

PART II

RUN THE RISK

How strength shapes identity

CHAPTER 4

Born and Made

At Christmastime in 1973, when Jan and Terry Todd were newly wed, they celebrated the holidays with Terry's family in Austin, Texas. One day during the visit, they went to the Texas Athletic Club, a grungy, heavy-duty gym in downtown Austin that had been almost a second home to Terry when he was younger.

Like practically all weight-training gyms in that era, the Texas Athletic Club was an overwhelmingly masculine space. On the day of this visit, Jan Todd was one of only two women there. She had not come to do any serious training, but to lift a few light weights, without much sense of purpose except for the hope of improving her posture. She was mainly there to keep her husband company while he worked out.

The other woman in the gym that day, though, was lifting heavy weights, relative to her own body weight. "She weighed about 115 very svelte pounds," Todd later wrote, "and she had worked up in the deadlift to 225 pounds."

Fascinated by this, Jan Todd watched the woman's every move. The woman stood at the midpoint of the length of a barbell, which was loaded with plates and resting on the floor. With her feet firmly planted and spaced roughly parallel to her shoulders, the woman bent the hinge of her hips, pushing her rear end backward. The woman gripped the bar, with arms straight down and shoulders retracted; and then she extended her hips, pulling the weight up as she stood erect. After a brief pause in this position, the barbell touching the front of her thighs, she reversed the movement, and she controlled the weight's descent back to the floor.

Watching these movements changed Jan Todd's life. More than forty years afterwards, she tells me about it—her first sight of a woman lifting heavy weights—and calls the chance encounter "actually something like a revelation." Having grown up in a culture where, Todd says, by and large "girls didn't try hard, and girls didn't do sport," she was astonished to see a woman exert great force by means of muscle.

"I liked the idea of the challenge the bigger weights represented," Todd has written. In deadlifting, she began to see a new kind of challenge that she as a woman could meet, even a new example of what a woman could be. Like many people of her generation, and some in successive generations, too, Todd had a vague sense that femininity and muscular strength were somehow incompatible. The feeling was more reflexive than considered. A notion based on absence of experience—and involving fears of physical danger in lifting heavy weights—that there might be something inappropriate about women training this way. But in the gym that day, a wild surmise welled up in Todd while she watched the other woman: "*She looks fine, so maybe that makes this okay.*"

Having decided it was okay for a woman to lift heavy weights, Todd began a regimen of training, and within the almost incredibly brief time of eighteen months, she earned a place in *The Guinness Book of World Records* for the heaviest deadlift by a woman.

"Breaking the record opened a new world for me because I realized that almost all of the fears we have and the barriers we set for ourselves as women are in our minds," she later wrote. "I felt so free."

For more than a decade of competitive powerlifting, Todd was one of the strongest women in the world, setting an example for others curious to know how lifting weights could change their lives.

For many of the curious, even in the twenty-first century, interest in weight training is infused with some fear. In 2017, after sociologists at Southern Illinois University interviewed members of seven collegiate women's athletic teams—cross-country, gymnastics, soccer, softball, tennis, track and field, and volleyball—they reported that most of the women began college with little experience of weight training and lots of hesitation about it. The sociologists, Rachel I. Roth and Bobbi A. Knapp, wrote that the athletes' "fears associated with weightlifting included the idea that it was intimidating and scary," the idea that it involved "a high risk of injury, fear of doing it wrong, and fear of getting bulky." These fears, the researchers observed, had not really changed in the previous couple of decades. In the mid-1990s, women had shared substantially the same concerns with the sociologist Shari L. Dworkin during her immersive ethnographic field research as a participant observer at a university gym in California.

Fears about lifting weights were even greater during Todd's competition years in the 1970s and 1980s. The physical risk of weight training was widely exaggerated by people who poorly understood it. The reputation of many elite lifters for taking performance-enhancing drugs added to the air of danger around heavy weights. Traditional gender norms were beginning to be more fluid and flexible than in earlier times, but women who lifted weights and built conspicuous amounts of muscle were nevertheless sometimes perceived as "gender outlaws," as sociologists who studied women's bodybuilding observed.

Jan Todd has run the gauntlet of all those fears and concerns. Through years of competition and decades of recreational training, Todd has sustained no major injuries while lifting weights. She claims never to have taken performance-enhancing drugs, and when drug testing became available in the sport of powerlifting, her claim was supported by test results. And she has helped to establish historical and scientific proof that muscular strength and access to knowledge about strength training are important aspects of social equality and opportunity for women.

By demonstrating how strong a woman can be, Todd has inspired all kinds of people to imagine themselves lifting weights and has shown how strength could shape their identities, as it shaped hers.

After that Christmastime visit to the Texas Athletic Club in 1973, Jan Todd says, "I asked Terry about using heavier weights."

Her husband Terry had some knowledge about using heavier weights. Nine years earlier, in 1964, Terry Todd won the first-ever U.S. national championship in the sport of powerlifting. At the time, he was a doctoral candidate at the University of Texas, working on a dissertation that would be titled "The History of Resistance Exercise and Its Role in United States Education." By 1966, when he received his PhD, he had set some fifteen national and world records in powerlifting. He was, for instance, the first man to squat 700 pounds. With this background, Terry was game to discuss Jan's question about heavy lifting and to engage her curiosity about "strength, size, and womanhood," as he later recalled.

Those topics came vividly to life, in their conversation that day in Austin, when Terry talked about Katie Sandwina, the vaudeville and circus strongwoman. Born Katharina Brumbach in 1884, she later took the stage name of Sandwina, perhaps borrowing some brand equity from Eugen Sandow, the most famous strongman of the time. Sandwina was a Barnum & Bailey center-ring attraction in the early twentieth century. She thrilled audiences with her beauty and strength. In 1911, one journalist wrote, "Lo! These eyes have beheld the Superwoman," after meeting Sandwina. "She is as majestic as the Sphinx, as pretty as a valentine, as sentimental as a German schoolgirl, and as wholesome as a great big slice of bread and butter."

Standing almost five-ten and weighing about 210 pounds, the strongwoman performed a circus act that involved lifting her husband Max—five-six, 160—and performing the manual of arms with him, tossing the man around as if he were a rifle. She hoisted a 600-pound canon onto one shoulder and carried it across the circus ring. She repeated these stunts almost daily, and sometimes twice daily, even while pregnant, until she gave birth to a son, at which point the great Sandwina reportedly took only a single day to rest before resuming her labors.

In 1911, Sandwina symbolized a scientific truth about women's bodies that was formally documented in 1974. Brand-new physiological research published that year showed that, with weight training, women could increase their leg strength—considered in proportion to their body weight—almost exactly as much as men. In proportion to lean body weight only, the same study found "the women actually exceeded the men in leg strength."

The physiologist who made these findings, Jack Wilmore, later observed that only a few women ever realized their full strength. "While there are rather substantial physiological differences" between untrained males and females, he wrote in 1979, the differences between highly trained male and female athletes in any given sport are much smaller. After analyzing earlier studies of male and female athletes, Wilmore concluded: "What once appeared to be dramatic biological differences in physiological function between the sexes may, in fact, be more related to cultural and social restrictions placed on the female as she attains puberty."

Weight training, Wilmore's own experience suggested, was not typically presented in a way that invited women to see the practice as being relevant to them. In 1971 at the University of California in Davis, when he had taught weight training to college students, only four females had shown up for the class he listed as "Weight Training I" in the course catalogue. When he changed the name to "Weight Training for Figure Control," class enrollment jumped to two hundred—a 5,000 percent gain—and in 1972, *Life* magazine ran a piece about California's weight-lifting co-eds under the title "The 'Weaker Sex' Comes On Strong."

As Terry Todd told Jan Todd stories such as these, Jan saw the woman deadlifting at the Texas Athletic Club as more than an anomaly. That woman's heavy lifting and Todd's chance meeting with her were part of a history, in which training is driven by force of example.

The force of example has an especially concrete, quantifiable aspect that captured Todd's imagination, too. Jan and Terry had the idea to consult a copy of *The Guinness Book of World Records*, where, Jan remembers, they found only a couple of records for women's lifting. The one that jumped out at her was the record for the "Two Hands Deadlift." In 1926, in France, Mademoiselle Jane de Vesley had lifted 392 pounds.

"I asked Terry if he thought I could break that record if I trained for it," Jan later wrote. "He answered simply, yes."

That day, her life changed. When she woke up in the morning, she had no thought of becoming a serious lifter, and when she went to bed that night, she wanted to set a world record.

But since childhood Todd had been preparing, without knowing she was preparing, to bank this turn.

"Show us your mouse!"

As young girls, Jan and her two sisters would beg their dad, and he would roll up his shirtsleeve and flex his biceps, make the muscle twitch, and the girls would giggle. Because when he flexed, he told them, teasing, it was like a little mouse squirming inside his arm.

James Suffolk was a big, tall man. He served in the Army in World War II. After he came home from Europe, he went to work in a steel mill and broke his leg on the job. In the hospital, one of the nurses who took care of him was Wilma Yerty; and she married him. They had three girls, Susie, Jan, and Linda, and the family spent the late 1950s on the move, James chasing work in the declining steel industry.

In 1961, after some of the girls' friends from school got sick, Todd remembers, "I had a little bit of flu, but no big deal, and then Susie became very sick on December 28. She and I shared a bedroom, we had twin beds in our bedroom, and I woke up in the middle of the night, and she was in convulsions."

Jan's parents called a doctor, but he would not make a house call. They called another doctor, and he would not come, either. Wilma and James wrapped Susie in a blanket and put her in the car, and James drove fast, speeding toward the hospital—but on the way, before they got there, Susie died.

"They didn't really recover from that," Todd says, remembering her parents' struggle through three more years of marriage, their separation, and divorce. "That was just too sad."

Before the marriage ended, the family moved to a cinder-block suburb of Tampa, Florida, where in 1963, Jan started sixth grade in the sixth school she had attended in as many years.

"The main thing I remember about growing up is that I always felt like I had a weight problem," she says. Compared to her classmates, she has written, "I was much larger in my bone structure," she recalls, "like a larger species of the same animal."

At Jan's new school, some of the girls in her class took dance and ballet lessons, and Jan, admiring the grace and the physical abilities of dancers, asked her father if she could join the other girls in taking lessons. She remembers his answer: "Why would you want to do something like that? Ballet dancers have such big, muscley calves." She also remembers his disdainful tone, which seemed to suggest that muscley calves were bad or disgusting.

Jan learned another lesson about the kind of strength that was supposedly appropriate for a woman in late November of 1963. In the days following the assassination of President John F. Kennedy, Jan watched events unfolding on the black-and-white TVs at school and at home. The parade with the caissons. The First Lady in her mourning veil, the long veil hanging from her forehead to her waist. This one woman, Jacqueline Kennedy, with her thin, small body, was strong enough to hold a whole nation together, Jan thought. And that seemed right to Jan. That was how women should be strong. Emotionally.

The next year, when Jan was in seventh grade, she began to learn a different kind of lesson about the strength of girls and women. Jan's mother Wilma helped her learn about sports. "She encouraged me to join a softball team," Todd once wrote, "and spent hours in the backyard playing catch with me. She also taught me to swim and then signed me up for the local swimming team. My mother had never played any sports—outside of her school gym classes—but she wanted me to have the chance."

By this time, Jan's parents' marriage was over. Her dad had moved out, and her mother needed to find a job. Wilma and the girls moved to a small apartment, and after paying the men who helped them move, Wilma sat down with Jan at their kitchen table, and they talked. Wilma was worried about money. "She said, 'I have forty dollars until payday,'" Todd tells me.

The daughter was determined to be strong for her mother. Todd says, "I started trying to do what I could do. I looked for things to do. Tried to have meals ready when she got home." Found babysitting jobs and took care of her younger sister Linda. Walked herself to and from school every day.

The walk was not much more than a mile each way, and she could have taken a shortcut, but the young Jan liked to take the longer way, especially in springtime. The long way passed a field where she could see and smell a great expanse of fragrant, beautiful pink phlox flowers blooming; and some days, the air above the field was full of butterflies, flying to pollinate the blooms.

When Jan Todd was growing up, she defined strength for herself, as every person does, in a complex process of invention and reception.

As a young woman who was bigger than she wished to be, she was sensitive to the connection between size and strength and confused by it. Being strong was good, important, and necessary, she believed. She also believed that, for her, in her female body, being big was somehow wrong. She was not sure quite what was wrong with it, but the way her father said the word *muscley* seemed to point out the problem. It didn't, of course. But it did indicate an actual problem.

Words like *strong* and *muscle* can be blunt instruments. Through personal associations, shaped by social conventions, they can limit people's understanding and experience of muscle. Some of those limits fall away, however, when roots of these words are uncovered, and old notions reveal new possibilities.

The English word *muscle* comes from the Latin word for mouse, *mūsculus*. For the better part of a millennium, in early versions of languages including Greek, German, Dutch, Icelandic, and Armenian, muscles and mice were named by variants of the same word—especially when referring to the muscles on the front of the upper arm—"apparently because of the resemblance of a flexing muscle to the movements made by a mouse," one dictionary speculates—which the young Jan Todd could have told you, when she begged to see her dad's "mouse." Yet later, the girl's sense of muscle as being playful and delightful got shut down, when Jan's father groused about ballerinas' "muscley calves."

As for *strong*, it developed from the word *strang*, which in early Germanic languages named the material that in English now goes by the name of string. Anthony Esposito, an etymologist at the *Oxford English Dictionary* who has studied *strong* and related words, tells me how *strang* spawned *strong*: "String is something that you bind things up with, it's made of material that is twisted together tightly. What seems to have been the original core sense of the word *strong* is this idea of tightness." The connotation of bodily power, he says, "is secondary to that, it's come later, it's developed out of that. Originally, *strong* seems to have had quite a specific literal sense of something tight. And as a result of being tight, it's hard. As a result of being hard, it's powerful."

The tight line from Germanic *strang* to English *strong* runs through a wider web of words for strength and associated concepts in many older languages. The line from *string* to *strong* vibrates, too, with the tone set by Aristotle's writings on anatomy and movement—his account of the gaseous substance called *pneuma* heating and cooling tiny sinews in the heart, the little strings that pulled on bigger strings that pulled on bones, to move us.

Various languages have associated strength with a range of other concepts as well, including healing, refreshment, comfort, and most consistently, with manliness. But the deepest roots of *strong* and all its cognates go back thousands of years before Old Germanic, to the parent language of hundreds of Asian and European tongues, a language scholars call Proto-Indo-European. Strength, as we now know it, arises partly from some of the earliest words that people used to name one type of action.

Some of the deepest roots of strength are intermingled with the concept of standing.

There has never been a time when Jan Todd liked to lie around. For as long as she can remember, she has preferred to be up and moving, often moving at high speed. Through a childhood busy with intense play and hard work—running, jumping, pulling, pushing, swimming, softball, riding her bike—she developed extensive and robust connections among her nerves and muscles.

Such early experience may have lifelong effects. Children who become involved in a variety of physical activities and sports at earlier ages grow into more active adolescents and adults. The opposite seems also to be true. Children who grow up with fewer opportunities to exercise and play "may be less likely to engage in more challenging activities later in life and more likely to experience the adverse consequences of a sedentary lifestyle," according to one review of pediatric exercise research. Though it's never too late for an individual to acquire new skills and change habits, the pattern of physical activity in most people's lives tends to be set early.

Some kinds of motor skills may be more likely than others to help children grow into more active adolescents. In a 2009 study of hundreds of young people in the Australian state of New South Wales, children's proficiency in walking and running had no connection to their overall activity as adolescents, but childhood motor skills demonstrating "object control proficiency," such as catching, throwing, and kicking, were strongly correlated with adolescent physical activity. In this study, girls tended to score lower on object control skills than boys did. Yet girls and boys with equal object control skills were equally likely to grow into highly active adolescents.

After childhood, across the lifespan, physical activity declines for all, and the steepest decline occurs between the ages of thirteen and eighteen. Activity declines in adolescence in many other species, too, from insects to monkeys, indicating that the mechanism of decline, in addition to being affected by early experiences, acquired skills, and habits, is at least partly biological.

In 2021, the international medical journal *The Lancet* published a review titled "Physical Activity Behaviors in Adolescence," by Esther M. F. van Sluijs at Cambridge University and colleagues. The researchers analyzed objective measures of physical activity in Europe and the United States collected by accelerometers such as step-counting devices. "These data show that men and boys are consistently more active than women and girls, and that there is a clear trend for decreasing physical activity with advancing age in early and late adolescence," the authors concluded. After this decline, they added, each person's level of physical activity tends to stabilize at a certain level and seems to track at that same level into young adulthood.

In her highly active childhood and adolescence, the young Jan Todd followed a track that was countercultural, as she has continued to do in many ways throughout her life.

"I was probably always stronger than a normal woman," she tells me. That was clear by the time she was a teenager. She had an aptitude for athletics, too, especially for events involving brief bursts of power. "When we did the high school physical fitness test, what I was best at were shorter distances," she adds. "When we swam, I was best at the 50 freestyle." But she could not complete even one pull-up, and she never spent enough time on the track or in the pool to find out how fast her sprints could be.

During high school, she never got a clear view of her own gifts, partly because her school gave more athletic opportunities to boys, who could play any of six team sports, than to girls, who had one choice: swimming.

But even if Jan had been lucky enough to meet a coach with the knowledge and desire to train her then, she probably would not have devoted herself to athletics at a young age. Like many of her peers, especially girls and women in households where money was scarce, she did not have the spare time or cash that would have allowed her to train up her physical gifts.

Jan Todd's potential as an athlete did not become clear until after she was married. Following that fateful visit to the Texas Athletic Club, and after Todd started training to beat the women's deadlift world record, her husband wrote a letter to a friend about his wife's gifts.

"She has good levers," he wrote. He meant that the proportions of her limbs—long arms and short legs, relative to her overall height—gave her unusually good leverage for deadlifting. In the deadlift, arms and legs are levers for moving the weight up from the floor, around the fulcrum of the hips. Todd's proportions gave her two advantages for this lift. Long arms let her hands reach the barbell with ease, and short legs kept the weight close to her body as she stood erect.

Another natural advantage for deadlifting was Todd's unusual handgrip strength. Not long after she married Terry, he showed her a trick that he had learned through years of practice—he bent an old-style hard metal bottle cap in half between the tips of his thumb and index finger, and he did this while keeping his fingers straight—and Jan, after witnessing Terry's feat, asked, "Can I try that?" And then she did it as easily as he had done, immediately and repeatedly.

Later, Terry would call Jan's bottle-cap-bending "one of the most unbelievable things that I've ever seen." He told me, "I don't have any explanation for it. She was just put together in such a way that her tendon insertions were where, when she called on the muscles to do what she wanted them to do, they were able to perform the feat because of the leverage advantages she had, because of how her body was put together. It demonstrates that strength can be manifested in many ways, and there is no way one man or woman can be stronger than all others, because of all the ways there are to test strength. And that's a truism."

The most important thing for Jan as a novice lifter, Terry believed, was gaining confidence to meet a challenge.

So from the beginning, Terry made sure Jan trained to succeed. In the gym he chose weights for Jan that she could move through the full range of motion of her joints, with real effort but decisively, and with some energy left to spare.

Jan Todd's early weight training, while building muscle and strength, also enhanced her efficacy—a term psychologists use to describe people's confidence and abilities to realize their desires. Efficacy is subjective: It depends on people's own perceptions of their abilities to make things happen. Efficacy is also objective: Proof of ability is in the doing. Psychologists have found that in athletics, the relationship between those factors is reciprocal. Most studies of efficacy concern its subjective aspect, called self-efficacy, and one review of research on the topic found that "self-efficacy is both a cause and effect of performance"—a less beautiful statement of Terry's philosophy of training to succeed.

When the first study of heavy weight training and self-efficacy in untrained adolescent young women was published, in 1988, it would show some of the same things that Jan Todd's experience suggested in the 1970s. In the 1988 experiment, getting stronger improved young women's "confidence level about their bodies, and about overall effectiveness in life as well." Since then, studies of all kinds of people have shown that increased self-efficacy—increased confidence in one's ability to act upon the world—is both a cause and an effect of all kinds of physical activities: fourth-grade girls playing on playgrounds; guys in their twenties pedaling stationary bikes; nonagenarians who are learning how to walk again, after falling and fracturing their hips.

Jan Todd was not short on self-efficacy, as shown by her personal history of achievement. By lifting weights, she built even more confidence in her own abilities and potential. Recalling the first months of Jan's training, Terry said, "The early days of a lifter's career are usually pretty pleasant and happy days, because you can continue to make pretty steady progress for a good while. Because we don't tap close to our limits just by what we do walking around."

Making this kind of progress—working to push the limits of your own potential strength—is what makes up the objective part of strength training efficacy, and Jan Todd was doing that work.

By February of 1974, after training for a couple of months, she could deadlift 295 pounds—the weight of an empty cast-iron bathtub.

By March, she could lift 315 pounds up from the floor, and by April, 325 pounds.

The pleasure of lifting, she tells me, was enhanced by visual proof of her progress. "What was really intriguing was that there were more plates on the bar," she says.

It was not during a workout at the gym, though, but on a trip to the grocery store, when Jan Todd first felt how lifting had changed her.

"I picked up a watermelon," she says, "and the watermelon wasn't heavy."

From month to month through 1974, Jan's early training took place in Macon, Georgia, where she and Terry then lived. At Mercer University, she studied for a master's degree, preparing to become an English teacher, and he was a professor and school reform advocate. All year, they both looked forward to Christmas, when they would return to Austin for the holidays with Terry's family—and return to the Texas Athletic Club for another workout. When they did, Jan Todd deadlifted 370 pounds—just 22 pounds shy of the record. Within a few months, she would be ready to reach her goal.

But in 1974, there were no powerlifting competitions for women. Making her achievement a public event required special arrangements. To that end, Terry placed a phone call to the organizer of a men's meet scheduled to happen in Chattanooga, Tennessee, and asked him to make room on the program for Jan's world record attempt. The man in Chattanooga wanted to hear about her progress.

In the first months of 1975, Jan had built up to a personal record deadlift of 385 pounds. To equal the 392-pound record, she would have to add 7 more pounds to her lift. The sticking point for Jan was the lockout, the last few degrees of rising to upright posture. "My upper body was my weak link," she recalls. "I would get up the heavy ones, and then start to wiggle, trying to hitch it up."

When Terry described the situation to the meet organizer, the man replied, "Why don't you just put her on some Deca-Durabolin?"

Terry chuckled diplomatically. "We'll see about that."

After the call, he asked Jan: "Can you *believe* that?"

Jan said, "What's Deca-Durabolin?"

Deca-Durabolin is the brand name for nandrolone, one of the most popular anabolic steroids; and when Terry told her this, Jan saw her own lifting in a new light.

"I was serious about setting the record—kind of. But I hadn't gotten to the point as a lifter where, if I missed the record, I would be crushed by that experience," she says. This was her first glimpse of the risks that many lifters take, to hurry up building muscle and strength. "The idea that I would take those kinds of risks? It wasn't in my conception."

Then came springtime—and on May 5, 1975, at Central YMCA in Chattanooga, Jan Todd set a new Guinness world record by deadlifting 394.5 pounds. Within a few days, she told a reporter that her next goal was to be the first woman to total more than 1,000 pounds in the three power lifts.

Todd stood five-seven and weighed 165 pounds when she broke the deadlift record. Her training had shifted her body composition—she could see more muscle in her legs now.

In the next phase of her training, Todd began to build even more muscle. At the same time, she says, she had to negotiate "how people around me—female people in particular—were opposed to what I was doing."

One woman close to Todd warned her, "Deadlifts are going to make your waist bigger."

Another, knowing that Todd wanted to have children, asked if high-force muscular exertions, the "squeezing" and the "pressure," would make pregnancy more difficult for her.

A third, as if speaking from the nineteenth century, voiced sincere and flabbergastingly naive concern that if Todd kept lifting heavy weights, her uterus might fall out.

None of these negative opinions about muscle size and strength had any empirical basis. None were based on any facts about strength and size and how they develop.

Strength, like other attributes of muscle including power and endurance, is a collaboration involving the nervous system and the muscles. *Neuromechanics of Human Movement*, a textbook by Roger Enoka, describes the collaboration. Enoka is a professor of integrative physiology at the University of Colorado, Boulder, where he directs the Neurophysiology of Movement Laboratory. He is also a professor of neurology and of medicine in the specialty of geriatrics at the University of Colorado, Denver. He is one of the world's leading experts on the basic element of engagement between neurons and muscle fibers, called the motor unit.

"Changes in muscle strength" such as take place in weight training, Enoka writes, "are generally attributed to adaptations in either contractile properties or muscle activation." Contractile properties are the structural traits of muscles allowing them to contract at various speeds. Muscle activation is the signaling by which neurons transfer electrical charges, called action potentials, to engage the muscle fibers, yielding force that makes us move. Getting stronger involves both structures and signals.

So when it comes to building strength, size matters, according to Enoka, and it matters a lot. "The principal factor related to contractile properties is muscle size," he writes. However, "differences in the cross-sectional area of muscle" account for only about 50 percent of the differences in strength among individuals. The remaining 50 percent of what makes some people stronger than others depends on factors including "the details of the training program, such as the types of muscle contractions included in the exercises, the muscle groups involved in the actions, the number of limbs involved in each exercise and the relative timing of their activity, and the amount of support provided by the surroundings."

Size isn't everything, in other words. Size is fundamental to strength, but that doesn't mean it's always primary. As Enoka writes, "it is possible to obtain an increase in strength without changing contractile properties, but not without an improvement in muscle activation." Muscles can get stronger without getting bigger, as long as they are sufficiently activated by the nervous system.

Activation isn't everything, either, though. In early stages of a strength training program, activation accounts for the first changes in the force-generating capacity of muscles. Then the situation changes. "The contribution of adaptations in muscle activation to strength gains declines as a training program progresses," Enoka adds, "and the contractile properties begin adjusting to the imposed demands by increasing the quantity" of contractile proteins, the molecules that muscle is made of. Increasing synthesis of those proteins is a major part of what makes muscles bigger. "An increase in contractile protein content, which is known as hypertrophy, often increases muscle size," he writes, "and can be detected as an increase in the cross-sectional area of muscle fibers and whole muscle."

The magnitude of those changes is different for different people. One of the largest studies investigating how resistance training affects muscular size and strength, by Monica J. Hubal and colleagues, discovered some disparate responses to the exercise. The study, published in 2005, involved nearly 600 young adults in their twenties and thirties who spent twelve weeks training their biceps and triceps at high intensity. Before and after their training, researchers measured one of their biceps muscles using magnetic resonance imaging machines. MRI results showed that, in those twelve weeks, the cross-sectional area of some people's muscles had grown 40 percent larger, or more—the biggest gain was 59 percent—while other people's muscles had gained little or no size: They grew by less than 5 percent. Both groups of outliers were very small, however; and both groups of outliers were about equally divided between male and female. Overall, average gains were 20 percent for men and 18 percent for women.

Gender made a smaller difference to size gains, in this study, than it made to gains of strength. The data, Hubal and her colleagues reported, "show definitively that women gain significantly more relative" strength than men, in response to biceps and triceps training. The average gain for men who took part in her study was 40 percent. The average gain for women was 64 percent. Research since then has supported Hubal's findings.

Sex and gender aside, variability in size depends on as many factors as variability in strength—factors ranging from your genetics to what you eat for breakfast to the lengths of the levers of your limbs and the sites of your tendons' insertions on your bones. It may even matter what you are *thinking about* when you lift weights. Research has shown that "attentional focus" on moving a load builds more strength, and attentional focus on contracting a muscle may build more size—and this is not the only strange quirk of neuromuscular engagement, as we will see. Variability in size also depends on the interactions among all the above factors, the permutations of which may be almost infinite, especially since the interactions alter with time.

Studies of twins suggest that "up to 90% of the variance in baseline muscle mass is hereditary," according to the textbook *Science and Development of Muscle Hypertrophy* by Brad Schoenfeld, a professor of exercise science at Lehman College in the Bronx who is a foremost researcher on training to build muscle. Yet the influence of genetics on people's potential to add muscle mass, Schoenfeld adds, "declines with advancing age."

There is no simple final answer to the question of which matters more in building strength—muscle size or muscle activation. Sometimes it's one, sometimes it's the other. The only answer that is always true is, *It depends*.

Many of the facts just mentioned were not yet clearly known when Jan Todd began training with Terry. As an experienced coach, however, Terry had a firm grasp on the underlying principle of those facts. It is basically the same principle that Philostratus evoked in the *Gymnasticus* when he wrote of *kairos*—doing the right thing at the right time. Building strength and muscle, as a long-term project, has always involved an intricate, multifactorial play of contingencies that are different for every person.

Respecting that reality, Terry had questions about how he would train Jan over an extended period, because Jan was the first woman he had ever coached. He knew the feats that Katie Sandwina and other strongwomen had performed in the past, but he didn't know what training methods they had used; and he was not well acquainted with any women in the present day who trained with very heavy weights. When Jan asked him questions about her ultimate potential, he said, "I really don't know. I just don't know how you'll react." He added, "The only way we could know is to try."

The most surprising thing they discovered together was Jan's capacity for work. Jan operated on a level beyond anything Terry had seen. "Jan could do more—maybe twice as much work as I could do," Terry said. "Prodigious amounts of work, and still come back. But you never want to give your athlete, if you're trying to help them, more work than they can recover from." In training, they found that Jan performed *better* if she did more work, more often, with more weight. "It was very, very abnormal," compared to other lifters Terry had known.

"There was no grand plan" guiding their improvisation in the gym, Jan remembers. "It was, Let's try this! And how about this!" Later, when Terry and Jan learned about periodized training methods from the Soviet Union, they began following those principles. They organized her training in blocks of time devoted to various goals including hypertrophy, strength, and power.

The first phase, devoted to building muscle mass, was a variation on Thomas DeLorme's proven formula. For most exercises, Jan did a couple of warm-up sets and then completed three sets of ten reps at a target weight that was, as she and Terry liked to say, "tough but achievable."

Her body changed. The transformation was challenging, Todd says, though she thought of it as temporary: "I didn't like when my hips and legs and thighs got larger, especially after I started squatting—but I *understood* it.

"What I was more worried about—it wasn't so much having muscle, but it was—I don't know how to say this exactly—it was more that I would still be perceived as *womanly*, if I can say that. The idea that I self-identify as a woman, and I'm in a relationship with my husband, and I still wore dresses and wore makeup and tried to do things a lot of women do, to say, *Look, we are slightly different here.*"

When she says this, Todd articulates a common challenge for many female athletes. "Physically active women and girls face an intriguing paradox: Western culture emphasizes a feminine ideal body and demeanor that contrasts with an athletic body and demeanor," as the sociologist Vikki Krane and colleagues wrote in 2004, in "Living the Paradox: Female Athletes Negotiate Femininity and Muscularity."

Jan Todd expands on this paradox in detail and with eloquence in "Historical and Social Considerations of Strength Training for Female Athletes," a 2020 paper coauthored with three colleagues: Jason Shurley, Victoria Felkar, and Lindsey Greviskes. The paradox of the female athlete means "participation in sport requires women to negotiate a series of complex contradictions," they write, and they list the contradictions:

> First, to be feminine and an athlete can be perceived as paradoxical. Athletic participation develops and promotes traditionally masculine qualities such as aggression, competitiveness, physical power, and

strength. Second, many sports require muscularity, yet many female athletes feel that they cannot be overly muscular such that they would defy traditional gender norms. Third, many women fear that lifting moderate to heavy weights will make them look "too big" and like female bodybuilders—despite the fact that it takes years of specific, intense training, and possibly drug use, to achieve the physique of a female bodybuilder. Fourth, athletes themselves often define the ideal feminine form in a contradictory fashion as one that is strong but thin, toned but shapely, and fit but sexy.

There has been progress, they observe. "Although athleticism and femininity are no longer viewed as inevitably in conflict, contemporary female athletes continue to feel tensions" between the two, such that muscular female athletes live in two cultures, "that of sport and that of society." Living in two cultures, muscular female athletes face "the challenge of navigating both successfully." Todd faced versions of this challenge with some of her friends and family, the women who worried that lifting would enlarge her waist and endanger her reproductive system.

She faced no such challenge with her husband. Terry Todd's support and example, she says, helped make her transformation possible: "Had I been single, I don't think I would have done it. I really don't. I don't think I would have found enough comfort zone to say, I'm going to gain weight just so I can lift weights." But for those years, she says, she decided, "Okay, I will accept this because these other goals, these other things, are really much more interesting to me."

She adds, "At a certain point, my concerns about my appearance fell away in the face of my own goals."

By 1977, Jan Todd had mastered the squat and the bench press, in addition to the deadlift, and in hundreds of training sessions, she says, Terry had learned to gauge her potential strength on any given day at least as well as she herself could: "He could tell, more than I could, what I should lift," or whether it was better not to lift at all.

About those judgment calls, Terry once said, "You can tell if someone is a bit off. There's something about how the weights are moving, they aren't quite as connected to the will of the person who was calling on them to move." He added, "It's as much an art as it is a science, and art comes in years and many, many days of careful watching."

The couple had moved to Nova Scotia, where Jan taught English at the New Germany Rural High School, and Terry taught at Dalhousie University. They lived on a 185-acre farm where, in their spare time, they grew organic vegetables, put up hay, and played with their English mastiff, Muffin, and her bullmastiff sidekick, known as Blotch. (Muffin and Blotch were the first of twenty-three mastiffs and bullmastiffs that became the couple's loyal companions over the years.) The Todds did not own a tractor or any large piece of motorized farm equipment. They did all their farmwork with draft horses.

Another major commitment, for Jan Todd, was coaching—a way of sharing the knowledge that she had gained in her first years of powerlifting. At the high school where she taught, she started a powerlifting team. She organized bake sales to raise the money to buy weights, and then she stored the weights in the back of her classroom. At the end of the school day, her students would move their desks to make space for the weight benches, and the team of more than twenty, most of them young women, all trained together right there.

But initially, when they started the powerlifting team, there was no competition for the young women to prepare for. There still had never been a dedicated women's powerlifting meet, or a women's bodybuilding meet, or a women's Olympic weightlifting competition.

Through the early 1970s, the few determined women who showed up at men's powerlifting events to compete with their male counterparts risked being subject to harassment. Bullies justified their abuse by claiming they were only following the sport's rules. Jan Todd remembers the trouble: "Because the rules were written for men, and all judges were men, weigh-ins were held in the nude," so men could compete in the lightest possible weight category. "When women began showing up to compete, some judges would not bend this rule, which was a clear form of sexual harassment."

She adds, "Another rule required competitors to wear jock straps, and there was no rule permitting the wearing of a bra until several years later," when Jan and Cindy Reinhoudt, another pioneer of women's powerlifting, drafted the first official rules for women's contests. "Not all the male judges were bullies, of course," Jan says, "but there were men who were not happy to see women trying to share the platform."

Those ugly circumstances were starting to change, incrementally, with bodybuilding as the beachhead for broader awareness of weight training in general. The book *Pumping Iron: The Art and Sport of Bodybuilding* by Charles Gaines, copiously illustrated with photographs by George Butler, had been published in the autumn of 1974, while Jan Todd was training with the goal of setting her first Guinness world record.

The author of *Pumping Iron* had a goal, too. Gaines wanted his book to bring bodybuilders out of what he called the "shadowy corners" of the public imagination, steeped in homophobia. According to *Pumping Iron*, the stereotype of bodybuilders was of "narcissistic coordinatively helpless muscleheads with suspect sexual preferences."

To some extent, *Pumping Iron* succeeded in its mission of humanizing bodybuilders. In Lubbock, Texas, the book review in the *Avalanche-Journal* newspaper reported that "the grotesque forms in the photos become more recognizable as men than as monsters, and their attitudes and motivations are revealed to be, if not conventional, at least human."

By the spring of 1975, *Pumping Iron* became a surprise bestseller—a couple of weeks before Jan Todd earned her first entry in *The Guinness Book of World Records*—and then *Pumping Iron* inspired a documentary film; but the main character in both book and film, Arnold Schwarzenegger, would not truly be a household name for some years yet. In January of 1977, at a New York party to celebrate the movie premiere of *Pumping Iron*, Jacqueline Kennedy Onassis appraised him coolly, telling a newspaper reporter, "Actually I don't think that kind of body is particularly erotic looking."

A few months later, the first powerlifting meet for women finally took place, in Nashua, New Hampshire. Todd says the All American Women's Open Powerlifting Championship drew twenty-seven competitors, and six of them were young women from her high school powerlifting team.

On April 17, 1977, at the YMCA in Nashua, Jan Todd's students watched their teacher win her weight class and become a United States women's powerlifting champion.

In Newfoundland two months later, Todd finally reached—and exceeded—her goal of becoming the first woman to lift a total of 1,000 pounds in the three powerlifts. Her exact total was 1,041.8, tallying the 424.4-pound squat, the 176.4-pound bench press, and the 441-pound deadlift. The squat and the deadlift were also new world records. Still, she ended up with a second-place finish in her weight class at the meet—because she was the only woman in the competition.

That summer, Terry Todd put the final touches on a 101-page book. *Inside Powerlifting* was the first book anyone had published about the sport. The concluding chapter, about the concept of the total, told the story of how Jan reached her 1,000-pound goal.

After describing what training helped Jan Todd do, Terry also wrote about what he thought training helped her to be. People often asked Jan why she trained and what she got out of it, Terry noted, and the question seemed to imply something incongruous about her lifting, as if weakness would be more appropriate for a woman. "Some people still preach that weakness is an aspect of femininity—that there is an iniquity in muscle itself, that there is a virtue in weakness, and a kind of assertive insolence in vigorous health," Terry reflected. "Meanwhile, Jan plans to keep on lifting, teaching through her life that you don't have to be weak to be a woman."

And then in tribute, Terry opened his voice and rang the bells for Jan, and for what Jan's lifting meant:

> Every time she shoulders a 100-pound sack of corn at the feedstore, every time she carries a bag of groceries under each arm, every time she lifts our 150-pound mastiff into the back of the pickup, every time she helps our neighbors do some logging, every time she changes a tire, and every time she goes to the gym and trains, she chips away a little more

of the wall of prejudice that has denied so many women access for so long to a crucial aspect of their human heritage, the aspect of strength. A wise and civilized nation should be concerned with helping men *and* women to attain the strength to endure and the skill to live not only in their bodies but through them. It takes strength to be gentle and compassionate, strength to love and be free. These are not, I know, scientific objectives but they do seem to be worthwhile.

Jan Todd was born to lift—born with rare, sensational genetic gifts. She was made by lifting, too—made for her own pleasure and purposes, which changed as her muscles changed.

Her muscles changed, and watermelons weren't heavy anymore. Her muscles changed, and world records were broken. Her muscles changed, and she came to understand herself differently. As she continued to act from that understanding, she inspired others to begin their own processes of discovery and renewal.

The circumstantial discovery of Todd's singular gifts raises questions about the untapped potentials that exist in all of us. It raises questions, too, about how women who share her ideals of femininity, or her ambivalence or fears concerning body size, can by training discover new dimensions of their own strength.

In the dusk of an evening in the middle of November in 2016, during one of my first visits with Jan Todd in Texas, where she and Terry eventually made their permanent home, she talked about what might hold back other women from discovering their potential strength the way she had discovered hers. She clipped a leash onto one of her dogs, a 150-pound bullmastiff named Monte, and took him for a walk, and I walked with them.

She spoke of the potential for great muscular strength that women have always carried within themselves, and I asked why, even today, many people don't see strength as inherent to women's identity.

"It's like with kids in school," she answered. Her giant dog strained at the leash, leaping at a bird, and Todd braced her shoulder to keep the creature close, not letting his distraction interrupt her: "If you always tell kids they're *stupid, stupid, stupid,* how can they ever end up being anything but stupid?"

Back in the late 1970s, Todd taught her high school students the opposite lesson, and when she took some of them to New Hampshire, to that first powerlifting meet for women, the students showed how well they had learned.

At the meet, a reporter approached the students for an interview and asked this question: *Why would girls want to lift weights?*

After an awkward moment, Jan Todd's students politely answered. They could not think of any reason girls should not be strong.

CHAPTER 5

Big and Small

Jan Todd's first world record, and her first powerlifting championship, launched her on a ten-year odyssey as an athletic pioneer.

On November 7, 1977, *Sports Illustrated* documented the first leg of her journey with an eight-page profile, a career-making story. "The Pleasure of Being the World's Strongest Woman" celebrated the athletic triumphs and social trailblazing of Todd's heavy lifting.

The shocking thing about the magazine's portrayal of Jan Todd was that—by mainstream stereotypes of the time—her strength was in no way shocking. The *Sports Illustrated* reporter Sarah Pileggi wrote that Todd was "just an attractive young woman with a body admirably adapted to its labor. Dress her up in a dirndl and a peasant blouse, give her half a dozen beer steins to hold and she could pass for a model in a Löwenbräu ad."

The magazine received letters for Todd and passed them on to her. Coaches thanked her for setting an example. She remembers one who described his trouble persuading female athletes to do much weight training "because they don't want to turn into Arnold Schwarzenegger. They don't want all those muscles. And now I can show them your picture."

Following *Sports Illustrated*, a lot of other magazines and newspapers made a spectacle of Todd's gender expression, the supposed paradox of her being beastly strong and also conventionally feminine. "I needed to see what the strongest woman in the world looked like," wrote the *Globe and Mail* reporter Nora McCabe. "Was she some kind of gallumphing elephant?" Then the telephone rang—a producer from *The*

Tonight Show—and "Good heavens!" Johnny Carson exclaimed, when Todd set another deadlift world record, on Groundhog Day of 1978, before a TV audience of some 14 million viewers. After she lifted 415 pounds—about half the weight of an Arabian horse—once, twice, and . . . *three times*, Carson was all keyed up, asking her, "How long have you been doing this?"

But when Todd began to answer, he interrupted: "The dumb question, of course, is *why?*"

But before she could speak, Carson interrupted again, answering his own question: "I suppose to be the best in the world at whatever you do," he opined, serene as his necktie's Windsor knot.

Todd gave serious thought to the why of weight training, and she liked to talk about reasons for lifting. Two things above all motivated her training. She wanted to be healthy, and she wanted to practice being the type of person she enjoyed being and had always strived to be: a person who could do hard things.

When she got a word in with Carson, Todd made her points about lifting, health, and character. Training "keeps me in pretty good shape," and it satisfies "a fascination in what is difficult," she said.

Jan Todd's reasons for lifting, as she named them in 1977—to keep fit and healthy, and to practice meeting adversity—may sound somewhat conventional today. Yet in the 1970s, these views were daring. Most people, including medical authorities, did not see weight training as a healthy exercise regimen, but a risky one, not least because so many lifters obviously used drugs that endangered their health. To be sure, Todd's weight training was risky, but in a different way. She bet that she could be a world-class strength athlete without using drugs. Against long odds, she won that bet and walked away with the added prizes of a new vocation and career.

Big muscles can inspire a primal fascination in almost anyone. But in the small world of weight training, during Jan Todd's time as a record-setting athlete, this common fascination was inflamed by esoteric, drug-fueled obsession. By the end of the 1970s, anabolic steroids all but defined muscle in the popular imagination.

Steroids are hormones. As defined by the endocrinologist Hans Selye, a hormone is a "chemical messenger-substance, made by an endocrine gland and secreted into the blood, to regulate and coordinate the functions of distant organs." Some steroids build up the body's tissues, and others tear them down. Steroids that build are anabolic. Steroids that break down are catabolic. When people talk about lifters taking steroids, they mean anabolic steroids.

Anabolics are synthetic versions of testosterone, a hormone that stimulates protein synthesis, including the building and repair of muscle tissue. Testosterone is the primary hormone associated with male sex characteristics; and with vanishingly rare individual exceptions, people who are born male produce much more testosterone than people who are born female. But for everyone, the hormone is so basic to bodily function and to personal behavior that it can be startling to consider how briefly people have known about the stuff.

In nineteenth-century France and Austria, a few scientists collected testicular extracts from guinea pigs, dogs, and bulls; they injected these substances into themselves; and they reported extraordinary effects, including increased strength and mental acuity. Yet chemists did not isolate the testicular hormone until 1935, when they also named it testosterone and learned to make it in a lab.

Easier access to the substance enabled experiments with synthetic testosterone treatments, as chronicled by Terry Todd's historical research. Cancer patients in Pennsylvania took it, and their pains went away. Testosterone was an elixir of youth for the aged and an aphrodisiac for pretty much everybody—although advocates for testosterone treatment focused mainly on how it affected men. In 1945, one enthusiast proclaimed: "It's chemical crutches. It's borrowed manhood. It's borrowed time. But, just the same, it's what makes bulls bulls."

Soon it also became basically what made weightlifters weightlifters, in the sport's top ranks. The Soviet Union had never won a medal in Olympic weightlifting until 1952, when suddenly it won seven. "I know they're taking the hormone stuff to increase their strength," grumbled Bob Hoffman, coach of the USA Olympic weightlifting team; and two years later in Vienna, in a bar one night during the World Weightlifting Championships, a Soviet team doctor fessed up to the U.S. team doctor, who went home to Baltimore and began his own experiments.

Meanwhile, in Southern California, the bodybuilder Bill Pearl, a former Mr. America and Mr. Universe, happened to see some bulked-up cows and horses ("beautiful animals," he would later reminisce) and asked a veterinarian what was different about those creatures. The veterinarian told him about steroids; in 1958 Pearl tried the drugs; and he kept winning competitions through a period of thirteen more years, until he was forty-one years old.

Steroids did not win bodybuilding titles, however, and steroids did not win Olympic medals. Steroids are not magic. People who take these drugs still have to train hard to build muscle and strength. What they don't have to do so much is rest. "The thing in today's world that's underestimated is the importance of recovery," says William Kraemer, a physiologist at the Ohio State University who is among the world's leading researchers on resistance training for athletics and for health. Athletes took anabolic drugs to speed the process of recovery from training. "You could make mistakes in training," Kraemer adds, "and the drugs kind of were the forgiveness factor. That's really what happened with anabolics."

In 1958, the first mass-produced anabolic steroid, Dianabol, was approved for sale by the U.S. Food and Drug Administration. Weightlifters and bodybuilders were not the only athletes who used such drugs. By the 1968 Olympic Games, anabolics were used in training for many kinds of competitions—Tom Waddell, a United States decathlete who was also a medical doctor, later told *The New York Times* that more than a third of his track-and-field team had used the drugs in the pre-Olympic training camp. But weightlifters were notably indiscreet. One American superheavyweight, looking forward to meeting a Soviet rival in the 1972 Munich games, bragged to a reporter, "We'll see which are better—his steroids or mine."

Blood tests to detect the use of some anabolic steroids were invented, and when the first Olympic drug tests were administered at the 1976 games, weightlifters accounted for every positive drug test but one. These tests were, like all drug tests so far devised, flawed and full of loopholes, and athletes determined to get around them have always done so.

As anabolics became more common, so did effective training regimens and more scientific nutritional programs for building muscle. With these developments, the body weight of bodybuilding champions increased—and the average lifespan in the same group decreased. By the 1970s, according to the historian John D. Fair, muscle size had become the most important factor in bodybuilding competitions. Arnold Schwarzenegger's mentor Joe Weider, who produced the most prestigious bodybuilding shows from the 1960s until his death in 2013, declared that "mass should take precedence over all else." Since the seventies, at the highest levels of professional bodybuilding, it has been an article of faith that it's not possible to build the mass required to win without taking anabolic steroids.

As early as the 1930s, scientists observed the potential of anabolic hormone therapies to cause adverse side effects, now known to include infertility, increased cardiovascular disease risk, liver dysfunction, tendinitis and tendon rupture, and sex-specific effects, including enlargement of the clitoris and shrinkage of the testicles. Many of these side effects are long-term consequences, associated with long-term use of anabolic drugs, though; and to many of the most talented young athletes, long-term consequence is abstract or irrelevant.

Drugs did not create this mindset. From ancient times, some elite athletes have valued performance higher than health or even life—as suggested by that Greek tale of Cleobis and Biton, who died after completing a feat of strength to help their mother. But drugs took the question wide, forcing it in many lives where it might not otherwise have come up.

Back in 1975, when the question first found its way to Jan Todd—when the man in Tennessee asked Terry Todd, *Why don't you just put her on some Deca-Durabolin?*—she was bewildered, but shook it off. To her, athletics beckoned as a path to vitality and fun. The other way repelled her.

She and steroids would meet again, but not for a while. Not until she made more progress toward answering a question that has shaped her life since childhood: the question of how strong she can be.

The Guinness Book of World Records took a shine to Jan Todd. She was paid to put on strength shows. As she remembers the towns she played—Las Vegas, London, Tokyo—her blue eyes brighten, and she conjures up a crowd, a stage, and a canvas tent on the north side of Portland, Oregon, at the Multnomah County Fair, in the summer of 1978.

"It felt like I was in the circus," she says. As many circus acts had done, Todd's strongwoman show played twice a day. Onstage, "me and my announcer and sidekick"—Terry—demonstrated the difference between powerlifting and Olympic weightlifting; but barbells did not excite the audience, she says, as much as feats involving "real objects"—elements of everyday life, handled with unexpected degrees of force. With her bare hands, she drove nails into wooden boards, bent a railroad spike, and folded bottle caps in half, before performing her grand finale: The Table Lift.

She invited children to come up onstage and climb onto a platform six feet wide and four feet deep. When the platform was crowded with squirming kids—up to fifteen or sixteen, totaling near 1,000 pounds—she asked them to be still. Standing in front of the platform, she hinged at the hips, flexed her knees to about a 45-degree angle, inched her rear end back, stiffened her trunk, and pressed her back up against the platform, while straightening her legs to raise the big load of little ones—and basked in the roar of the crowd.

After the fair, Todd continued her world-record-setting spree. Since 1975, she had repeatedly bested her own deadlift record. She also held the women's overall world record in the squat—480 pounds—and in the powerlifting total—1,127 pounds.

While traveling around setting records, she told audiences and reporters how lifting changed her. Her comments stand out in newspaper clippings in her scrapbook from these days. She told a paper in the West, "The main thing it has done for me is that I don't have a sense of limitations in my life, that there are things I can't do." She told a paper in the East, "I didn't have great expectations for myself. Now I've lost my fear."

Jan Todd's experience confirmed her reasons for lifting—to be healthy, to build character, and to improve her athletic performance. Yet medical and scientific research to back her up was scanty during Todd's competition years.

That was about to change; and to appreciate the change, it's worth taking time for a close look at some reasons that reputable research about lifting weights was so scarce for so long. The ancient Greek aphorism of Hippocrates that "Everything in excess is opposed to nature," like the ancient Roman argument of Galen that heavy flesh would wreck the body and suffocate the soul, echoed in medical discourse of the nineteenth and twentieth centuries. In 1902, *The Journal of the American Medical Association* published an editorial, "Discrimination in Physical Culture Methods," making what then was a common medical argument that "large muscles in this day and generation serve little purpose indeed." Large muscles not only served little purpose, the editorial said; they could be a kind of ailment or affliction:

> In developing an exaggerated muscular system it is possible to sap the vitality. Men of experience in physical training not rarely see individuals with large muscles well used and yet with poor general health. And large muscles, once obtained, must be lived up to or else, like other unused organs, by their very existence they will render one more prone to disease.

Around the same time, and also following ancient authorities including Galen, doctors and scientists generally affirmed that movement had a kind of sanctity, or an importance verging on sanctity. *A Text Book of Physiology* by Michael Foster, first published in 1877 and then for decades widely considered the definitive work in English on that subject, called muscular movement "the most important means to the chief ends of animal life." Showing muscle's importance, Foster said "the greater part of the animal body is a collection of muscular machines" for producing locomotion, supporting sensation, and enabling voice and speech—but nothing could be moved or sensed or said, Foster added, unless the "will," operating in the brain and through the nervous system, did the job of *"putting this machinery into action."*

By this orthodox scientific view of muscular movement, bodily exercise was central to health. Therefore doctors needed to know about exercise and ought to follow exercise regimens themselves, too. That was the message John B. Hamilton, the second surgeon general of the United States, delivered to an audience of medical students in 1891. "I do not know that I can on this occasion render you a better service," Hamilton said, "than to direct your attention to the necessity of caring for your body not less than for your minds."

Hamilton exhorted future doctors to "set the example" of being "strong, healthy, vigorous," and to that end he related his understanding of ancient, medieval, and modern thought on "systematic training of the body," framing the history of exercise as a resource from which they might build regimens of their own. "The body cannot become a perfectly working machine without culture or training, any more than the mind can become intellectually expert without that drill which we call education," Hamilton said. *The Journal of the American Medical Association* published the text of his speech under this urgent headline: PHYSICAL CULTURE A NECESSITY.

Since the *Journal*'s founding in 1883, exercise and physical culture had been promoted in its pages, occasionally but emphatically. Articles described exercise as an aspect of medical treatment for diabetes, heart disease, and obesity, among other ailments—with "exercise" often meaning light exercise. Yet heavy exertion was examined, too. In 1885, the *Journal* published two reports about M. J. Oertel, a physician in Munich who "tried the effect of mountain climbing in restoring the equilibrium" of the circulatory and respiratory systems and "reducing superfluous fat."

Heavy exertion might in some cases also be associated with longevity, according to a substantial 1892 *Journal* report on "The Influence of Bodily Exercises upon Length of Life" by J. Madison Taylor, a prominent Philadelphia neurologist. Taylor compiled "brief histories of a few old athletes, now in good condition," and some of them were heavy athletes—such as Scotland's all-time most honored sportsman, the wrestler and Highland Games champion Donald Dinnie, then fifty-eight years old. Taylor's article said Donald Dinnie was "still wrestling, lifting and throwing weights."

One surprising fact in the article, a stark anomaly, concerned the muscles of a much older athlete, Major Knox Holmes, an eighty-three-year-old bicyclist in England. In a twelve-week training period, Holmes had "in a striking manner 'developed muscle' in external and internal vasti and rectus muscles" on the front of his thighs.

Muscle-building in an octogenarian was reported by *The Journal of the American Medical Association* in 1892 with a truly arresting statement, the sort of claim that, to later generations, would indicate potential for a scientific paradigm shift: "This refutes physiological doctrine that muscle is not newly developed after three score and ten," J. Madison Taylor wrote.

What were doctors to make of new evidence about muscle-building? One more effect of Galen's legacy, mentioned earlier, may have made it hard for them to know quite what to think. The organ of voluntary movement was now established as a locus of anxiety. Muscle was entrenched in that ancient rhetorical double bind, in which nothing was said to be more important than muscular movement, but almost nothing could be worse than having too much fleshy muscle.

Anxiety about muscle was baked into the culture of many colleges and universities, where medical doctors were in charge of new physical education programs. With broad mandates to improve students' health for the good of society, and ultimately for the glory of God—most universities then professed to be religious institutions—many early physical educators had almost utopian ambitions. Their objectives, according to the historian Roberta J. Park, "came close to embracing total human perfection," on standards owing at least as much to Classical Greece as to Christianity. Many physical educators, as James Whorton has written, "were virtually unable to discuss their work without describing it as nothing less than a crusade to establish within American society the Greek ideal of a sound mind in a sound body."

Some physical educators pursued their Greek ideals by means of anthropometry, the study of bodily proportions. The chief apostle of anthropometry was probably Dudley Allen Sargent, the medical doctor who spent forty years, 1879 to 1919, running Harvard's gymnasium. In his autobiography, Sargent described the physical examination he liked to make of incoming freshmen. He took a medical history; administered some tests of strength; assessed conditions of the spine, muscles, and skin; and took measurements. "Every dimension of the body, apparently significant or insignificant, was given," Sargent wrote, and that was no exaggeration. For each individual student, he recorded more than forty lengths, widths, weights, and girths. After comparing those measurements to averages of what he'd found in other exams, averages from which Sargent had extrapolated sets of ideal proportions, scaled to the individual's height, weight, and age, he "prescribed an order of appropriate exercises, specifying the amount of work and the adjustment of the apparatus used." The apparatus was an elaborate set of thirty-six resistance-exercise "developing machines" of Sargent's own design, involving lots of ropes and pulleys. He made similar examinations of students at the Society for the Collegiate Instruction of Women, later known as Radcliffe College, and he gave them similar prescriptions.

The prescriptions were meant to help students build muscular symmetry, which Sargent believed would make for highly functioning bodies. He knew a college crew team coach who told athletes that "to learn to row was to row," but this training philosophy, Sargent observed, produced "unbalanced development" in the oarsmen. Not only in athletics, but in all activities of life, Sargent "found men accustomed to some special form of labor over-developed in the muscles which they used, and under-developed in those which they did not call into play"—an observation that foreshadows Charles Stocking's concept of the sport-specific paradox.

Sargent came to think "the over-development of any faculty meant the dwarfing of another." He thought imbalance should be avoided because he believed the body's many faculties, as well as the whole body and the mind, were all thoroughly integrated. In his words:

> The object of muscular exercise is not to develop muscle only, but to increase the functional capacity of the organs of respiration, circulation and nutrition; not to gain in physical endurance merely, but to augment the working power of the brain; not to attain bodily health and beauty alone, but to break up morbid mental tendencies, to dispel the gloomy shadows of despondency, and to insure serenity of spirit.

For many years, Sargent tirelessly repeated this holistic, perfectionistic philosophy of physical education, both in his work with Harvard and Radcliffe undergraduates and with students at the normal school he founded, where thousands of people from all over the world—China, England, France, Japan—studied to become physical educators.

Yet symmetry always turned Sargent's head; and as a perfectionist, he may never have been more pleased by any body than by Eugen Sandow, the Prussian strongman whose fame was built on the premise of being a Greek ideal made flesh, like a classical statue come to life. In 1893, Sargent examined and measured Sandow in the many ways he often examined and measured college freshmen, and afterwards told newspaper reporters, "Sandow is the most wonderful specimen of man I have ever seen," a man with "the characteristics of Apollo, Hercules, and the ideal athlete." Later, a life-size sculpture of Sandow was installed in Harvard's gym, to exemplify the symmetrical type of physique Sargent hoped undergraduates would build.

Not all doctors were as ensorcelled by Eugen Sandow as Dudley Allen Sargent. A few months after Sargent examined Sandow, *The Journal of the American Medical Association* sent a Chicago physician, G. Frank Lydston, to meet the strongman. The doctor admitted he was "compelled to confess that I have rarely examined so perfect a type of good health as this man presents." Regardless, he thought Sandow's example should be a warning, not an inspiration. "Athletics for health, should be the motto of the man who trains. Athletics for big muscles and competitive feats of strength and endurance are pernicious, illogical and dangerous," Lydston wrote. And he recited a litany of dangers of "athletics for big muscles," including heart failure, stroke, and emphysema, repeating rhetoric that had circulated at least since Galen railed in Rome against the supposedly paralyzing, crippling effects of athletic training to build mass.

On some level, G. Frank Lydston must really have believed what he wrote about big muscles, and so must have many of his colleagues who read the *Journal*. Yet all these doctors were at least as eager to position themselves on what they thought was the winning side of an argument as they were to find and tell the truth about training to build strength and muscle. Moreover, some of what those orthodox modern doctors cited as dependable, ancient knowledge—such as that litany of the dangers of big muscles—may originally have been articulated not as propositional truths but as rhetorical dirty tricks.

This much is certain: Toward the end of the nineteenth century, practically everybody whose job had anything to do with health and strength was in the business of slinging hype, even if much of the hype was in earnest. On paper, the conversation among doctors and physical educators about medicine and gymnastics and muscle, including the never-ending gabble about how to build enough muscle without building too much, quivers with tail-chasing redundancy that can make it exhausting to read. Sometimes all the anxiety about minds and bodies and measurements must have made everyone wish that someone would just let rip with a joke and have done with it—and in 1895, Mark Twain was pleased to oblige.

Twain had made mistakes with money and gone bankrupt; and to pay his creditors he set out on a round-the-world lecture tour, at the age of sixty and in poor health. But after the first leg of the tour—twenty-three lectures in twenty-two cities—he was feeling better, and he wrote about that in the *San Francisco Examiner*:

> Lecturing is gymnastics, chest-expander, medicine, mind healer, blues destroyer, all in one. I am twice as well as I was when I started out. I have gained nine pounds in twenty-eight days, and expect to weigh six hundred before January. I haven't had a blue day in all the twenty-eight.

My wife and daughter are accumulating health and strength and flesh nearly as fast as I am. When we reach home two years hence, we think we can exhibit as freaks.

Eugen Sandow, by contrast, could not have been much more industriously serious about the business of physical culture. He was the first performing strongman to build himself into something like a multi-platform, vertically integrated lifestyle brand—Sandow's career is the only near precedent for Arnold Schwarzenegger's. Among Sandow's many enterprises, one stands out for actively promoting exercise as a treatment or cure for common afflictions.

In London, where Sandow spent most of his adult life, he opened Sandow's Curative Institute in 1897, offering services "invaluable to sufferers from Insomnia; Neurotic Persons, or those whose nervous system is run down; all who desire a graceful carriage and symmetrical form; those whose sedentary occupations have caused sluggish liver or constipation; athletes in every brand of Sport." Serving a clientele of men and women, Sandow's Curative Institute was the flagship of a chain that had two more locations in London, and one each in Manchester and Liverpool. As the historian Dominic G. Morais has written, in a study of Sandow's marketing techniques, the apparatus of healing at Sandow's Curative Institutes "consisted solely of dumbbells."

By the turn of the century, some physicians felt the ground of their field shifting. Exemplars of physical culture such as Sandow were encroaching on their turf, gaining status as authorities on health. Rumblings of the shift were registered in headlines of *The Journal of the American Medical Association*.

"Physical Culture and Medicine" was the title of another *Journal* editorial, published in 1901; and this one in retrospect seems to mark a watershed. Exercise and medicine were parting ways. "At the present time perhaps more than at any former period in our modern civilization the public has grasped the idea of physical culture and its advantages," the editorial began. "The notion that symmetrical development of the body is a safeguard against disease and a cure for existing disorders is widespread," and "is being exploited in all quarters as the grand cure-all to the disparagement of drugs and doctors." For the damage to medicine's authority on healthy regimens, the editorial said, doctors had no one but themselves to blame: "while we have prescribed exercise, massage, etc., in a general way we have left the details and even the oversight of them to others as unworthy of our attention."

Now it was "high time," the editorial continued, "that the medical profession gave more attention to the physical education and development of the patients," to disabuse patients of "the notion that the practice of medicine is merely the giving of drugs." Doctors should be prescribing every method of physical training that is "scientifically and therapeutically of value," and should assist patients in finding qualified help to administer those prescriptions. Also of the essence, "there should be thorough instruction given in all departments of physical culture in medical schools."

Yet medical schools and the medical profession as a whole made none of these changes—most of which probably seemed impossible due to another transformation, which had been occurring in the realm of physical culture.

The very meaning of the term *physical culture* had changed, and the shift had driven a wedge between exercise and medicine, in the ten years between 1892, when the *Journal* published the headline PHYSICAL CULTURE A NECESSITY, and 1902, when it published "Discrimination in Physical Culture Methods," the piece that said "large muscles in this day and generation serve little purpose indeed."

Earlier, the term *physical culture* had been a common noun phrase, a generic and somewhat dignifying umbrella term for exercise, massage, and other practices and regimens for health guided by people with or without medical degrees. Now the generic term had been rewired such that, in the public imagination, *physical culture* increasingly registered as a proper noun, the title of a popular magazine that had been published in New York since 1899. *Physical Culture* magazine was America's first successful mass-market fitness publication written for ordinary people; and *Physical Culture* magazine probably was the trigger for *The Journal of the American Medical Association* to urge "discrimination in physical culture methods."

That editorial lamented "expensive advertising which is done nowadays by the propagators of various physical culture systems," beginning with a "get-strong-quick" phase, hooking people "on the task of building their muscular system after the model of the athletes" pictured in those magazine ads. This was only the beginning of a bitter, protracted medical controversy about muscle-building and about the salutary powers of exercise and diet more broadly that *Physical Culture* magazine provoked.

Much later, in 1950, after hostilities died down a bit, *The New Yorker* magazine published a three-part profile of the editor and publisher of *Physical Culture*, Bernarr Macfadden. The profile said Macfadden was "the recipient of as much organized abuse as any man in the history of the country," and then listed Macfadden's "liveliest tormentors." The American Medical Association was first on the list.

Bernarr Macfadden was a self-educated health reformer, born dirt poor in rural Missouri in 1868. In his mid-twenties, at the 1893 World's Columbian Exposition in Chicago, he saw Eugen Sandow perform onstage, and it changed his life. Inspired by Sandow's resemblance to Greek statues, Macfadden trained in hopes of building that kind of body. In the process, he patched together an approach to exercise, diet, and health that in due course made him much richer than Sandow, even if, in most people's eyes, he never would be as beautiful.

Macfadden believed, in the summary of historian Donald J. Mrozek, "all ailments were really variations on the one and only disease besetting humankind, impurity of the blood. Specific ailments which doctors called diseases were Nature's way of removing the impurities. Thus, Macfadden railed against the 'medical trust' and opposed the use of drugs. Even the term he gave to his method of cure—'physcultopathy,' or healing by use of physical culture—hinted at his distance from conventional medicine." In short, as the *New Yorker* profile said, Macfadden taught that "anybody could cure himself of anything by exercise and diet," making doctors unnecessary.

It's impossible to know how many lives were damaged or destroyed when readers emulated Macfadden's fanatical antipathy to doctors. In 1925, the coroner of Lake County, Indiana, reported that a hotel cook, twenty-eight years old, who "was not ill but thought he was a little too heavy," had started fasting after meeting Macfadden and buying a copy of his book *Fasting for Health*. Two months after the young man stopped eating, he was hospitalized, and he "died from starvation by fasting." Many others, though, did not follow Macfadden's teachings so strictly but drew from them more flexible guidelines, in the spirit of *Physical Culture* magazine's larger message.

Physical Culture's message was that readers could change their lives by changing their bodies—and survival and success would be the outcomes of self-reliant health. Macfadden's unsystematic system of physical culture promoted, in addition to fasting, walking, fresh air, and abstinence from alcohol and tobacco—but not from sex. Macfadden had such a candid, open, positive view of sex that *Time* magazine sometimes called him by the nickname "Body-Love."

Muscle was a keynote of Macfadden's sprawling message. The term *physical culture* "means muscular strength," Macfadden taught. "Strength or power of some kind is what runs the human machine," he wrote, adding that the person "who is looking for health, but does not want muscles, will search in vain."

The magazine had a tagline, "Weakness a Crime / Don't Be a Criminal," that hit home with readers; and Macfadden leveraged the success of *Physical Culture* to build a publishing empire—eventually more than three dozen magazines and newspapers, plus a conglomerate of other businesses, including restaurants, schools, hotels, and sanitariums; and he produced a series of Physical Culture Exhibitions, which included physique competitions, at New York's Madison Square Garden. (Charles Atlas got his start at one of these events, by winning the title of "World's Most Perfectly Developed Man.")

Macfadden also dabbled in politics, both national and international, apparently unconstrained by ideology. He was involved with Franklin Delano Roosevelt, and also with Benito Mussolini, both in the same year. Consistency was not of much concern in his publications, either. Popular science and fringe quackery, determinist eugenics and progressive self-empowerment: Macfadden printed it all, in articles by writers including the muckraker Upton Sinclair, the millionaire Cornelius Vanderbilt II, the feminist Charlotte Perkins Gilman, the boxer John L. Sullivan, Harvard University president Charles Eliot, and Nobel Prize–winning playwright George Bernard Shaw, as well as Winston Churchill and Mahatma Gandhi.

Macfadden persuaded an enormous number of people, including giants in their fields, to provide and find positive value in his publications. He convinced them all, in one sense or another, to buy what he was selling.

The rise of Bernarr Macfadden and of his *Physical Culture* magazine happened in a period when the medical profession was wracked by struggles for legitimacy. In the United States, these struggles played out simultaneously on several fronts, involving among other matters the public licensing of doctors, certification of drugs and treatments, and regulation of medical education and training.

On the education front, some of today's prerequisites for practicing medicine were then radical innovations. In 1893, the year Macfadden first laid eyes on Eugen Sandow, a new medical school opened in Baltimore, Maryland, at Johns Hopkins University, the first medical school ever to require all entering students to have college degrees. Johns Hopkins was also the first medical school where science and research were integrated with clinical practice in a hospital setting.

Hopkins remained exceptional, however, and the general state of medical education was chaotic until the American Medical Association and the Carnegie Foundation issued a report in 1910 that branded 120 of the country's 155 medical schools "worse than useless" and said they should be closed. Then medical training changed. The process of becoming a doctor grew lengthier, more challenging, and more expensive; and practicing physicians became more effective, thanks also to scientific breakthroughs in bacteriology, immunology, and pharmaceuticals.

By the 1920s, the American Medical Association would spare no effort to protect the prestige of the new class of highly educated, strictly licensed medical doctors. Government officials also took steps to protect doctors, because of medicine's power to improve public health. As the historian Paul Starr has written, scientific medicine and government together "took up the war against 'quackery' as a great cause."

The Journal of the American Medical Association published frontline reports about that war, as did the AMA's magazine for general readers, *Hygeia*, named for the ancient Greek goddess of health. In 1925, the editor of both publications, Morris Fishbein, assessed the opposition in a book, *The Medical Follies*. Fishbein ridiculed Macfadden and his ilk—"The Big Muscle Boys," he called them.

"Turning the pages of a periodical like *Physical Culture*, one might become impressed with the notion that the chief goal of man is muscle," for the pointless exertion of "strength for strength's sake," Fishbein wrote. "And not just ordinary strength, but the kind of strength that bends crowbars between the teeth," he added, "while giving huge grunts to the accompaniment of an orchestra."

After the mid-1920s, according to one of Macfadden's biographers, Robert Adams, "the AMA seemed to realize that lavishing negative publicity on Macfadden was like pouring kerosene on a campfire, and adopted a policy of public silence." Yet in the pages of the *Journal*, animosity still flickered sometimes, especially when Macfadden suggested changing the health care system.

In 1936, Macfadden wrote in one of his magazines that "healers, whether osteopaths, chiropractors, or whatever they be—should be paid by the government." He also speculated that if market competition were eliminated from health care, "all would work together for the purpose of making the patient healthy and strong." The *Journal* answered with an editorial: "O.K., Mr. Bernarr Macfadden! Why not use part of the magnificent Macfadden fortune, erected by the sale of physical culture hokum to the suckers who purchased the Macfadden courses, and the 'dumbbells' who purchased the dumb-bells," to provide free health care for everyone. "Maybe, after all, Mr. Macfadden, muscle and not mind or morals is your special field!"

In 1939, Macfadden again argued that "doctors of all kinds should be paid by the government," prompting another *Journal* editorial. This one noted that, coincidentally, "the Medical Section of the Communist Party" recently had urged all American communists "to demand salaries for all physicians who have given freely of their service in the past to public institutions and for the care of the indigent sick."

In 1940, Macfadden made a suggestion about military medical care. He wanted the U.S. Army to let practitioners of alternative medicine treat soldiers, and to measure the effectiveness of their techniques for comparison with standard medical treatments. Again the *Journal* shot back. In one more anti-Macfadden editorial, the year before the United States entered World War II, the *Journal* exulted that layers of right-minded authority would protect the Army troops from unscientific treatments: "Fortunately for the American soldier the medical corps is under the direction of competent officials, and the work of that department is, in turn, controlled by the general staff."

Macfadden's nemesis Morris Fishbein was fired from his job as editor of *The Journal of the American Medical Association* in 1949; and after that, for a brief time, the *Journal* published some articles about weight training for health. The first one, in 1950, concerned another of those astounding quirks of the neuromuscular system: cross-education.

The name *cross-education* refers to the phenomenon in which, as the *Journal* article stated, "vigorous unilateral exercise elicits associated movements or increased tone in symmetrically disposed muscle groups." Because of cross-education, if you have a musculoskeletal injury on one side of your body, and you continue progressive resistance training with the corresponding muscles on the other side of your body, you may substantially preserve the muscles and their powers on the injured side. After mentioning some promising potential applications of this discovery for rehabilitating people with, for example, hemiplegia— paralysis on one side—the *Journal* item concluded: "More research in this field certainly seems indicated."

A few months later, the *Journal* published a detailed report on progressive resistance exercise as a "form of treatment," in this case for polio. The neurologist Sedgwick Mead summarized the findings of Thomas DeLorme and colleagues. Mead also described how, in his own practice, he had applied DeLorme's technique, and he wrote: "I believe it is by far the most important contribution yet made to the treatment of this disease."

The next year, the *Journal* published a positive review of the book *Progressive Resistance Exercise*, about DeLorme's weight training prescriptions. And by the end of the 1950s, the *Journal* would publish two more articles about progressive resistance training for health. But this cluster of articles was like a cloudburst in a desert. After 1959, as far as I have been able to find, the *Journal* made no more mention of the technique for almost forty years.

A long history of medical controversy about athletic training, which had culminated in clashes between Macfadden and the AMA, meant that by the middle of the twentieth century, there were not many credible high-profile physicians who considered physical culture a necessity, much less weight training.

When weight training registered in scientific and medical discourse, in those decades, it was often dismissed as callow and jejune. "The waddling gait and breathlessness of a muscle-bound weight-lifter are salutary warnings of the danger of over-specialisation" in athletic training, "and of what occurs when muscular development has been carried to excess," the British neurologist Roger Bannister wrote in 1955—not long after he became the first person to run a mile in less than four minutes.

Little was generally known about how muscular development affected health, because cardiovascular concerns took up the cultural oxygen. Doctors who dismissed "the waddling gait and breathlessness of a muscle-bound weightlifter" may have been churlish, but they were not *only* being churlish. They were trying to relieve a medical and social crisis with life-and-death stakes. The cardiologist Eugene Braunwald, a professor at Harvard Medical School, has written that acute myocardial infarction—heart attack—"was described as a clinical pathologic entity only as recently as 1912," but within a few decades was acknowledged as "the single most common cause of death in the United States." By then, "cardiovascular disease accounted for more than half of all deaths, not only in the United States, but also in the remainder of the industrialized world." Therefore, addressing the crisis of cardiovascular disease was the top research priority for medically oriented exercise science in that era.

During the 1960s, in many of the world's wealthier countries, the tide began to turn in the fight against cardiovascular disease. Hospitals created coronary care units; new technologies for treatment, including portable automatic defibrillators, were invented; and the number of deaths caused by heart attacks declined. As care and treatment changed, physicians and public health experts put more energy into preventing heart disease and promoting health. In 1974, a report by the Canadian government minister of health and welfare, Marc Lalonde, drew global attention to lifestyle factors that affect heart disease. The main lifestyle factors discussed in the Lalonde report were drinking, smoking, drugs, diet, and exercise.

The most popular book about exercise through most of the 1970s was *Aerobics* by Kenneth H. Cooper, who was then a physician in the United States Air Force. After its first edition in 1968, *Aerobics* was translated into forty-one languages (and braille) and repackaged in sequels: *The New Aerobics* (1970), *Aerobics for Women* (1972), *The Aerobics Way* (1977), *The Aerobics Program for Total Well-Being* (1982), and so forth. Cooper's nineteen books have reportedly sold more than 30 million copies.

Aerobics promoted a scientific rationale for exercise that builds cardiovascular fitness, and the book belittled the "muscular fitness" of weight training for how it "concentrates on only one system in the body, one of the least important ones." As mentioned earlier, *Aerobics* was the book that said lifting weights was "like putting a lovely new coat of paint on an automobile that really needs an engine overhaul."

By the time those words were published, the message that weight training was superficial and secondary was common sense, the obvious deduction any rational person would have made, based on muscle's checkered reputation. The cardiovascular disease crisis provided a virtuous practical justification for ancient medical prejudice against athletes with big, meaty bodies, and prejudice against the kind of training that was associated with that kind of body. Yet the prejudice was still only prejudice, now more insidious than ever.

Scientific and medical publications about weight training, including those by sports medicine doctors, were too often inept. The first official statement about weight training by the largest group of self-identified sports medicine specialists concerned the use of anabolic steroids—and the statement concluded, absurdly, that the drugs had no effect on athletic performance. This was in 1977, when many elite athletes had been flaunting the performance-enhancing effects of anabolics for well over twenty years.

Sales of *Aerobics* notwithstanding, and outside the small world of highly educated people who paid attention to things like Canadian government reports on health, even cardiovascular exercise was not very widely practiced, valued, or understood in the 1970s.

Results of the first study to show a connection between vigorous physical activity and health, a study of London transport workers, had begun to be published in 1953. On London's iconic red double-decker buses, the drivers, who spent all day sitting, were twice as likely to die of heart disease as the ticket-takers, who spent all day walking up and down stairs.

Through the 1960s, more data confirmed the value of physical activity for cardiovascular health, says William L. Haskell, emeritus professor of cardiovascular medicine at Stanford University School of Medicine. The story of that research, as Haskell tells it, starts in 1964, when the first English-language physical activity questionnaire combined with direct observation of research subjects showed that more active people are less likely to develop heart disease—even if they smoke cigarettes. Clearly, the study of exercise had the potential to protect public health.

To be taken more seriously, the field of exercise research needed a makeover, according to Michael J. Joyner, a Mayo Clinic professor of anesthesiology who runs an exercise physiology lab there. University physical education departments, in which exercise research typically took place, were rebaptized with more scientific-sounding names. They became departments of exercise science, movement science, exercise physiology, and kinesiology.

As the names changed, much else did, too, in a sweeping process of educational reform, somewhat comparable to the nineteenth-century medical education reforms mentioned earlier. In the United States, in the 1960s and '70s, coordinated efforts of the Big Ten and Pacific Coast Conferences and Pennsylvania State University to reform their physical education departments set standards for others to follow. The new departments had more stringent entrance requirements and tougher curricula, emphasizing the principles of biochemistry.

Since the study of exercise also had to pull its weight in the economy of higher education, Michael Joyner adds, some of these academic departments were reoriented to support the NCAA sports teams and to win the government and corporate research grant money that became financially indispensable to universities. Much scientific knowledge about muscle atrophy, for instance, comes from research funded by the National Aeronautics and Space Administration, to prepare astronauts to operate in zero gravity.

Yet for people who didn't play sports and never rode in spaceships and wondered if they should exercise—and if so what kind and how much—there was no solid consensus medical guidance about what to do.

Some of that uncertainty resolved in October of 1978, when the American College of Sports Medicine published its "Position Statement on the Recommended Quantity and Quality of Exercise for Developing and Maintaining Fitness in Healthy Adults." The document suggested that healthy adults should do 15 to 60 minutes of aerobic activity three to five times a week, at medium to high intensity. This formula became the standard answer for anyone who wanted to know how much exercise is enough; and it became the basic template for exercise guidelines in many countries.

Weight training was not recommended by the 1978 document, and was mentioned only in a brief, two-sentence paragraph, which said that lifting weights had no effect on aerobic fitness—which later proved to be untrue.

The medical doctors and physiologists who wrote these early exercise guidelines were part of the same group who had denied that anabolic steroids affected athletic performance. They wrote little about weight training because they knew little about weight training. At this point, not one powerful voice in any leading scientific or medical institution was publicly asking if the average person might have good reasons for lifting.

Even to raise that question, weight training needed some leverage. And so another struggle for legitimacy began.

"They had to create it as a science," Jan Todd says. "We had good ole boys and practitioners and knowledgeable folks, but there was no science of strength. What was happening in the late seventies and early eighties was, we were trying to invent science for strength."

In those days, the value of weight training was widely questioned even in athletics. Many coaches took a dim view of lifting, or even opposed it, on the basis of stubborn misconceptions. Voicing some of those skepticisms about heavy training that dated back to ancient Rome, they claimed that lifting weights used up energy that athletes should save for competitions. They claimed that lifting put an athlete's heart under excessive strain. They claimed that building muscle reduced flexibility—the myth that lifting weights makes people "musclebound." Even more problematic were their claims that the development of muscles in women might turn them into men or turn heterosexual women into lesbians.

A minority of coaches had a different experience of lifting. At the University of Nebraska in Lincoln, for example, assistant football coach Boyd Epley organized systematic weight training programs for the Cornhuskers, and the team won more games. When Epley invited like-minded coaches to Lincoln in July of 1978, about seventy-five men showed up. To promote the use of strength training in athletics, they formed a group, the National Strength Coaches' Association, and they started putting out a newsletter.

Jan Todd says the newsletter was "working through old mythologies," publishing articles "trying to justify squatting," for instance, "and letting people know that squatting will not hurt you," when done properly. Over a couple of years, the rough-hewn newsletter built a following; and in the 1980s, this group—now known as the

National Strength and Conditioning Association—began to run a new leg in the relay race of physical culture. Taking the baton from older magazines such as Bernarr Macfadden's *Physical Culture*; and *Strength and Health*, which inspired Thomas DeLorme's interest in strength training; and *Muscle Builder*, which helped launch Arnold Schwarzenegger's career, the group of strength coaches turned their newsletter into a peer-reviewed academic journal.

The National Strength and Conditioning Association Journal published increasingly reputable scholarship, starting with collaborations between strength coaches and their university colleagues in those burgeoning exercise science departments. They tested old training practices and reformulated them in scientific terms. They turned a set of traditional practices into a more systematic body of knowledge.

Among the most valuable publications were a series of position papers modeled after the American College of Sports Medicine's position statements, reviewing and summarizing research about aspects of weight training that were controversial or poorly understood, such as strength training for women.

"Strength Training for Female Athletes: A Position Paper" was published in two parts in 1989. The paper's purpose was to answer a question that had been unanswerable when Jan Todd started training with her husband Terry in the early 1970s: Should males and females lift weights in different ways?

A committee of six women searched for answers to that question. Jan Todd was one of them. The others were Meg Ritchie, who as head strength coach at the University of Arizona was the first woman to hold that position in the NCAA's Division I; Lori Gilstrap, a part-time strength coach at Georgia Tech; Lynne Stoessel and Denise Gaiter, both graduate students in exercise physiology; and Jean Barrett Holloway, who taught strength training at UCLA Extension School and had conducted the first study of heavy weight training and self-efficacy in adolescent girls.

Research gathered by this committee contradicted and corrected common assumptions about female muscle. The research showed that women's relative strength (relative to body composition) was similar to men's; women's relative potential for hypertrophy (relative to initial muscle mass) was the same as men's; and normal menstruation did not impede athletic performance. Such evidence led to an all-embracing conclusion. "Due to similar physiological responses," the committee wrote, "it appears that males and females should train in the same basic way," using similar programs and exercises.

Almost thirty years later, when Jan Todd collaborated with the sport historian Jason Shurley on a review of women's strength training research since 1989, they found the position paper's major findings and main conclusion had held firm.

By this time, athletics, medicine, and physiology had all developed distinct branches of research on weight training as it affects each field's central concern: sport performance; health, disease, and physical function; and biology. But the trunk of that tree is athletic weight training, since the science of lifting was invented by people who liked doing it.

The tree's root system is intricate, broad, and deep, reaching to sources as far-flung as Romantic philosophy, archaeology, and the circus. Another of Jan Todd's big lifts, at least as impressive as her world-record-setting feats, is her historical research to excavate some of those roots and map their connections—work that was inspired as accidentally as her barbell training.

———

On Memorial Day weekend in 1979, Terry Todd asked Jan to do him a favor. Terry purchased a collection of books, letters, documents, posters, photographs, and objects pertaining to the history of physical culture—a gargantuan private archive built up over seventy years by a former circus strongman, Ottley Coulter, who lived in rural Pennsylvania—and Terry wanted Jan to retrieve the purchase for him, which would mean packing the whole collection into almost four hundred boxes.

"When packing happens," Jan quips, "Terry has other places he needs to be"—but she didn't mind running the errand; and when she arrived in Pennsylvania, she worked fast.

She was less focused on the contents of the boxes than on her next public appearance, the next week in Dayton, Ohio, where she would set another world record, this for the "man-woman deadlift." Todd gripped one side of a barbell, the men's powerlifting world champion Larry Pacifico gripped the other side, and the shaft of iron sagged as, together, they pulled 1,105 pounds.

After that, months passed before she took a peek into a few of the boxes she had packed up in Pennsylvania; and when she did, she was surprised. She was surprised to find some help that she needed, in the pages of a book. *Les Rois de la Force* (*The Kings of Strength*) by Edmond Desbonnet, published in 1911, is a lavishly illustrated "history of all strongmen from ancient times to the present day," a pictorial roll call of "kings" that includes both male and female strength athletes and performers. Turning the pages, Todd came to a photograph of Katie Sandwina, the performing strongwoman she had heard about from Terry on the day she set the goal of setting her first world record.

Kathi Brumbach dite Sandwina et son mari

The picture of Sandwina in this book struck her. Even after all Todd had accomplished, even though *The Guinness Book of World Records* had been calling her the "strongest woman in the world" for the last five years, she still craved the reassurance of an example, a sense of being less alone in this one aspect of lifting: adapting her ideals of beauty to the muscular requirements of her athletic life.

Todd's experience of being a woman with big muscles still rankled sometimes, she says, partly because of the unfairness, or the unevenness, of an aesthetic standard. "When a man changes his body, if a football player or a track athlete puts on muscle, that's not considered a negative," she tells me. "Guys don't agonize as much about how their bodies adapt to sport as women did and probably still do. And the culprit has always been muscle. If you build a body with more muscle, then people look at you with suspicion."

Versions of this quandary grew more common in the twenty-first century as weight training grew more common among female athletes. Sociologists have found that "sportswomen live a paradox of dual and dueling identities," and as Jan Todd and her colleagues wrote in 2020, many female athletes "perceive their bodies differently, depending on the context."

This phenomenon has been documented in many sports and settings, including soccer in New Zealand, wrestling in Norway, and track and field in Canada. In 2005, after two years of ethnographic research as a participant observer with a women's soccer team at a private university in the Rocky Mountains, Molly George concluded that "a woman athlete literally and figuratively embodies two physiques, the performance body and the appearance body."

The report of her research, "Making Sense of Muscle," noted that many of the college athletes thought they were less attractive with larger muscles, so they chose, at times, to subvert their own weight training. They did extra aerobic exercise, skipped lifting sessions, or lifted lighter weights, in hopes of slowing or counteracting muscle growth. Off the field, to conceal their muscles, they wore baggy clothes or avoided wearing shorts. "My legs bulge out and it makes me so self-conscious," one player told George.

Other members of the team "seemed unafraid of muscle and intentionally pushed the constructed boundaries set on size," George wrote. "These women were more concerned with their performance in competition and devoted time to maximizing their muscle through weightlifting." One such woman, "an impressively tall and heavy forward, used her physique to hold the ball under pressure and knock other players down. Because of this, she received the nickname 'Big Bird' and a reputation for being one of the most feared players on the team."

The player's name was Sharon, and among her teammates "she was adored for her size, precisely because it helped the team win games." Asked how she felt about being called Big Bird, Sharon told the sociologist that it "didn't bother her and that she was proud of what her body allowed her to do on the field."

Sharon's perspective sounds like the view that, a couple of generations earlier, Jan Todd ultimately took of her own muscles—the shift Todd made when her concerns about appearance, as she put it, "fell away in the face of my own goals." Similar transformations of bodies and of beliefs about bodies happen among female athletes in cross-country, volleyball, gymnastics, swimming, basketball, tennis, rugby, and ice hockey, according to Vikki Krane and her colleagues, who wrote that they "found as much pride as consternation in athletic bodies." The sociologists added that transformation happens in different ways for different people doing different things: "In general, the tennis players were more concerned with portraying a traditionally feminine appearance and the rugby and hockey players most pushed the boundaries of femininity."

The lift that probably built the most muscle for Jan Todd was also the lift in which she moved the most weight, the deep knee bend called the squat.

Terry Todd called the squat "the single greatest exercise in the world for gaining bodyweight" and "a cornerstone exercise in any decent program of weight training for athletics," because it stimulates the body's biggest, strongest muscles. "It works the hips, thighs, and lower back *as a unit* and, because of this, it is indispensable as a developer of athletic power."

Squatting is basic to ambulatory power, too. "Done no doubt by people all over the world as a daily part of their lives once they began walking upright," as Terry noted, the squat movement has two parts, descent and ascent. When you sit down in a chair, you make the descent. When you stand up from that chair, you make the ascent.

Jan Todd never really thought about the practical usefulness of squatting, though. She liked squats because she was the best at squatting. "They were the biggest pounds I lifted," she says. "I was a hundred pounds more than any woman in America, or any woman we knew of. That was pretty cool."

When Todd talks about her athletic achievements, she also talks about anabolic drugs and her choice to abstain from using them.

At the top women's powerlifting meets in the late 1970s and early 1980s, she remembers, so many competitors used anabolics that, from one year to another, they were liable to look like wholly different people.

Individuals' choices to take steroids had consequences for others, too. As female lifters' bodies changed—"they stopped looking like slightly larger gymnasts," Todd says—so did the image of women's lifting. At one of the first and last women's powerlifting meets covered by a major television network, Todd remembers an NBC sports announcer telling her, "This is just going to be really hard to sell in Iowa."

Even in the 1970s, image mattered less to Todd than the effects of the anabolic drugs on health. She avoided the drugs because, at that time, she hoped to have children, and she was concerned that "you could have children who would be different because of something you were taking."

If powerlifting had been her whole identity, she says, she might have made different choices. She knew that anabolics could have helped her lift even heavier weights. But she preferred another strategy.

She says, "My only hope was to fight the drugstore with the grocery store."

In December 1980, Todd started a new training cycle. If all went according to plan, this cycle would bring her to peak strength in the first week of February, the highlight of the year for women's powerlifting, the week of the national championship—where, it seemed reasonable to expect, she could set another world record in the squat, a record that might stand for a while. And that would be her last day as a heavyweight.

In seven years of lifting, Todd had gained a total of about 65 pounds. Along the way, she made peace with the weight gain, but a struggling peace. "I'm not going to think about body weight," she would tell herself, "because I do this for a specific purpose. To see how strong I could be." Now she was ready for a change. She decided, "I had carried the weight long enough."

She chose not to compete during this training cycle, to keep her powder dry for the championship. Then on impulse, the week before nationals, she entered a minor-league regional powerlifting competition, the West Georgia Invitational.

On the morning of January 31, 1981, Todd walked into the event at a gymnasium in Columbus, Georgia, and found a quiet place to prepare herself. She wrapped her knees with elastic bandages for support, while Terry and a friend rallied her confidence: *You've got this. Don't worry. Think of what you've done in training.*

She needed the pep talk. Being the best at heavy squats didn't mean Jan Todd didn't dread them. She says she dreaded the physical danger, "the weight being on top of you," together with what she calls the "emotional burden" of the feat. "You have to be stupid enough not to be afraid of it," she jokes. "There are *reasons* to be afraid." If she moved with poor form, she could lose control of the weight, pitch forward or backward. In competitions, she had seen those things happen to powerlifters, a few times resulting in awful injuries: snapped hamstrings, detached kneecaps.

Squatting also made her vulnerable to the crowd's gaze. "No one looks *good* squatting five hundred pounds," she says. "Your eyes pop, your face is red, puffy, double- or triple-chinned. You've got a wide stance, so your legs are spread and your butt is sticking way out. Really not your prettiest moment."

Todd stuffed the fear of injury way down in her duffel bag, and she embraced the fear of looking her worst when doing her best. She thought, *It's okay to be strong, it's okay to be a woman and do this, I know I'm not going to be attractive as I'm doing this.*

Her husband supplied a different type of reassurance. "Terry would always say, *Go back to the numbers,*" she remembers, and "the realism of that appealed to me." In a workout with Terry the week before, she had squatted 540 pounds, exactly equaling the women's world record squat at that time—the record was her own—and she did that in the workout not once but *twice*—and it wasn't even very hard. *Remember the 540 you squatted twice,* Terry said as Jan got ready. *There's no way that you can't do 541,* and probably more.

With the adrenaline of competition pushing Todd to near all-out effort, Terry thought she could reasonably expect to squat 565 or 570. She remembers him telling her, "I think you might make 575."

In powerlifting competitions, the three lifts always take place in the same order: the squat comes first, then the bench press, then the deadlift.

For her first attempt at the squat, Todd lifted 451 pounds—and she made it look so easy, people in the audience laughed out loud.

For her second attempt, the bar was loaded with 545.6 pounds—5.6 pounds more than her previous world record.

"I remember it like a ritual," she says, about what happened next.

She stepped under the bar and positioned it across the top of her back—just below the vertebra at the base of the neck, the one that protrudes when people bow their heads. She had a callus there, from long practice squatting, a thin length of thick callus from the barbell bearing down on her upper back.

She stood up straight to raise the bar from the rack: "I have a big memory of *This is really heavy on my back, Oh my God*—okay, this is really *heavy.*"

Continuing the ritual, she stepped back from the rack the same way she always did. Left foot first, then right. Then scooted her feet laterally, a bit wider than shoulder-width, toes angled 45 degrees from heels. Describing these moves, Todd shifts into second person and present tense, as if this squat is still happening now, and she's coaching herself through it: "Be careful not to jiggle the bar too much—do all this in slow motion."

On the descent, her hips and knees bent as she lowered the weight quickly, easily, to the bottom of the movement. With her legs being so big, she knew she'd squatted deep enough when she felt the backs of her thighs touch her calves.

On the ascent, something unusual happened. She felt the bar drift slightly forward, and she adjusted to stay under the weight; she *had* to stay under it, to keep pushing it up. With that much iron on the bar, even tiny shifts can make big differences in torque. As she drove the weight up, she also had to wrench it back—which made the rising a true fight, a grind.

At the sticking point, she says, "a third of the way up, it would just get tremendously hard. Legs and glutes trying to push it on up. I would internalize that: *Come on come on keep pushing don't give up don't give up.* And then you pray that they grab it as soon as possible"—*they* are the spotters, a strong pair, one near each end of the barbell, ready to catch the weight if you faltered or fell.

Driving up, she did not think about her muscles. She did not think about the weight.

She was all inside her will. "Just: *Push. Keep going.*"

And she did. Pushed the weight all the way up. Stood straight. Saw the signals from the meet's three judges that her lift was good: three white lightbulbs lit up; it would have been red lights for a miss.

But what had made this lift extra demanding? The tiny shift that made the big difference, she discovered, involved a shoelace.

At the bottom of Todd's descent, one of her shoelaces had come undone. In the slightly loosened shoe, her foot shifted slightly forward. That minuscule shift of a shoelace weighing a fraction of an ounce made the 545.6-pound weight on her back drift forward and made her drift forward with it. This threw off her balance enough to make the lift demand much more effort than it would have if she had been wearing two tightly laced shoes.

Because the meet had just started—and because she had her eye on the prize of nationals the next weekend, the peak she'd been training for—she took a pass on making her third squat attempt. That second squat had taken a lot out of her, and it was good enough for the time being—and, incidentally, a new world record. There seemed no reason to spend more energy on squatting that day.

But she did not hold back in the deadlift. There she set another new world record by pulling 479.5 pounds. Adding that number to her best squat and bench press at the meet in Columbus yielded a total of 1,229 pounds in the three lifts—still another new world record. "That was it. I could have tried to do more" in the squat; and she probably could have done more, she believes. "I was saving it for the next week."

Two days later, Todd was at home by herself, unloading the dishwasher. She reached for the silverware. Reaching, she did not notice that a knife had been loaded in the utensil drawer, sharp side up.

She felt the sharpness. Slicing her right hand.

She pulled back her hand, looked: saw blood. Blood was spurting—"little arterial spurts."

She grabbed towels to wrap her hand and drove to the hospital. The doctor who stitched her up said the knife had nicked the tendon on the base of her first finger. That tendon connected the finger's bottom bone, the proximal phalanx, to the muscles in her forearm that gripped the barbell.

And she did grip the barbell when she went to the women's powerlifting national championship the next week. She remembers her performance, extraordinary by any standard but her own—475 squat, 204 bench, 440 deadlift—as being "all subpar."

It was the week before—at the little meet she almost didn't go to, not the big event she planned for—when Jan Todd came closest to finding out how much weight she could lift.

Her voice is even, all pride and all humility, when she says, "January 31. That's the strongest I ever was."

That's the strongest she ever was, in an absolute sense. Todd never lifted that many pounds again.

Yet her heaviest world-record squat was still far from the upper limit of her strength in relative terms. The amount of weight that she could lift, relative to her body weight, kept rising.

After setting her final world record in the heavyweight class, Todd took some time for active rest, and with the goal of losing weight, she changed her training. She rode her bicycle for an hour or two a day. She tended her vegetable garden. She ran—short distances, then longer ones. She maintained her strength, too, by lifting heavy weights, but she also did more reps with lighter weights, careful to minimize loss of muscle while she was losing fat. In a year, her body weight dropped from 230 pounds to 150.

She became a powerlifting coach at the sport's highest levels, teaching others what she had learned. She led both the U.S. women's team and the U.S. men's team at their respective powerlifting world championships, twice—and she was, to the best of her knowledge, the first woman ever to coach a national men's powerlifting team from any country.

She also helped organize and administer the sport. This work made her more finely attuned to the influence of drugs on competition. In addition to her long-standing concern about health, she says, "the ethical issues became more clear to me." She received letters from competitive female powerlifters who did not take steroids, asking her, *Why aren't we doing drug testing for women?* and asserting, *It's unfair for me to go to a meet and be beat by someone who I'm sure is on drugs.*

Todd came to share this conviction, and her duty as a leader seemed clear. "You wanted to protect the sport," she says, "because you wanted to protect the people."

The only Catholic monk in powerlifting, as far as Jan Todd knew, was Brother Bennett Bishop. He taught physical education and ran a powerlifting club in Bay St. Louis, Mississippi, at St. Stanislaus College, a boarding school and ministry of the Brothers of the Sacred Heart, in which he lived.

"A lot of good people teach weightlifting," Brother Bennett once said, "But as a Brother, I also teach religious truths and help in the moral formation of the young people in my club. That's what being a Brother is all about."

The monk's calling to teach moral formation was activated on a large scale in 1979, after the governing body of U.S. powerlifting mandated drug testing and the athletes revolted. A few top-ranked American lifters broke away and formed their own group, declaring that no international organization, involving foreigners, should "dictate to the U.S. lifters." But this was jingoism. Powerlifters in the breakaway faction were largely motivated by desire to take the drugs they wanted to take, judging by advertisements for their meets: "Don't want testing? We won't have any." One leader of the group wrote a booklet called *Anabolic Steroids: What Kind and How Many*. Another, in a meeting where Jan Todd was present, hailed steroids as "the salvation of mankind."

Brother Bennett didn't think so. He was tall, with a thick beard, and he spoke plainly. "Drugs offend the concept of fairness," he said. He believed that "athletic competitions are becoming more and more chemical competitions." Trying to change this, he also believed, was his duty. "If we are to have respect for others, we must first have respect for ourselves," he said. "A different world cannot be made by indifferent people."

Brother Bennett set up an alternate powerlifting federation, with help from Todd and others. They called it the American Drug Free Powerlifting Federation. Drug testing was mandatory for their meets. At the first one, in the summer of 1982, in Mobile, Alabama, Jan Todd was the first athlete, male or female, to set a world record. At a body weight of 149 pounds, she deadlifted 446 pounds—a feat of relative strength significantly greater than her heaviest world-record squat. The load on that squat had been a little more than twice her body weight. The load on this deadlift was more than three times her body weight.

Todd's risk had been rewarded. Her victory suggested that the grocery store *could* beat the drugstore. Was there a chance that Todd's personal drug-free lifting triumph might help change the culture of weight training as a whole?

That same summer, the up-and-coming archetype of the drugstore-style physique scored a victory, too. The movie *Conan the Barbarian* was released, putting Arnold Schwarzenegger on course to join pop culture's pantheon; and from that height, he would set the tone for how most of the world saw muscle and weight training for a long time to come.

Todd's own view of muscle and weight training continued to evolve. She decided that she wanted to enroll in graduate school at the University of Texas at Austin, to work toward a doctorate in American studies by researching and writing about early performing strongwomen such as Katie Sandwina. In 1983, the Todds moved to Texas and settled there for good. Terry worked at the university lecturing in sport history, and together the couple taught undergraduate courses in weight training. They also coached the university powerlifting team—which they founded—through more than ten team titles at the national championships.

In addition, the Todds began to donate to the university their vast collection of physical culture materials, more than 100,000 books, pamphlets, magazines, instructional posters, images, films, letters, manuscripts, and other artifacts—amounting to what must be one of the most extensive records of pursuits of muscular strength that any library has ever held. In the following years, Jan would draw heavily on this collection in her academic research.

All the while, throughout the 1980s, the Todds worked to publicize the dangers of anabolics and the value of drug testing in strength sports. Terry wrote some lengthy, lucid articles on the topic, including "The Steroid Predicament" for *Sports Illustrated*. (He had met the editors when the magazine published Jan's profile.) Jan wrote columns for big newspapers like *USA Today* and niche publications like *Strength Training for Beauty*. To popularize lifting as a healthy regimen for the average person, Jan and Terry together wrote a book, *Lift Your Way to Youthful Fitness*, published by Little, Brown in 1985, trying to do for resistance exercise what *The Complete Book of Running* did for aerobic exercise—but with an emphasis on how lifting weights could help slow the processes of aging.

In this, the Todds were prescient. Academic research on training to build stronger muscles in old age was in its early stages. The next year, in 1986, the Scottish geriatrician Archie Young published a seminal paper, "Exercise Physiology in Geriatric Practice." Young's work with older patients at an Oxford rehabilitation center and at London's Royal Free Hospital had yielded evidence that strength training at low levels of intensity could partially "reverse the effects of immobilisation" for old people and "readily produce a 10 to 20% improvement in strength and aerobic power, effectively postponing" functional impairment for a decade or two.

In the 1980s, Archie Young's findings about exercise for older and younger people bore structural resemblance to what Jack Wilmore had found about exercise for women and men in the 1970s. Beneath some absolute differences in measured ability between groups, both researchers found proportional similarities in adaptive potential. As Young wrote, "the effects of physical training are much the same, proportionally, in old age as in youth."

Young also drew attention to scandalous ignorance among medical students regarding exercise for older people. With a couple of colleagues, he surveyed British final-year medical students in June of 1980, and found that 57 percent of those future doctors were unaware that "physical training reduces the elderly person's heart-rate response to exercise." Young hoped his own research might help change that situation.

But change would require collaboration between physiologists and physicians, Young knew—and he declared the need was "urgent," because the older population was rapidly growing and their risk of disability was high. His research, a foretaste of more dramatic findings soon to be made by colleagues overseas, was a milestone in the long process of winning legitimacy for strength training as a safe and healthy lifelong practice.

By the spring of 1990, the struggle to win legitimacy for strength training—a struggle that had called forth Jan and Terry Todd's relentless labors—met with some success.

That April, the American College of Sports Medicine revised its influential exercise guidelines. The biggest change was a new recommendation that all adults should do "resistance/strength training"—at least twice a week, performing at least eight to ten exercises that would, collectively, involve all the major muscle groups.

This did not mean that everybody had to pick up a barbell right away, the guidelines clarified: "Although resistance training equipment may provide a better graduated and quantitative stimulus for overload than traditional calisthenic exercises, calisthenics and other resistance types of exercise can still be effective in improving and maintaining strength." But the guidelines affirmed heavy weight training in the most explicit terms: "Muscular strength is best developed by using heavy weights (that require maximum or nearly maximum tension development) with few repetitions."

In that sentence, the tacit knowledge handed down by generations of serious lifters gained unprecedented sanction from medical and scientific authority.

No one knew the characters who had defined that tacit knowledge better than Jan and Terry Todd knew them. And thanks to the Todds, during the same season when the new exercise guidelines validated knowledge won by trial and error in the gym, many eminent figures in physical culture started to receive more careful consideration from historians, too.

In early 1990, the Todds began publishing a peer-reviewed academic journal called *Iron Game History: The Journal of Physical Culture*. The journal's intended readers, according to the first issue, were academics and others "who, for one reason or another, enjoy reading about the days before anabolic steroids changed the field of physical culture so dramatically." In an editorial, Terry expressed hope that "stories of pre-1960 athletes will serve as guideposts" for young people "as they search for strength *and* health."

The best way forward for people who loved lifting weights, the Todds believed, led through the past. Instead of focusing on what contemporary steroid users did wrong, *Iron Game History* sought to show what lifters in earlier times did right.

For Jan Todd, the hard work of creating this new journal was also a pleasure. As she once told me, "I like the oldest things best."

CHAPTER 6

Old and New

In the early 1990s, Jan Todd kept a lot of old things in her book-crammed office, next to the indoor archery range in the basement of the University of Texas women's gymnasium, near the center of the Austin campus. Much of the material she worked with had come from those hundreds of boxes that, a few years earlier, she had retrieved from the house of Ottley Coulter, the former circus strongman who had amassed a private archive of the history of physical culture.

For Todd, one of the most precious treasures in Coulter's collection was a scrapbook of clippings with pictures of performing strongwomen, clippings gathered early in the twentieth century. She considered those women to be her "athletic foremothers," as she later wrote, and she was grateful for how their strength helped make hers possible.

Todd was now in her early forties, an academic star on the rise. One of her first papers, "Bernarr Macfadden: Reformer of Feminine Form," was a revisionary study of the controversial publisher. Todd wrote about Macfadden's magazine *Woman's Physical Development*, the first English-language women's fitness magazine, launched in 1900. The magazine taught readers that "all beauty has its roots in a physical, active life," Todd wrote; and her essay was honored by the North American Society for Sport History as the year's best work by a graduate student.

Her passion was for narrative history, her research largely driven by straightforward questions about her forebears: What were the early strongwomen's lives like? What kind of training did they do, following what lessons, from what teachers? What did the women think about and what did they feel while they were doing these things?

Todd's many questions about the strongwomen expressed her identification with them, too. She wanted to know if training changed their views of their own bodies and their sense of what was possible in their lives, as training had changed her.

Though history preserved few details about strongwomen's training techniques, promotional stories about them survived in plenty, and in these, Todd found a curious pattern that, she wrote, "seemed to contradict everything I knew about womanhood in Victorian America."

According to historians, the ideal woman of the Victorian era was fragile and submissive—she knew her place was in the home. Yet strongwomen of the time were celebrated, not stigmatized, for defying such conventions. Writers glorified the big, robust bodies, vigorous feats, and adventurous spirits of Katie Sandwina, her peers, and their predecessors, while also describing these women as conventionally feminine and beautiful. The contradiction made Todd ask, especially about Sandwina, "How could she be celebrated as this great beauty? Why would crowds go and watch her lift weights? What is it that allows that to happen?"

Todd's prizewinning initial research on turn-of-the-century women's fitness deepened her interest in earlier times, when the athletic foremothers of her athletic foremothers were young women. "To understand the strongwomen and why they found the cultural acceptance they did, and why those bigger, more muscular bodies were beautiful and attractive," she tells me, "I had to write about what happened *first*."

It's challenging for historians to write about what they can't read about, however, and there were big blank spaces in the story of exercise. "No one," Todd found, "had ever written a book dealing solely with the early history of women's exercise in America."

The few scholars who had even engaged with the topic—all women; not one man had written about it—based their research mainly on medical texts, not the more plebeian popular exercise and health publications, which "were not in libraries—until, kind of, *we* come along," Todd says, meaning herself and Terry. "That was kind of the point," she adds, of donating their voluminous physical culture collection, including what they had acquired from Ottley Coulter, to the University of Texas. She raises her voice a notch, as if talking through a door: "It's like, '*Hey*. There is this enormous other world out here of stuff that *normal people do. All* the time. It's *exercise*.'"

In this exercise keep the body as erect as possible.

No. 23. Having the arms perpendicular over the head, perform the same exercise as in the last number, with right hand, left hand, then alternately and simultaneously.

Figure 21.   Figure 22.

No. 24. Placing the feet in the position of *Fig.* 21, raise the arms with great force from the hanging position to that seen in *Fig.* 21. On the next beat bring the arms to the position seen in *Fig.* 22; on the next

Figure 23.

to that seen in *Fig.* 23; on the next beat sweep back

81

Todd read her way through piles of old magazines, pamphlets, and books. At first it was hard to see past the long skirts and bloomers, but then her vision adjusted. "I began to realize: '*Wait*, they have dumbbells in their hands,'" she recalls. "I had stupidly thought: the strongwomen, that's where women's lifting begins." Yet other women, earlier, had followed lifting regimens, too. Todd painstakingly collected evidence of their training, and the focus of her research emerged.

She focused on a certain kind of training: exercise done with purpose, to meet a goal and change a life. She called this purposive exercise. She described it as being "always rational," a type of program to help people "meet specific physiological and philosophical goals"—to change their whole selves, in the process "creating a new vision of the body."

Physical Culture and the Body Beautiful: Purposive Exercise in the Lives of American Women, 1800–1870, published in 1998, was Todd's first academic book. As she wrote in the volume's introduction, new visions of the body were emergent hot commodities in the nineteenth century.

Early European settlers in North America had been too busy churning butter and digging fence-post holes to think much about exercise. *Life* was exercise. After the American Revolution separated the colonies from England, citizens built a new society in the United States, including rudimentary systems of education and health. Exercise became a topic of interest and debate, especially in Boston.

In the first decades of the 1800s, Boston's civic leaders undertook a self-improvement strategy for Massachusetts. The strategy was built on education. Some of Boston's best and brightest went to study in the states of Germany, where the university curriculum was more demanding than what they'd known at home. These young Americans abroad thrived on the challenge, and they thought it helped explain other things they admired about this foreign country. They admired German society as a system of decentralized republican power, grounded in knowledge and culture—a "republic of letters," in the phrase of one Bostonian—and they dreamed of remaking Massachusetts on the German model.

Following the European example would entail education reform, inspired by innovative German schools called gymnasia. Since the late eighteenth century, these prestigious schools—at first mainly for boys—had experimented with a new type of education: preparing young men to navigate the modern world on principles derived from ideas about ancient Greece.

These were some of the only European schools since antiquity to put vigorous outdoor exercise in the core curriculum. Jumping, running, climbing—and some rudimentary weight training, such as lifting sandbags and holding them with arms outstretched for as long as possible. Initially, Germans called programs of such exercises Gymnastik, inspired by the vocabulary of Greek athletics. Later, as the system evolved, it was renamed Turnen, to make it sound more German. By either name, gymnastics fostered a worldview that was complicated, to say the least—at once rigidly hierarchical and radically egalitarian.

Doing gymnastics, boys were supposed to learn that they, as members of the German Volk, inherited a physical vigor making them superior to other peoples. (Superior especially to their main rivals, the French.) Doing gymnastics was also supposed to turn a boy into a citizen who, in the words of a renowned German educator, would "feel, think, and act with the state, through it, and for it, and in it; he will be one with it and the people in life, woe, and love."

Equality within superiority—meaning, equality with peers and superiority to outsiders: the social philosophy of gymnastics appealed to elite young Anglo-American Protestant men. By 1825, a group of them—Boston's first gymnastics boosters—had hired one of Germany's leading gymnasium heads, who had moved to Cambridge; and in 1826, he built Harvard's first outdoor exercise gym, a hodgepodge of monkey bars, ladders, ropes, and platforms.

In the same year, one of those boosters also established the first physical education program in North America. The program, Jan Todd discovered, consisted of gymnastics lessons at a school for girls—and not, as historians previously believed, at a nearby school for boys.

Todd found a report on this educational experiment written by the principal of that school, and, reading between the lines, she saw that the girls worked out hard. The principal, William B. Fowle, observed that "many hands have been blistered, and perhaps a little hardened" with calluses, from gymnastics, and that "many weak and feeble children have at least doubled their strength." Wishing there could be broad public access to such workouts, he wrote, "I hope the day is not far distant when gymnasiums for women will be as common as churches in Boston."

The only drawback of the regimen, he thought, was the name. Fowle's report was titled "Gymnastic* Exercise for Females," with the footnote to the asterisk suggesting it might be best to avoid the "coarseness" of the word *gymnastic* in connection with girls' exercise. Apparently, Fowle assumed his readers knew enough of ancient Greek language and culture to know that young males in Greece—unlike young females in Boston—did their exercise in the nude.

Also in 1826, across the Charles River from Cambridge, the city of Boston built its own outdoor gymnasium, thanks in large part to the work of another gymnastics booster, John Collins Warren. He was a well-known physician, a founding member of the Massachusetts General Hospital, and his father had founded Harvard Medical School; but neither his dignity nor his age—he was then forty-eight—kept him from regular exercise on the bars where young people trained, too. To the end of his life in 1866, he continued to extol the benefits of gymnastic training.

Warren grew up in one of the first generations to know the term *physical education*, according to the historian Jack Berryman, who writes that Warren was in the vanguard of physicians committed to "teaching the 'Laws of Health,'" a newly fashionable term for "instruction about how one's physical body worked." Warren called it "a general law, that health may be preserved to a late period of life by the use of those things, which are friendly, and the avoidance of those which are noxious. Most diseases are the consequences of violations of the laws of nature, sometimes the result of ignorance, more frequently of inattention."

Even the heaviest gymnastic exercise was gentle, compared to many common medical treatments of the time. Doctors were as liable to bleed, purge, and blister their patients as they are to write prescriptions for pills today. "Medical treatment at the hands of regular physicians," as Jan Todd wrote, was "a gamble, at best."

In medical circumstance, women gambled against especially long odds. Many doctors presumed that female bodies were inherently unstable. In Todd's words: "Hysteria, menstruation, neurasthenia, and other so-called common female disorders were associated with women's lack of bodily control and exemplified their physical and emotional instability."

The historian Patricia A. Vertinsky, in her study of nineteenth-century women's experiences of exercise and medicine, *The Eternally Wounded Woman*, writes that an adolescent girl's first experience of menstruation "was considered an illness to be weathered only with particular care. For the next thirty years of life's pilgrimage, women were advised to treat themselves as invalids once a month, curtailing both physical and mental activity" in order to minimize risk of "accidents, disease, and loss of fertility." Yet if "vigorous activity was frowned upon and periods of rest strongly encouraged," Vertinsky continues, "regulated healthy exercise was definitely indicated"—and this was especially true for older women, during and after the menopausal transition. "Indeed," Vertinsky writes, "at the age of forty to fifty and beyond, the need for appropriate exercise was seen to be greatest."

As to how much and what kinds of exercise were appropriate, establishment physicians and popular health movements offered various answers. Vegetarians, hydropaths, homeopaths, hypnotists, mesmerists, and exercise teachers promoted their regimens with religious fervor, Jan Todd wrote: They "preached that health was a moral duty and that every man—and every woman—must take responsibility for his or her own physical destiny."

Much of their teaching has since been debunked. For instance, one of the most popular exercise teachers, Diocletian Lewis, claimed that lifting very light weights was always, for everyone, sufficient to make "performances in the business or pleasures of life become most effective," as he wrote in *The Atlantic Monthly* in 1862. Lewis taught that two-pound dumbbells "are heavy enough for any man"—meaning for any person. Yet even the lightest exercise, for people who led sedentary lives, did produce results.

Diverse types of training had diverse effects, but all had one thing in common. In Todd's view, exercise regimens "encouraged women to have a new relationship to their bodies—to view them as trainable, and, more importantly, controllable."

The regimens were shaped by old philosophical ideals, Todd found.

According to one ideal, expressed in the notions about exercise and gender mentioned above, women's bodies were so different from men's bodies, they required a whole different form of exercise—gentle and mild activities, such as walking or light calisthenics. His and hers regimens were rationalized by reproductive roles. "Women should not be strong like men, but for them, so that their sons may be strong," wrote the French philosopher Jean-Jacques Rousseau, in 1762; and Todd rephrased his repressive message: "The more attractive women were," she wrote, "the more men would wish to mate with them, and the stronger the family would be."

Yet Todd contended that Rousseau's words had some paradoxically empowering effects for women, too. By raising the topic of gender-appropriate exercise regimens and affirming the importance of these regimens in female lives, Rousseau influenced others, including his detractors, "to think seriously about women's physical nature, women's education, and women's need for physical exercise."

For instance, in 1792 the British writer Mary Wollstonecraft published *A Vindication of the Rights of Woman*, which "attacked Rousseau's fecund domestics and introduced an alternate ideal" of physical and intellectual equality between the sexes. "She believed that woman's subordinate role," Todd wrote, "resulted not from real, physiological inadequacies, but from woman's limited educational opportunities and the low expectations of her culture."

Big and strong was beautiful, Mary Wollstonecraft believed. She predicted that "we should see dignified beauty and true grace" if girls and women were allowed to build up their bodies through exercise.

"Not relaxed beauty, it is true, or the graces of helplessness; but such as appears to make us respect the human body as a majestic pile, fit to receive a noble inhabitant, in the relics of antiquity." That last line, Todd pointed out, was a sidelong reference to statues of Venus, the Roman name for the Greek goddess of sexuality and reproduction, Aphrodite. "Large-limbed, broad-waisted, and somewhat athletic in

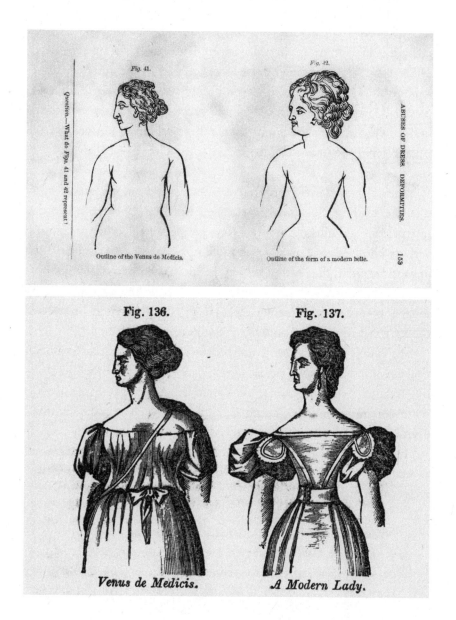

appearance," the Venus was featured in Victorian women's exercise books to illustrate a standard of female health and beauty.

Wollstonecraft "initiated an ideological movement that argued women had as much right to physical strength, muscularity, and robust health as men did," Todd wrote. The movement's "new physical model for American womanhood" was "based on size, strength, and substance." Todd called this ideal "Majestic Womanhood."

Not only in America, but also in Britain and Ireland, as Todd showed, and to an extent in other progressive societies of Europe, ideals of Majestic Womanhood informed or inspired some intensely challenging exercise regimens, including body-weight resistance training, that made women's muscles bigger and stronger. Largely overlooked or underappreciated by other historians, these regimens jumped off the page to Jan Todd because of her experience of lifting weights.

When she studied early training regimens, Todd put her whole body into the reading. She imagined herself into the nineteenth-century instructions and illustrations, based on her personal history of training and her knowledge of exercise physiology. She evaluated the challenge of each regimen, the level of skill and experience it required. She pinpointed the muscles involved in the movements. She even made reasonable conjectures about what people of past generations may have thought and felt as their muscles contracted and relaxed.

She raised the hems of crinolined titles as *A Treatise on Gymnastic Exercises, Or Calisthenics, For the Use of Young Ladies*, published in Dublin, Ireland, in 1828, which recommended exercises that, in Todd's words, "required levels of strength and fitness which would be unusual even in an athletic, twentieth-century female."

To understand what this book meant to teach its readers—addressed as students or pupils—and what it expected of them, consider its description of proper pull-up form:

> The pupil will take her position under the bar, raised a foot or more above her head. Having placed her hands thereon, will raise the body gradually by the strength of the arms from the ground, the palms of the

hands turned from the body, and the toes pointed to the ground, that the knees may be properly extended. The body to be raised until the chest be on a level even with the bar, and the exercise to be repeated several times without putting the foot to the ground, and descending gradually by the exertion of the arms alone.

If Todd read this with both eyebrows raised, she unpacked it with incisive authority. "Chinning movements such as this require a high level of upper body strength," she told her readers, with a mini-lesson in biomechanics. The form recommended by the *Treatise*, "placing the palms away from the body," she emphasized, "actually made the exercise more difficult for women, since this position decreases the biceps' ability to help in carrying the load." The move is so exceptionally challenging, Todd drilled down on it in a footnote about her experience as a teacher of weight training. She estimated that "less than five percent of the college-aged women in my classes were capable of performing overhand (palms forward) chins, even after a twelve-week course of strength training."

The pull-up was no anomaly in the *Treatise*, but one of several suggested tough moves, including also the dip between parallel bars. "The dip is another difficult exercise for women," Todd wrote, "because of the relatively small size of the muscles in the female arm and shoulder."

Right here, Todd's scholarship all but merged with her coaching. These lines from her weight training class lecture notes could almost be crib notes for her dissertation: "Dips—Hard for women but perhaps the best overall assistance exercise for bench and upper body. Actually works triceps and pecs. More they lean into the movement, the more they're working the pecs—Straighter up they are, the more they're working triceps."

For the *Treatise* to mention such an advanced move, Todd knew, the book's first readers must have "trained regularly, seriously, and with strength as a physiological goal. Simple calisthenics movements would not be enough to allow an average woman to perform either chins or dips."

What do you do if you strain a muscle? Sprain a joint? Dislocate a bone? What happens if you get too hot while exercising? Or if you're one of those people who always sweats like crazy?

The *Treatise* spoke to all these problems and concerns, indicating that some women who read this book surely experienced all these issues. One way to avoid such problems, the book said, was to execute each movement with "determined firmness and presence of mind." Doubt or hesitation in gymnastic exercise, the book warned, "is always liable to expose the performer to danger, which determination and courage will in every instance prevent." This was one of the first lessons Jan and Terry Todd taught in their University of Texas weight training classes, too: *Always lift with confidence.*

Every move involved in every exercise in the *Treatise* was therefore practice in virtues that, for women of the time, were expressly countercultural.

As practice in courage and determination, gymnastics could help

women realize their potential in other parts of life, according to the *Treatise*. Advocates for highly challenging exercise regimens, Jan Todd observed, "viewed women's physical potential from a much more egalitarian perspective" than teachers of gentle calisthenics, by and large. The *Treatise* trumpeted that women's gymnastics would "produce 'souls of fire, in iron hearts.'"

The author of this book, J. A. Beaujeu, directed physical training for hundreds of soldiers' children, including orphans of war—both girls and boys—at the Royal Hibernian Military School in Dublin; and according to the school doctor, Beaujeu's gymnastics lessons improved those students' lives. In 1827, in the Dublin *Morning Register*, the doctor wrote that gymnastics had produced especially "beneficial results" among the school's "female department." Before the young women learned gymnastics, he wrote, "we had constantly a majority of females in the infirmary," but after learning gymnastics, the female students were "much improved in carriage, health, and appearance."

Beaujeu also owned a gym in the city that offered classes for adults, both men and women; and he led the women's classes with his wife, Madame Beaujeu, who had been one of his students.

Dublin's early adopters of gymnastics, like Boston's a couple of years earlier, included leading citizens: government officials, doctors, and scientists. Among them was William Rowan Hamilton, the astronomer and mathematician whose research changed the fields of algebra, optics, and dynamics and contributed to the development of quantum physics. As a young man, Hamilton regularly worked out at Beaujeu's gym, sometimes accompanied by a friend who eventually became his biographer, and whose recollection of the scientist's workouts depicted implied intimate links between Hamilton's mental and physical feats:

> . . . and pleasantly can we recall, as having been fellow-votaries in the pursuit, his vigorous prosecution of gymnastics at the academy of M. Beaujeu, where we have seen him as earnest about circles, of which in his own person he flew along the circumference, or about the ascent of perpendicular poles and slanting rope-ladders, and the swinging between parallel bars, as ever he has been in exploring the mysteries of ink-drawn curves and right lines, or in ascending by the ladder of algebra to the specular heights of science.

In September of 1828—the year the *Treatise* was published—Beaujeu and his wife advertised major expansions of their gymnastics class offerings. Beaujeu opened a second gym, this one outdoors, "an extensive uncovered Establishment for the instruction of Gymnastic Exercises, for young Gentlemen." The same advertisement announced: "Mrs. Beaujeu will open her School for Calisthenic Exercises, for young Ladies"—a school of her own, apparently—"on the 1st of October."

But within the month, J. A. Beaujeu was dead—killed in a freak gym accident, a fall from the parallel bars. As another Dublin newspaper reported, "he lost his balance, the bar not having been properly fixed in the groove adapted for it, and fell to the ground—in the fall he injured his spine so severely that he terminated his existence." The death made news as far away as Liverpool, across the Irish Sea, where a story in the *Mercury* newspaper cautioned readers: "The weak and timid-minded will lay hold of this accident in order to denounce gymnastic exercises. Nothing can be more absurd than such conduct."

Yet by now Dublin had enough committed gymnasts that Beaujeu's establishment stayed open. A German prince who frequented the place not long after Beaujeu's death wrote a letter about working out there among a group notable for including "much older people—men of sixty." He wrote: "I constantly found men of this age, who played their part very well among the young ones." The prince also wrote that he had seen, among the young ones, a man, "the arch of whose breast, after an uninterrupted practice of three months, had increased seven inches; the muscles of his arms and thighs had at the same time enlarged to three times their volume, and were hard as iron."

Despite Beaujeu's fatal accident, his widow carried on the work they had begun together and held to her vocation as a gymnastics teacher, in a career of determined creative adaptation to constantly changing circumstances. The year after her husband died, her "Ladies Gymnasium" was open for business. She advertised it in 1829 as her own "Academy" in Dublin "for the Practice of the Kalisthenic Exercises," movements that "met with the approbation of the most eminent Medical Practitioners."

Then, for reasons unknown, she moved to the United States, where in 1837 she taught calisthenics at a gymnasium in New York City—one of the city's first gyms, operated by the English boxer William Fuller.

By 1841, she had moved again, this time to Boston, where she became the proprietor of another gymnasium of her own, another gym for women. Within a couple of years, she returned to the New York City area, where she taught calisthenics at several girls' schools in Brooklyn. And then by 1846, she had reportedly opened yet one more gym of her own, this one in Manhattan.

In 1855, when she was probably well into middle age, she was still teaching gymnastics and calisthenics, "Classes for Ladies and Misses" on Tuesdays, Thursdays, and Saturdays, at the Brooklyn Gymnastic and Calisthenic Institution owned by Joseph B. Jones, a physician who was an early promoter of and participant in the new sport of modern baseball.

She was by this time known professionally as Madame Beaujeu Hawley, or sometimes as Mrs. Hawley—she seems to have been remarried by the 1840s. By whatever name, she was in Jan Todd's words a "direct link" between European and American gymnastics (like the Germans who had come to Boston a few years earlier), and "one of America's first exercise entrepreneurs of either sex."

With all her gyms and jobs, this early exercise entrepreneur was at the forefront of an eclectic new global physical culture with some overtly political dimensions, a culture that grew out of and in response to German gymnastics during the nineteenth century. As the historian Allan Guttmann has shown, German gymnastic exercises were also adopted and adapted in networks of private clubs, national recreational federations, or state-sponsored physical education programs across Western Europe, in Belgium, Holland, Italy, and Switzerland. In Eastern Europe, "Czechs, Slovaks, Poles, and Hungarians appropriated German gymnastics and used it," Guttmann writes, "in their struggle for national liberation." In the Bulgarian capital of Sofia, he notes, gymnasts "performed their gyrations in an abandoned mosque that had been erected on the site of a Byzantine church." And German gymnastics influenced training regimens for military troops in many countries, including Denmark, France, Spain, and the United Kingdom.

Madame Beaujeu's repeated resettlements exemplify this great gymnastic migration that reshaped physical culture throughout Europe and in the Americas—not just in Boston, Brooklyn, and Manhattan, but also in Baltimore, Cincinnati, Philadelphia, and other North American cities, as well as in Latin and South America, especially in Argentina, Brazil, and Chile.

In a movement of such breadth and scale, the woman who brought her expertise from Ireland to the United States stands out in retrospect for her accomplishments as an educator and entrepreneur over a career spanning many years and touching many other lives. The power of her achievement is edged with poignance, too, for practically every personal detail about her seems to have been lost. Jan Todd traveled to the great libraries of Boston and New York in a comprehensive search for information about the individual who was known at various times as Madame Beaujeu, Madame Beaujeu Hawley, and Mrs. Hawley, but was not able to discover even her first name.

In the middle of the nineteenth century, when conspicuous muscular size and strength were sometimes points of conflict in debates concerning proper medical care of the body, those debates played out in the context of escalating feuds between university-educated "regular" doctors and "irregular" practitioners, such as the self-taught exercise instructors. Patricia Vertinsky writes: "Regular doctors became increasingly vocal in their desire to defend public health against the irregulars, 'an enemy more subtle than disease' who could 'never do honor to the profession.'"

These feuds were early stages in the struggle for legitimacy that would consume so much energy in medicine and physical culture around the turn of the twentieth century. Struggling for legitimacy was in fact the American Medical Association's original reason for being. To consolidate their own authority by setting and enforcing standards for medical education, the regulars founded the AMA in 1847.

In the same year, Elizabeth Blackwell enrolled at Geneva Medical College in Geneva, New York. Blackwell went on to become the first woman to graduate from medical school in the United States, in 1849, and the first woman to be registered as a physician in Great Britain, in 1859. She argued that vigorous muscular exercise was important for everyone, male and female. In Elizabeth Blackwell's view, the *"great object* of the muscular system" was "to furnish a varied and powerful instrument for the expression of the soul," making muscular exercise the most important aspect of early education, "the grand necessity which every thing else should aid."

While Elizabeth Blackwell, like John Collins Warren in Boston, embraced and promoted gymnastics, other doctors were much more cautious. They believed "the body had finite resources and that these were used up during a person's lifetime," Jan Todd wrote. The heavier the exercise, the more likely regular doctors were to distance themselves from it. Many regular doctors were leery of activities such as lifting heavy weights because, Todd wrote, they believed "the exertions of exercise and the enlarged muscles caused by that exercise used up the body's vitality, drained the brain of intelligence, and hastened one's death."

Victorian physiology sounds fanciful today, but they did not just make it up—they inherited it from ancient Greece and Rome. The Victorian view of muscle as a "powerful instrument for the expression of the soul" and of muscular movements being fueled by a vital substance carried weight of ancient authority.

After the fourth century BC, when Aristotle attributed movement to the soul's perceptions of pleasure and pain, warming and cooling that substance called *pneuma*—the substance in the heart that, he believed, expanded and contracted as its temperature rose and fell, transferring force to pull and release those small strings of sinews in the heart that in turn pulled bigger strings of sinews attached to people's bones—his view was endorsed, but with major changes, by Galen in the second century AD. As we've seen, Galen believed that a special kind of pneuma, the "power from the brain" called psychic pneuma—the part of the soul responsible for cognition and for voluntary motion—infused the nerves with psychic power, or *dunamis*, that flowed to muscles and tendons, which pulled and released people's bones, moving bodies around.

Galen's view seems to have predominated, at least among the highly educated whose writings have survived, for more than a thousand years. In 1543, however, a twenty-eight-year-old professor of surgery from Brussels named Andreas Vesalius published a landmark book about anatomy, *De Humani Corporis Fabrica*, raising serious questions about some of Galen's teachings on bodily structures.

In the course of his dissections, Vesalius came to believe that what moves people is the fleshy part of muscle—the part that Aristotle and Galen thought was only padding, only a kind of cushion or insulation. Vesalius wrote: "I consider this flesh to be the particular organ of motion and not merely the stuffing and support of the fibres." The year Vesalius published this new theory of movement, Copernicus published his theory that the Earth revolves around the sun.

Nevertheless, until the nineteenth century, scientists kept seeking to understand the circulation of psychic pneuma—called *spiritus animalis* in Latin, and "animal spirits" in English—which they also conflated with the material psychic power that caused movement. They called this power "the more subtle part of the blood," "the vapour of the blood," and "the juice of the nerves."

But all these names *"are mere words, meaning nothing,"* wailed the Danish anatomist Nicolaus Steno, in a 1667 treatise on muscle. Steno investigated the mechanics of contractions of the fleshy part of muscle, and he established that the tension of a whole muscle was created by the tensions of all the individual fibers in that muscle. Steno's research also gave him "grave doubts" concerning the reality of animal spirits, doubts based in part on the wild variety of speculations among his peers as to how the spirits worked.

Maybe the spirits were like wine, some speculated—maybe a bit like how "wine restores exhausted powers," spirits could give muscles something like liquid courage. Or maybe spirits converged with the blood "like the explosion of gunpowder," firing human movement as cannons shoot cannonballs. Or maybe spirits rushed into muscles and collided with little particles of a second substance, making "effervescence," so that muscles in motion were always inwardly bubbling and being inflated, and then at rest shrinking back to their normal size.

A sip, a shot, a fizz, a swelling. Many similes for the spirits had the same premise: A substance fills up a muscle. Because—as anyone can see, looking at any flexing biceps—and as practically everyone had been seeing, since the third century BC when doctors of the Alexandrian school declared that flexing muscles were filling up with pneuma—that sure is what it looks like.

More than a century after Vesalius died, Isaac Newton still believed that nervous action was caused by some kind of ether; and for a century more, as the disciplines of physics and chemistry matured, the old animal spirits kept taking on new names, such as "animal electricity" or "vital force."

Then in 1841, a twenty-year-old medical student in Berlin, Hermann von Helmholtz, dropped some little pieces of boiled meat in flasks, and he looked through a microscope to see what happened as they began to rot. Noticing similarities between conditions that help living muscle thrive and conditions that make dead muscle decompose, he became "convinced that the notion of a vital force should be abolished" from all of physiology, including the study of muscle function, as the neurophysiologist John Farquhar Fulton later wrote. This conviction drove Helmholtz to establish a basic principle of modern science, the concept of the conservation of energy, which states that energy can be neither created nor destroyed.

Yet it took a long time for faith in animal spirits to wane, such that doctors, scientists, and laypeople would cease to think of movement as a willed whooshing of a mysterious substance through the body that filled up muscles like balloons.

The tradition of those beliefs in an immanent vital principle, identified with some of history's most celebrated intellectuals, helps to account for the fears of nineteenth-century doctors that heavy exercise would tax the body's reserves of vitality. In that era, one academically trained doctor in the United States stood out for his teaching that lifting heavy weights was good for health—George Barker Windship, a diminutive man whose message grew from his own early training to build muscle and grow stronger.

George Barker Windship graduated from Harvard Medical School in 1857. The Roxbury, Massachusetts, native had a small frame. He stood not quite five-seven, and he never weighed more than 150 pounds. In his freshman year of college, when he was much smaller than that, Windship was teased and picked on, and in hopes of making his tormentors back off, he learned gymnastics. By his senior year, by Windship's own account, he was known as "the strongest man in my class."

After his college graduation, during a visit to Rochester, New York, Windship happened to see a crowd gathered around a strength-testing machine on the city's main street. But when it came Windship's turn to test his mettle on the apparatus, he impressed no one, least of all himself. For all his gymnastic training, Windship determined that he was still missing something, "the strength of the truckman and the porter," a quality he called *"main strength."* This name for extraordinary muscular maneuvers goes back at least to the seventeenth century. In 1719, Daniel Defoe wrote that Robinson Crusoe heaved a rowboat up on the beach "by main strength." In the mid-nineteenth century, when George Barker Windship came of age, boxers with main strength were the ones who dominated the ring by alternately standing their ground and unleashing overwhelming force.

Wanting to build that kind of strength, Windship rigged up his own version of the Rochester lifting machine in the backyard of his family home in Roxbury and began his new workouts. With his feet astride a hole in the ground containing a barrel full of gravel and rocks, he would bend his knees and reach down for a handle attached to a rope affixed to the barrel. Then he would straighten his body to pull the handle up, raising the weight of the barrel.

While training these partial lifts, Windship "became fascinated with the great weights he could hoist," Jan Todd wrote, although "the weight moved no more than a few inches." Training this lift for about five years, he almost tripled the amount of weight he could move, from 420 pounds to more than 1,200.

During that time, Windship later reflected, he experienced a sense of renewal like what he had found in gymnastics as an undergraduate: "I discovered that with every day's development of my strength there was an increase of my ability to resist and overcome all fleshly ailments, pains, and infirmities," including "nervousness, headache, indigestion, rush of blood to the head, and a weak circulation." Now the young doctor connected his early experience of gymnastics with his new regimen of lifting heavy weights, and he inferred that what made him stronger made him healthier.

George Barker Windship condensed this lesson of his life into a catchphrase: *"Strength is Health."*

To spread the word, he arranged to deliver a public lecture on physical culture and to perform some feats of strength, onstage at Mercantile

Hall in Boston, on May 30, 1859. "But from the moment of the public announcement of my lecture," Windship later wrote, "my appetite for food, for meat particularly, began to fail me." He lost 9 pounds—now the budding strongman weighed 134—and when the big day arrived, he could barely eat a bite of food, from breakfast through dinner. When Windship finally began his lecture, his stage fright was so intense that, about ten minutes into the talk, he fainted.

But ten days later, he tried again, and he triumphed. "Without any artificial assistance," the *Boston Medical and Surgical Journal* reported, "Dr. Windship lifted *nine hundred and twenty-nine* pounds." He even "sustained and lifted the weight of his body by his little finger, and by a single hand"—translation: he did chin-ups with his pinkie (he did three of those) and one-handed (he did twelve of those).

This Thursday Ev'ng,
JUNE 23d, 1859,
— AT —
CITY HALL!
CAMBRIDGE,
Dr. WINDSHIP,
WILL
LECTURE
ON PHYSICAL CULTURE,

And Illustrate by FEATS OF STRENGTH; one of which will be the

LIFTING WITH HIS HANDS 929 LBS.,
The Greatest Weight on Record.

TICKETS 25 CENTS.
Doors open at 7½ o'clock.   -   Lecture to commence at 8.

[*From the Boston Transcript, June 9th.*]

There was a fine audience, among which were many of our most distinguished citizens, at the Music Hall, last night. The lecture was *one of the greatest successes we have ever witnessed*. The style was eloquent and vigorous. Dr. Windship has all the qualifications of the consummate elocutionist. He was frequently interrupted by the most animating bursts of applause.

The lecturer performed some remarkable feats (after the lecture). Mr. C. J. B. Moulton, city sealer of weights and measures, being present, weighed before the audience the iron discs, lifted by Dr. W. *by the hands alone*, and certified that the whole amount lifted by the lecturer was 929 pounds, a feat it will be found difficult to parallel.

A repeat performance of his lecture, and of his 929-pound lifting feat, was advertised to take place weeks later at City Hall in Cambridge—and soon he was lecturing all over the United States, billed sometimes as "The Roxbury Hercules." In Chicago, he organized what Jan Todd has called "the first true weightlifting competition ever held in the United States," the month after Abraham Lincoln's presidential inauguration. *The Atlantic Monthly* published Windship's essay "Autobiographical Sketches of a Strength-Seeker" in the same issue as Ralph Waldo Emerson's essay "Old Age." (The next month's issue featured a new poem called "Battle-Hymn of the Republic.")

Windship's regimen was based on lifting heavy weights in short workouts, no more than half an hour a day; but the system was also more supple than that. "And what is my method?" he asked theatrically, in an essay called "Physical Culture," and he answered:

> It is doing the right thing, in the right way, at the right time. It is the obtaining a sufficiency without going to excess. It is the using my own discretion about everything, without blindly following another's precept or another's example, or even tying myself down to the rules of my own devising. It is the carrying out of what may seem to be the expediency of the moment. It is the doing what may seem best under the circumstances.

Windship's description of his method seems to echo that ancient Greek concept of *kairos*—the principle of athletic training as the practice of doing the right thing at the right time—as mentioned by Philostratus in the *Gymnasticus*. Yet Windship was not consciously emulating that ancient idea, as the text of the *Gymnasticus* had only recently been rediscovered and had barely begun to be studied by a few classicists in Europe when Windship wrote his essay.

Windship also taught that all kinds of people, including the oldest people, can handle a challenging workout. No less a medical authority than John Collins Warren, Windship related, "tells us of a member of the legal profession who practiced gymnastics for the first time and with the happiest results, when nearly seventy years old." Windship understood, though, why many were wary of heavy workouts such as weight training, or why they gave up soon after starting. As the *Boston Medical and Surgical Journal* summarized his remarks: "It is because beginners undertake too much *at first*, and are too eager to attain and surpass the prowess and strength of accomplished gymnasts. If they would eschew this ill-advised ambition, and be content to progress gradually, the exercises would be at once delightful and beneficial."

A correspondent for *Harper's New Monthly Magazine* recast the doctor's message: "You are to keep doing, and are never to overdo. But by small beginnings, gradually increasing from day to day, a remarkable muscular condition may be produced by any man."

By 1865, when Windship patented the first of several of his exercise equipment designs—for a plate-loaded dumbbell, adjustable from 8 pounds to 101 pounds, which he advertised for the price of $16— roughly equal to the monthly salary of a private in the Union Army that year—he was known for his ability to lift 2,600 pounds on his own heavy partial lifting machine (which he kept in one corner of his medical office), and he wanted to build up to 3,000 pounds. He had somehow decided that 3,000 pounds was the heaviest weight his body could support.

Within a couple more years, he opened a "combination gym and medical office" on Washington Street in Boston, which "may have been America's first sports medicine facility," Jan Todd wrote. Windship's hybrid general practice/sports medicine facility was a brick-and-mortar version of his catchphrase that strength is health. The corollary of that central idea, according to one of Windship's more observant contemporaries, Thomas Wentworth Higginson, who wrote a number of long, detailed articles about physical culture for *The Atlantic Monthly*, was that "every increase in muscular development is an actual protection against disease."

Jan Todd also found that Windship advertised his medical "office and gymnasium" as having a "Separate Apartment for Ladies"—more evidence of women's early involvement in heavy exercise that no scholars had previously found, or found worthy of notice. Yet "no evidence has been found to suggest precisely what women did at Windship's gym or how many women were members," Todd wrote.

George Barker Windship, in an 1862 photo on his carte de visite. Jan Todd calls it "one of America's first physique photographs."

Among the women who preferred other types of exercise and equipment was one of Windship's own first cousins, the writer Louisa May Alcott, who chose to train with a very different kind of teacher—the one mentioned earlier who taught that no one ever needed to lift anything heavier than a two-pound dumbbell.

While Windship was becoming a physical culture lecture star, Alcott—who had recently caught her first big break as a writer, selling a story to *The Atlantic Monthly*—was becoming a star pupil of one of his rivals, Diocletian Lewis, in an evening gymnastics class in Concord, Massachusetts.

We can still catch sight of her workout, in the fictionalized version of it she later presented in a story: a class where "unsuspected muscles were suddenly developed," and "dumb-bells flew about till a pair of white arms looked like the sails of a windmill." Alcott's gymnastics class was a whole field of these windmills, a motley mass of townsfolk exerting themselves together, in an ecstatic event, "a sort of heathen revival," in which "even the ministers and deacons turned Musselmen; old ladies tossed bean-bags till their caps were awry, and winter-roses blossomed on their cheeks; school-children proved the worth of the old proverb, 'An ounce of prevention is worth a pound of cure,' by getting their backs ready before the burdens came," and "pale girls grew blithe and strong."

More subtly, Alcott's early experience of gymnastics continued to surface in her writing for many years. The narrator of *Little Women*, published serially in 1868 and 1869, observes—in a manner suggesting she thinks readers will agree—that muscular exercise is part of what makes women well: "Want of exercise robs them of cheerfulness, and too much devotion to that idol of American women, the teapot, makes them feel as if they were all nerve and no muscle."

Historical research, Jan Todd has said, is "like seeing a wild animal flitting through the woods—you get a glimpse, and then the animal moves on."

George Barker Windship also had another rival, David P. Butler, who shared his belief that lifting heavy weights was good for health and who built a much bigger business than Windship did. Butler fabricated and patented several contraptions similar to Windship's heavy partial lifting machine, and, spinning Windship's "Strength is Health" catchphrase, promoted the Butler Health Lift. He wrote a book, *Butler's System of Physical Training: The Lifting Cure*, and he franchised the apparatus to establishments that became, effectively, America's first gym chain.

Jan Todd tracked the rise of Butler's business. "By 1871, the Butler Health-Lift Company had five different branches in New York City, four of them on Broadway," with more branches in Boston, Providence, and San Francisco, she wrote. Butler was a sophisticated salesman: He segmented the market into tony studios and "Low-Rate" rooms, and knock-offs popped up in Cincinnati and Chicago.

The historical record is not reticent, at all, about how women experienced the Butler Health Lift—or, at any rate, about how it was marketed to them. "Contrary to what may commonly be supposed," Butler wrote, "we have been unusually successful in treating female diseases and weaknesses, owing, probably, to the fact that women and girls are unaccustomed to the use of nearly or quite all their power."

"Ladies' and Gentlemen's Parlors of the Health Lift Company, 46 East Fourteenth Street." Illustration from The Daily Graphic, New York City, 1873.

Since females did less heavy manual labor than males, they stood to gain more strength than men from lifting regimens, according to *Butler's System*. The book also said that "after a few months of this preparation, continued during the period of pregnancy, labor has, in all cases, been comparatively easy and of short duration, and the children have, in every instance, been healthy and strong."

One of the simplest, grandest claims for Butler's system was made in a pamphlet, published for the company in 1870. The pamphlet said the Health Lift machine was made "for the calling out simultaneously, as nearly as is conceivable, all the voluntary muscular forces of the body, in a strictly harmonious mutual relation with each other—as in attuning an instrument—and in the best possible position for their individual action, viz., that instinctively assumed by man when he feels his fullest powers and 'stands up like a man.'"

The pamphlet's title is *Exercise a Medicine*.

Because George Barker Windship had been the most prominent early advocate for heavy lifting as a healthy practice, he remained a public emblem of the regimen; and his sudden death in 1876, at the age of forty-two, showed how, in weight training, the force of example may not only ratify regimens but also undermine them. Windship died at home, of "apoplexy," probably meaning a massive stroke.

"Weightlifters everywhere were suddenly concerned. If 'strength is health,'" Jan Todd wrote, "then why was Windship dead at forty-two? Didn't his death prove that lifting was dangerous?"

Likely yes, said the *New York Times* report of Windship's passing. "Dr. Windship probably carried lifting too far," the newspaper scolded, just like "all athletes who forget that caution is the parent of safety."

Windship's inspiring example became a cautionary tale. The year after he died, the Butler Health Lift studio in Boston went out of business, and then Butler's other studios closed, too. For decades, partisans of lighter exercise alluded to Windship and to the Health Lift when they warned against heavy lifting.

Later, people just forgot Windship, forgot almost every fact about him. After Dudley Allen Sargent retired as Harvard's head of physical education in 1919, he wrote his memoirs, which mentioned some traditions of physical culture that had influenced him. Among them was heavy lifting as practiced by George Barker Windship, whose "convictions came from his own amusing training that began with his lifting a calf," Sargent wrote. He added that Windship "continued to lift and gain strength as the calf gained weight, until one day he was lifting a full grown ox. His feat, widely advertised, became proof of the efficacy of lifting weights."

Dudley Sargent's confabulation of George Barker Windship lifting Milo of Croton's calf is one of those weird double exposures populating the panorama of physical culture, in which episodes of some lives are always welling up into and reshaping the lives of others. As the gods' gifts to the *Iliad*'s warriors and to the early Olympic *stadion* victors reshaped Charles Stocking's life. As the spectacle of Katie Sandwina's beautiful strength reshaped Jan Todd's life.

In physical culture, the force of example drives metamorphoses that superficially respect a rule of precedence, but at some deep level these transformations seem not to require—and seem somehow actively to resist—anyone's understanding or even awareness. As if all the human bodies—old and new, big and small, male and female, yours and mine—really were the same bodies, all through history, while also being radically different from one another and indelibly singular. As our selves.

Jan Todd says that lifting weights helped prepare her to be an academic historian. "I think the main thing I learned from training was the importance of discipline and regularity," she tells me. "I had no expectations when I started that I was going to lift 500 pounds. But I discovered: Go to the gym, take that small bite of work every day, and that small bite eventually becomes a feast"—and the same thing happened to Todd when she went to the library every day.

She adds that, for doing research, "Lifting also taught me patience. You can't expect change or transformation, whether intellectual or muscular, to happen immediately. You have to give it time for adaptation to occur. You've got to build the frame and build the muscle so that you can support the work."

Publishing her book about women's exercise in the Victorian era, *Physical Culture and the Body Beautiful*, advanced Jan Todd's academic career with broad-spectrum force. In 1998, she condensed the whole history of women and weight training into a long lecture to the national meeting of the American College of Sports Medicine, and it was such a hit, she went on the road, giving the talk to many more groups of doctors and trainers.

Her timing could not have been better. At the turn of the twenty-first century, more doctors were becoming more open to exploring how progressive resistance exercise could improve and maintain health. Medical dogma that weight training was dangerous for health had been largely based on cardiac concerns. But in the year 2000, the American Heart Association issued a science advisory on the "Benefits, Rationale, Safety, and Prescription" of "Resistance Exercise in Individuals With and Without Cardiovascular Disease," including those recovering from heart attacks or cardiac surgery. Whereas previous guidelines said such patients "should avoid resistance training for at least 4 to 6 months," the new document endorsed the possibility of resuming sooner—as soon as three weeks—starting with light weights. By 2000, even Kenneth Cooper, the author of *Aerobics* who in 1968 had denigrated muscular fitness as being no more important than the paint job on a car, was following a regimen of weight training, as he began the eighth decade of his life.

Todd's ongoing historical research, published in *Iron Game History* and other journals, explored myriad topics, from "The Legacy of Pudgy Stockton," whose gymnastic feats on Santa Monica's Muscle Beach, captured by 1940s newsreels, made her the most muscular woman ever seen by many World War II–era moviegoers—to "The History of Cardinal Farnese's 'Weary Hercules,'" about the massive statuary that Philostratus mentioned in the *Gymnasticus*. When the bodybuilding impresario Joe Weider said, about the Weary Hercules, "What he has is what we want," he said it in an interview with Jan Todd; and ten years later, Todd's paper helped inspire Charles Stocking's research on the *Gymnasticus*.

Starting in 2001, Todd collaborated for more than a decade with Roger Farrar, a University of Texas exercise physiologist, on what they called "Project Firepower," a strength training program for female firefighters in Austin. Also in 2001, when Arnold Schwarzenegger decided to produce an elite strongman competition, he asked Jan and Terry Todd to collaborate in designing and running the event, and they said yes.

In her sixties, Jan Todd served a term as president of the North American Society for Sport History; and now, in her early seventies, she continues teaching at the University of Texas, as chair of the Department of Kinesiology and Health Education and professor of women's and gender studies, while holding a fellowship in sports history. She guides the research of doctoral candidates on modern physical culture's most illustrious figures, such as Jack and Elaine LaLanne, who starred in one of the first syndicated exercise television programs with an international audience—and on its foundational institutions such as the Milo Bar-Bell Company, North America's first commercial barbell manufacturer, established in Philadelphia in 1902.

Todd's most dynamic and fruitful achievement may prove to be her creation, with Terry, of the H.J. Lutcher Stark Center and Archive for Physical Culture and Sports, one of the world's largest collections of material pertaining to those fields. A sprawling complex in the north end of the Darrell K. Royal–Texas Memorial Stadium, home to the Longhorn football team, the Stark Center opened in 2009. Looming over the entrance is an exact replica of the Weary Hercules, more than ten feet tall, like the one the first readers of Philostratus saw when they went to the baths in ancient Rome. The center is both a library and a museum, with eleven galleries and a staff of six. One of the chief librarians for the United States Olympic Committee in Colorado Springs, Cindy Slater, left that job to establish this library.

Eleven thousand linear feet of shelving hold the center's archival collections. Here are almost all the physical culture magazines ever published in the English language, including America's first periodical devoted to the topic of health, earnestly titled *The Journal of Health*, published from 1829 by "an association of Philadelphia physicians"—physicians who were interested in exercise. The center also holds the archives of some of the most popular magazines about weight training, including *Strength*, the first American muscle magazine, published by the Milo Bar-Bell Company starting in 1914; and *Strength and Health*, published from 1932 to 1986. Here, too, are training logs and other papers and ephemera, including trophies, uniforms, barbells, and dumbbells, that belonged to lifters such as Pudgy Stockton—her actual first name was Abbye—and Tommy Kono, who first picked up a weight at the age of thirteen in 1943, when he and his family were being held in a California internment camp for Japanese Americans, and nine years later he won his first Olympic gold medal for weightlifting. But the heart of the holdings comes from the Todds' personal collection, which they built on the cornerstone of Ottley Coulter's collection and brought to the university when they moved to Texas in 1983.

For the history of strength, the Stark Center has been a defining institution. So many leading figures in physical culture have had such a facility with and tolerance for falsehood, and a soft spot for sham science, that the field never knew a history made of much but legend until Jan and Terry Todd came along.

Their work has changed the history of physical culture in general, and the history of weight training in particular. Their writings did much to ground physical culture in facts, and they set standards for telling stories based on diligent research about people and practices that had not previously been considered worthy of much serious attention.

Historical research on physical culture remains and always will be a particular kind of struggle. This is clear when Jan Todd talks about the main force of example in her own life, Katie Sandwina, whom she has been researching for decades. "I truly believe she is a construction," Todd says, and she does not say this lightly.

Todd has reconstructed the process of Sandwina's construction in some detail. How physicians invited by the circus examined Sandwina, with reporters present, and declared her "a perfect woman by all the accepted standards." How reporters handled the delicate matter of

Sandwina's muscles: "No horrid lumps of muscle, dears," one wrote, "just a little ripple under the skin, like mice playing in a mattress." How the strongwoman became a political symbol, a reason to support women's suffrage: When she lifted her husband overhead, "Sandwina's performances were certainly about power," Todd has written, quoting one reporter's avowal that "the 'female Hercules' was a living argument in favor of equal franchise."

"What is most fascinating about Sandwina," to Jan Todd, is how the strongwoman "suddenly created a new identity" when she debuted with Barnum & Bailey, and how her example was edited, framed, and sculpted to help the people in her audience explore their identities, too.

Terry Todd died in 2018, a few days after suffering a heart attack at the age of eighty.

He used to joke that his tombstone ought to be inscribed, "He Liked Big Things," but that wasn't really a joke. Terry did love big things, and big people, he really loved them; and so does Jan Todd. Much of their work, in coaching and scholarship, has focused on strength athletes at the highest levels, mostly big people who were born strong. But if not for Jan and Terry's example, many people who were not so colossal but were dedicated to weight training and determined to do it without drugs, and who wanted to understand the cultural and historical context of their training—people like Charles Stocking—would have had a much harder time finding their ways.

Charles Stocking had admired Jan Todd's athletic and academic achievements for twenty years when he finally met her in the summer of 2022. At the time, Stocking was a professor of classics at Western University in Ontario, and he visited the Stark Center to do research concerning some connections between ancient and modern athletics. Stocking joined the University of Texas faculty the next year, with his joint appointment in classics and kinesiology. He will also be Jan Todd's successor as director of the Stark Center, where his expertise in ancient physical culture, including athletics, medicine, and health, will further enlarge that institution's scope.

For women especially, Jan Todd's example has shown possibilities of transforming their bodies for the sake of action, possibilities of becoming strong and muscular, and possibilities of lifting weights as practice in being the kind of person who is unafraid to do hard things, both in the gym and out of it. That was not her intention, but it happened. By changing her body, she helped along some changes in the world. By lifting weights, Jan Todd lifted women.

"The thing that I find very hopeful," Todd says, "is talking with the young women that are at the university now, who have totally grown up in an era where not only they but their mothers did sport. And so, unlike my generation, they all know it's okay for girls to do sport. They are much more willing to understand that it's okay to try hard. They don't necessarily want to be powerlifters or lift weird things like I did, but they can kind of understand."

Of Jan Todd's many publications, one of the most frequently cited by other scholars is her definitive history of "Thomas L. DeLorme and the Science of Progressive Resistance Exercise," coauthored with Jason Shurley and Terry. In that paper, the authors show DeLorme solving problems of how to make patients' muscles grow bigger and stronger, faster, after surgeries. They show DeLorme discovering that solving those problems is only half the battle—and then they show him finding an effective way to put solutions into words, to make his technique more accessible and comprehensible to patients needing help.

In 1945, in his first published paper on this technique, DeLorme called it "heavy-resistance exercises." He soon found that *heavy* was too hard a word for most people to swallow. The word *heavy*, DeLorme later wrote, "bears false implications." Talk of "heavy-resistance exercises" seemed to imply that the quality of strength was fixed, not relative, and that was incorrect.

Resistance is "heavy," DeLorme wrote, only when the patient's actual "amount of muscle power" is considered. "For instance, a few ounces would be as heavy for a severely weakened muscle" in a person who had survived polio "as 50 or 100 pounds would be for a normal muscle."

No matter how he tried to redefine the word *heavy*, DeLorme found that his medical colleagues were not comfortable prescribing "heavy weightlifting." He discussed this problem with his wife Eleanor, who thought DeLorme should give the technique a new name. She suggested "progressive resistance exercise."

In the history of strength training, this turn of phrase proved to be a pivotal event. The shift in language from challenge to invitation—changing "heavy-resistance exercises" to "progressive resistance exercises"—made it easier for more people to imagine themselves lifting weights. Unquestionably, it was a decisive step toward showing how lifting can help everyone—even people who are sick or old or frail—to enliven their days.

No one knew how Thomas DeLorme had come up with that new name for heavy weight lifting until many years after his death, when Jan and Terry Todd made a telephone call to Eleanor DeLorme and asked her about it.

Some years after that, I spent a few days reading old copies of the pulp magazine *Strength and Health* from the 1930s, including the years when Thomas DeLorme was in high school and college and was learning to lift weights, and I came to a phrase that caught my eye.

In the August 1935 issue, an article called "An American Apollo" describes a former "weakling" who "transformed himself" through "the progressive resistance movements."

The phrase was so similar to the one that Thomas and Eleanor DeLorme later suggested they had made up, I wondered what to think of it. Continuing to read the magazine, I came to another article, with similar language in its headline: FACTS IN PROGRESSIVE TRAINING. And then, in other issues of *Strength and Health* published around this time, during DeLorme's impressionable early years of learning to lift, I ran across more variations on the phrase: "progressive methods," "progressive exercises," "progressive system with barbells and dumb bells." There is a report on weight training and competitions involving a large group of readers in Atlanta—four hundred of them—who called themselves the "Jewish Progressive Club." There are advertisements promoting mail-order weight lifting lessons in a "double progressive system" based on the principle of "progressive resistance."

Strength and Health not only foreshadows the name of Thomas DeLorme's lifting technique, it also suggests his basic lifting protocol, as well as his strength-testing technique.

The magazine's August 1935 advice column on training begins with this question from a reader: "What do you consider the ideal number of repetitions to use in training?" The magazine's answer to that question is complex, suggesting a cyclical weekly training protocol in which readers would "employ a weight at the beginning of the week which you can properly perform ten repetitions with," before shifting in the middle of the week to a weight "which will permit fifteen repetitions," and throttling up at the end of the week to work out with heavier weights, "the heaviest weight that you can use for five movements." Here, though, is the bottom line:

> Great strength can be developed by handling very heavy poundages as little as two or three repetitions. But you also wish muscular size and shape which is the principal purpose of the night on which ten movements are made in each exercise with eighty percent of your limit poundage.

The next month's advice column describes how to determine someone's "limit poundage" for ten repetitions:

> We gave him a light weight in the exercises and asked that he perform a set rate of ten repetitions and we added to this weight as long as he could correctly perform ten repetitions. The body is developed very quickly in this manner.

After Thomas DeLorme graduated from medical school, sets of ten repetitions at a high level of intensity relative to maximal strength, determined by the strength-testing technique that had been described in *Strength and Health* magazine, became the basis of the lifting protocols he published in medical journals.

Subsequently, protocols based on his description of the "repetition maximum" were validated by decades of medical and exercise physiology research. They became a basis of resistance training guidelines for the public, adopted and promoted by national and international medical and health organizations. They have been adapted in training by generations of athletes, including Jan Todd, Arnold Schwarzenegger, and Charles Stocking.

They are what DeLorme made of the advice he found by reading a muscle magazine when he was a teenager.

There may be some chance that DeLorme intentionally suppressed this source of his signature ideas. Or, after reading about progressive resistance training in *Strength and Health*, he may have followed the magazine's advice for so long that he forgot where it came from.

If Thomas DeLorme did remember the source, more especially of the strength-testing technique and lifting protocol, he certainly would have had good reason to hide it. Jan Todd says, "He's got to be self-conscious that he's a weight-lifting doctor, and therefore his credentials may be more suspect." When medical colleagues heard that DeLorme was treating U.S. Army soldiers with "heavy weightlifting"—one form of the alternative medical treatments that, just a few years earlier, *The Journal of the American Medical Association* dismissed as almost beneath contempt, after the reviled figurehead of physical culture Bernarr Macfadden had said the military ought to be providing just that kind of nontraditional treatment—DeLorme's more conventionally minded colleagues might have thought he was, as Jan Todd puts it, "in the realm of being a crackpot like Macfadden."

DeLorme could not afford to be associated with a crackpot like Macfadden, Todd adds, because "he was trying to give weight-lifting scientific terminology." He was "trying to put it all on a scientific level so that it would pass muster."

What matters most, Todd points out, "is not where the words came from," but DeLorme's determination to run the risk of applying his knowledge and experience of heavy weight training in his work at the hospital, in rigorous experiments that tested lifting weights as a medical treatment; and the fact that his experiments "demonstrated that it worked," at a time when medical orthodoxy said he was doing the wrong thing.

She says, "I think it would have been really easy for him not to have had that courage."

Thomas DeLorme's translation of *Strength and Health* into strength and health shows medical and athletic knowledge about muscle are inextricable. It also illustrates, clearly and winningly, the social and historical nature of strength: how knowledge of strength and training develops through acts of sharing between mentors and students, friends and strangers, in romance and family, and among competitors. It shows strength as something people create together, by giving and receiving in surprising ways, out of whatever resources they have at hand and even at great risk. Because the prize of life is worth it.

PART III

GAIN THE PRIZE
How muscle is a matter of life and death

CHAPTER 7

Heavy and Light

The first "Geriatric Olympics" took place in 1974 at the Hebrew Rehabilitation Center for Aged, a nursing home in Boston, Massachusetts. Competitors in this sporting event were frail. They were also hardy.

Some survived the Holocaust. Others fled the Cossacks. They all lived through the Great Depression, and few had experienced leisure before they retired. To help them make up for lost time, nursing home staff organized activities galore—shuffleboard, ceramics, gardening, a klezmer band—but the Geriatric Olympics was recreation on another level. After a torch-lighting ceremony, the games began. Relay races and horseshoe-throwing; basketball, soccer, and volleyballoon; arm wrestling, "and even a contest to see who could translate the most Yiddish words."

If much of this happened in wheelchairs, and if even the gold medalists led mostly sedentary lives, these annual rites at least gave this place—affectionately nicknamed "Hebrew Rehab" by its residents and staff—some rudimentary sense of a physical culture. For a nursing home, this was unusual in the mid-1970s. So was another element of life here. As a form of civic service, Hebrew Rehab residents volunteered in droves as research subjects in scientific studies at Tufts, Harvard, and other local universities.

They were not all willing to do just anything, though. Asked to bequeath his brain for posthumous dissection, one man balked: "Can you imagine when I'm in heaven, walking around, and they say, *There goes Ed, without a brain!*"

Beginning in the late 1980s, a series of experiments run by a young doctor named Maria Fiatarone, a faculty member in geriatric medicine at both Tufts and Harvard, gave a different kind of opportunity to that man who refused to leave his brain to science. Asked if he would commit to a regimen of lifting heavy weights, the man said yes; and scores of other Hebrew Rehab residents, male and female, started lifting, too. Their workouts produced unprecedented proof that high-intensity progressive resistance training can strengthen and build muscle even for the oldest people, with life-changing effects. Hebrew Rehab residents who lifted weights gained power to function more independently, and to live with more autonomy and dignity, into their last years.

After Fiatarone's research began to reveal this new view of muscle, in subsequent years she organized more studies, exploring how weight training can affect the human body from head to toe, and can help prevent or treat many diseases and disabilities throughout the lifespan: insomnia, depression, diabetes, heart failure, vascular disease, kidney failure, stroke, breast cancer, hip fracture, osteoporosis, and arthritis; and neurological conditions including multiple sclerosis, Charcot-Marie-Tooth disease, attention-deficit/hyperactivity disorder, mild cognitive impairment, and dementia.

Maria Fiatarone's decision to study high-intensity strength training for the oldest people was at least as daring as Thomas DeLorme's original experiment with heavy lifting for wounded soldiers. When Fiatarone began her research, she, too, confronted skepticism and prejudice. Even as she first proved that progressive resistance exercise is critical to long-term health for all, weight training was widely scorned, slighted, and dismissed as hedonistic and extreme, tainted by the anabolic steroid abuse of a few.

Fiatarone was not distracted or deterred. Her studies of weight training, built upon and inspired by others' research, helped establish a new perspective on their subject: For young and old alike, muscle-strengthening exercise can be a locus of regeneration, offering chances for growth and development; for service and self-improvement; and for healing and renewal.

The doctor might not have made her discoveries about muscle, though, if she had not joined forces with an adventurous group of colleagues and patients in Boston.

———

What happens to all the grandmothers and the grandfathers? What does time do to human bodies? How much of the body's powers and abilities does a person have to lose? When does the losing have to start, and how fast does it have to proceed? What can people do to slow or stop the loss? Seeking answers to those questions as they pertained to every aspect of the body was the mission of Irwin Rosenberg's laboratory at Tufts University. The Human Nutrition Research Center on Aging was a hive of two dozen physicians, dietitians, and physiologists, supported by money from the U.S. Department of Agriculture.

The importance of strength and muscle to daily life in older age is easy to quantify, says Walter Frontera, a physician who worked in the Tufts lab in the 1980s. If your strength is 100 and you need 70 to rise from a chair, Frontera says, you will use 70 percent of your maximal force to stand up. But if you can build your strength to 200, then using 70 to stand up from that chair becomes a whole different experience, involving half the effort—35 percent of maximal force.

The gain in strength Frontera describes—the idea that older people could increase their strength from 100 to 200—seemed ludicrous to most doctors and scientists for most of history. In the early 1980s, Frontera says, settled scientific views of strength in older age followed a neural paradigm, where any possible improvement in older people's muscular strength was attributed entirely to neural activation. Experts thought skeletal muscle could not grow or be regenerated during a human lifetime. They thought that as people aged, their muscles gradually lost capacity to adapt, so that by age fifty or sixty, it was unsafe to lift any weight that felt heavy to lift.

Walter Frontera wondered about that.

Frontera's interest in muscle started on the basketball court. He learned the game while growing up in Puerto Rico. He kept playing through college at the University of Puerto Rico in San Juan, then went to medical school, where he specialized in physical medicine and rehabilitation. In that specialty, which treats biomechanical disorders and injuries by physical means (such as physiotherapy, heat, electricity, and manipulation) instead of surgery or medication, Thomas DeLorme's strength training studies are considered classics. In the time since DeLorme had first prescribed heavy resistance exercises to hasten rehab from orthopedic surgeries during World War II, practitioners of physical medicine had broadly adopted the technique. Outside that specialty, few doctors seemed very interested in progressive resistance training as a form of medical treatment. In 1983, when Frontera started work at Tufts, he noticed that the lab's exercise studies were focused only on aerobics; and he asked his new bosses if he could do some research on strength conditioning, too.

Frontera got the go-ahead to recruit twelve healthy but sedentary older men, aged sixty to seventy-two, for a study of workouts with heavy weights—equal to 80 percent of each individual's one-repetition maximum strength—three times a week for twelve weeks. Scientists had never studied what happened when people of that age lifted weights that heavy, as far as anyone at Tufts knew. This cohort lifted weights about twice as heavy as their age peers had done in previous studies.

Their hard work produced phenomenal results. In twelve weeks, the strength of the men's quadriceps doubled. The strength of their hamstrings tripled. These gains of strength involved mental effort and increased neuromuscular engagement, certainly. But the neural paradigm did not fully account for them. As the men's strength grew, the contractile tissue in their muscles also grew. According to computed tomography scans—CT scans—the size of their whole muscles grew by more than 10 percent. When Frontera looked through a microscope at muscle tissue biopsies taken from the men's legs, he saw growth also at the level of individual muscle fibers.

Conventional wisdom about muscle and aging had been wrong. With effort, older people could make the same relative gains of strength and muscle as younger people could make.

The muscles in these twelve men's legs helped to clarify the role that muscle plays in every person's life. Walter Frontera's experiment showed possibilities of growth and renewal, where most research had traced slopes of loss.

From birth, through the typical course of growth, most people build muscle up to a peak at the age of thirty or so. General physical activity is a major determinant of peak muscle mass. The most active people have, on average, almost 2 kilograms more peak muscle mass than the least active people.

Then comes decline, when people lose muscle at a rate of at least 3 to 5 percent per decade. Some research has found a faster rate of loss in early adulthood and middle age—6 percent per decade. Frontera explains the variability in these numbers to me: "Cross-sectional studies that compare groups of persons of different ages at the same time tend to underestimate the magnitude of the loss when compared with longitudinal studies that follow the same individuals over time." Longitudinal studies of muscle loss, he says, "are a better reflection of the aging process than cross-sectional studies."

By the time we are older, in our sixties and seventies, the rate of loss can accelerate to roughly 1 percent per year, or 10 percent per decade. (For women the loss may be a bit slower on average—0.6 to 0.8 percent per year—and for men it may be considerably higher on average. In a twelve-year longitudinal study of a small group of older men, Frontera and colleagues found an average rate of loss of 1.4 percent.) Studies all over the world—Australia, India, Italy, Japan, the United Kingdom, the United States—have found significant age-related muscle loss. Studies show that general physical activity such as walking ceases to build muscle in young adulthood, though, and general physical activity does not protect against the muscle loss that happens with age. The most active eighty-year-olds have, on average, only about 1 kilogram more muscle mass than the least active people.

Changes in muscle fibers help to explain changes in whole muscle mass—but these changes are numerous and complex, and many of the details are in dispute. Some widely accepted views of how aging muscles change are based in part on measurements of the muscle fibers that have traditionally (but not universally) been associated with muscular strength and power—type II fibers. Measurements show that these fibers shrink in healthy older people, and the number of these fibers also strikingly declines. These changes cause functional impairment, many scientists say, on the following logic: Because most people tend to stop moving heavy weights and making quick movements as they get older, they also stop stimulating type II fibers. Without stimulation, motor neurons in the spinal cord that send signals to those fibers may be lost.

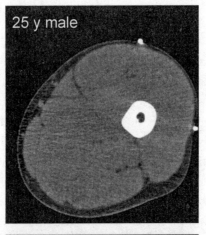

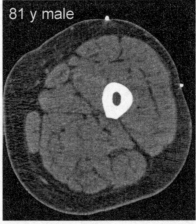

With age, our leg muscles can become marbled with fat, as shown in these cross-sectional CT scans of the upper thigh.

While muscle fibers change, the whole muscle's composition also changes. This change, visible in CT scans, is not in dispute: Connective tissue and fat seep in among the muscle fibers and interfere with fibers' ability to contract. As a result, older muscles are not as muscular as they used to be. Where muscle fibers make up 70 percent of the volume of a twenty-five-year-old's muscle, they comprise just 50 percent of an eighty-one-year-old's muscle.

With progressive resistance training, the losses can be slowed. Some of the losses can even be stopped for many years, or for decades. Walter Frontera's experiment was a giant step in learning how to stop that loss.

When Frontera wrote up his study in 1987, it set the Tufts lab buzzing. A new theory about problems of older age was emerging. William J. Evans, Walter Frontera's immediate boss, was also impressed by the strength training research that Archie Young was doing in the United Kingdom. (Evans told me that Young was "really the first to use strength training to improve functional status in elderly people.") Considering the new findings, Evans began to think the main problems of aging had to do with body composition.

Disease and disability in older age, Evans posited, were mainly caused by one imbalance. Too much fat and too little muscle. Many doctors counseled aging patients to lose weight, but Evans thought that might not be the best advice. In his view, a more nuanced and constructive type of change might better suit more people: *fat loss and muscle gain.*

Losing fat and gaining muscle reduces risk of many chronic diseases, including diabetes, because more muscle increases insulin sensitivity; and heart disease, because more muscle helps maintain high levels of good cholesterol; and cardiovascular disease, because more muscle consumes more oxygen, increasing aerobic capacity.

Losing fat and gaining muscle reduces risk of functional disability, too, since function is largely determined by how much fat and muscle a person has, and where that fat and muscle are located. People with more fat in combination with less muscle—a lot of weight to carry, but not much strength to carry it—have a hard time moving around, especially if the fat is concentrated around the middle of the body and if it invades the muscle tissue itself.

If body composition was key to the problems of older age, and if older men could gain muscle and strength by lifting heavy weights, then maybe, Evans and his colleagues speculated, progressive resistance training could help a lot of people.

Around that time, a new doctor came to work at the lab, with a double-barreled academic appointment at both Tufts and Harvard. Maria Fiatarone took these jobs immediately after a fellowship in geriatric medicine, a newly minted medical specialty focused on caring for older people.

From childhood, Fiatarone had been serious and bold-hearted—when she was in elementary school, her hero was Albert Schweitzer—and she had launched into adulthood with guarded mirth. When her university magazine asked graduating students to pen their own obituaries for a mock-heroic end-of-school feature, she wrote: "Maria Fiatarone, a fencer, graduated in French in 1976. She completed medical school and as a physician went to practice in West Africa, where she was known far and wide for her humanitarian activities. Her most noted achievement was 'training medical aid volunteers for the great famine of 2008.'" Fiatarone was not accepted to medical school straight out of college, though, so she spent a year developing her talent for fencing. In Oakland, California, she trained with an Olympic gold medalist from Romania, Ion Drîmbă, while she also worked at a food kitchen, and in her spare time made a second round of med school applications. When the University of California at San Diego medical school invited her to visit the campus, and the admissions officer who interviewed her turned out to be a fencer, she knew that was the place for her.

Fiatarone's training as an athlete did not involve any experience of lifting weights. In medical school, she learned basic facts about muscle contraction, but nothing about weight training. Not one lecture in any of her medical school classes focused on exercise of any kind—and this was typical in that era, when a survey of medical schools found that only 16 percent of them taught students about exercise.

But now, working alongside colleagues eager for more research on strength training, she jumped right in. She learned how to lift weights and how to teach others to do the same. She learned the intricate, recursive structure of structures that make up human skeletal muscle—the bundles of bundles, each muscle fiber containing hundreds or thousands of myofibrils (meaning "muscle strings"), arranged side by side and made up of sarcomeres, the basic cellular unit of muscle, laid end to end. She learned how those structures operate—how, inside each sarcomere, overlapping arrays of filaments of two molecules, actin and myosin, latch onto each other and pull, generating force that, combined with other sarcomeres' pulling force, propels every unassisted step taken by all the legs that ever walk on planet Earth.

Fiatarone was interested in sarcomeres because when muscle fibers changed, the whole person around them changed, too. The more she learned of progressive resistance training, the more conscious she was of how few had the benefit of this knowledge. Walter Frontera's impressive study of high-intensity training in older men made Fiatarone impatient. Men in their sixties weren't *that* old. They were young-old, disease-free, and physically functional. Was it such a big deal if guys like that lifted weights and built bigger, stronger muscles? If scientists wanted to know how exercise could ease the experience of aging, why didn't they focus on improving lives that most needed improving? Why not take this new knowledge about strength training and test it among the oldest people—the ones who could barely get around, the sick and frail in places like the Hebrew Rehabilitation Center? If there was any chance that lifting weights might improve the quality of life for people who were barely hanging on, why not give it a try?

One day, after visiting Hebrew Rehab, Maria Fiatarone asked one of her bosses, "Don't you think these guys need it more?"

Seven hundred and twelve people lived at Hebrew Rehab. Their average age was eighty-eight, and three-quarters of them were women. Every resident had multiple medical conditions. Almost half required help to engage in the essential activities of daily life: getting out of bed, going to the bathroom, bathing, walking, eating.

So in 1988, when Maria Fiatarone designed a proof-of-concept pilot study of high-intensity strength training for the oldest people, Hebrew Rehab was an ideal source of research subjects. In terrible shape, game to work hard, with solid faith that their efforts mattered: To Fiatarone, they were perfect.

Of the ten residents she chose for the study—six women and four men—eight had a history of falls. Seven used canes or walkers to get around. Most were on at least four medications. Most had at least four chronic diseases. The most common diseases in this group were osteoarthritis, coronary artery disease, hypertension, and osteoporosis.

Their workout program would be simple. It would consist of one exercise. The exercise was the knee extension. Picture someone sitting on the edge of a straight-backed chair. She contracts her quadriceps to extend her knees, lifting her lower legs. At the top of the motion, both legs extend straight out from her lap. Then she bends her knees to lower her shins and feet back down. The lift comprising that whole movement, up and down, is called the knee extension.

Two things make the knee extension a good exercise for ninety-year-olds. You can do it sitting down, and it strengthens muscles that help you stand up.

High-intensity strength training is safe and feasible for nonagenarians: This was Fiatarone's hypothesis. The hypothesis was so ambitious that even the first step toward proving it—maximal strength tests for the oldest people—lay in unknown territory. Western medical literature contained no record of anyone having ever done a one-rep max test with a frail ninety-year-old. Medical orthodoxy said lifting heavy weights was inherently risky for older people's hearts. "We didn't know if that was true or not, and we were quite conservative," she remembers.

On the day of the tests, the ten got on their marks. One by one, each was led into the physiotherapy room, carefully positioned on a chair, and fitted with a blood pressure cuff around the arm. Buttons were unbuttoned, electrodes taped onto loose skin and connected to the continuous-monitoring electrocardiograph, and the stylus scratched the heart's electrical activity on a scroll of graph paper unwinding as legs lifted weights and lowered them, building up to one-rep maxes. The knee extension apparatus was a cable-pulley machine with a weight stack that Fiatarone and her colleagues adapted for their purpose. They prepared little sandbags to attach to the weight stack so that small increments of weight, as small as half a pound, could be added.

Following the line of research established by Thomas DeLorme, doctors and scientists had refined the definition of a one-repetition maximum, or 1-RM, as the heaviest weight a person can lift through a full range of motion and with perfect form. The one-rep max test is not meant to involve grunting effort. It is a way to assess maximum strength within a margin of safety. Accordingly, the challenge of these tests at Hebrew Rehab ramped up very slowly.

Each person started by lifting a fairly light weight, relative to the person's capacity. Some of the frailest people lifted only a fraction of a kilogram at first—as little as half a pound, or one pound.

After a rest period of about thirty seconds, Fiatarone increased the weight by what she judged to be an appropriate increment. If the person could lift that weight, then more weight was added for the person's next effort, which took place after another thirty seconds or so. Blood pressure was monitored every three to four minutes.

On average, the ten maxed out at lifting less than 20 pounds. One person lifted almost 40. One person lifted about 5.

A battery of other baseline measurements, too, were made of these ten participants. Height and weight, and diet analysis. Body composition, assessed by CT scan—to quantify the muscle, bone, and fat in each person's legs to the nearest one-hundredth of a square centimeter. Functional performance, timed with a stopwatch: standing up from a chair without using one's arms for leverage; normal walking speed over a distance of six meters; and another six-meter walk with the more challenging and more deliberate tandem gait—placing the heel of one foot directly in front of the toe of the other, like in a roadside sobriety test.

Analyzing all these data against the nonagenarians' one-rep maxes, Fiatarone found associations in the data as notable for what they did not suggest as for what they did. Strength was not correlated with biological sex. Strength was not correlated with length of time spent in the nursing home. Strength was not correlated with how much food a person ate. Strength *was* significantly associated with size of muscles, with dietary intake of certain nutrients (vitamin B_6, magnesium, potassium), with walking speed, and with quickness in standing up from a chair.

The smaller their muscles, the weaker people were. The weaker they were, the slower they moved.

Each of the ten residents then started coming to the physiotherapy room for three strength training sessions per week. One month into the two-month program, an eighty-six-year-old man dropped out—an old hernia repair was bothering him—and then there were nine.

In each training session, the lifters completed three sets of eight repetitions, a routine much like Thomas DeLorme's standard prescription. They raised and lowered the weights very slowly: three or four seconds up, three or four seconds down. Between each of the three sets, they rested for a minute or two.

In week one, they lifted at 50 percent of their 1-RMs. In week two, the load increased to 80 percent of 1-RM. Their muscles adapted quickly, so the 1-RM test was repeated every other week, to keep the load at 80 percent of maximum. Every rep of every set was monitored closely by Fiatarone and her colleagues, and the ECG and blood pressure monitoring continued throughout most of those two months.

There were no cardiovascular complications among the nine. They lifted and lowered the weight so slowly that their blood pressure and pulse rates barely changed. Four of the nine occasionally reported some discomfort in the hips or knees, but not to the point where anyone had to take a pain reliever or miss a training session. Everyone was able to do what was asked of them. And everyone did it. The study's attendance rate was 98.8 percent.

These people were dedicated, and they got stronger. The smallest strength gain was 61 percent. The largest was 374 percent. The average gain was 174 percent. The average increase in walking speed as measured by tandem gait was almost 50 percent. The magnitude of improvements for men and women were the same. After eight weeks of lifting weights, no one's improvements reached a plateau, much less approached limits of potential. Their muscles had not only grown stronger, but also bigger. The ninety-year-olds' muscles grew by almost the same amount that a younger person's muscles would grow in response to a similar lifting program.

Maria Fiatarone's preliminary findings widened the scope of Walter Frontera's insight, revealing a fact about human strength so contrary to prior knowledge, many came to see it as a paradigm shift in understanding the process of aging: In the neuromuscular collaboration called strength, muscle remains a vital partner all through life.

What was happening at Hebrew Rehab got under people's skin. The Tufts lab director Irwin Rosenberg was a gastroenterologist, so he was used to thinking mostly about patients' guts, but he found himself giving more thought to muscle. On rounds at the hospital, he would stop at patients' beds and look at their bodies. So many older people, he noticed, had tiny withered legs.

Those skinny legs meant something. No one was talking about it. Rosenberg decided to change that.

"Now, let me stop here to make a sort of plea," he implored the audience, halfway through his speech concluding a medical conference in Albuquerque, New Mexico, in October 1988. He pled with his fellow physicians to remember some things that they already knew.

No aspect of aging, he said, more decisively determines a person's power to perform the most basic activities of everyday life—walking, eating, even breathing—than the loss of muscle mass. As men and women age, they experience the loss of muscle in different ways. During menopause, which accelerates the losses of muscle and strength, women also lose bone mass, a loss well known as osteoporosis. Every doctor knows how to measure bone density, how to tell the difference between strong bones and brittle ones, when to tell patients their options for reducing risk of bone fracture, and what to suggest as options for treatment. But doctors have no such clear standards and practices concerning muscle mass.

Patients, Rosenberg told his colleagues at the Albuquerque gathering, were suffering for the lapse. He asked, about age-related loss of muscle, "Why have we not given it more attention?" His question hung in the room.

"Perhaps it needs a name derived from the Greek," he said. "I'll suggest a couple: *sarcomalacia* or *sarcopenia*," both from *sarx*, that ancient word for "flesh." *Sarcomalacia* would mean sickness or weakness of flesh—*malakía* specifically denotes softness. *Sarcopenia* suggests a deficiency or lack of flesh—*penía* means "poverty" or "need."

In suggesting the Greek, Rosenberg was tongue-in-cheek, playing on the fact that sometimes doctors pay no attention to basic human problems that do not have fancy-sounding scientific names. Treating disease involves a public relations game where good names help to assuage bad experiences. A victor in that game, Rosenberg thought, was osteoporosis—with a name, from *osteon* for "bone," and *poros* for "little hole," that enabled doctors to help their patients keep their bones strong; and more generally it drew attention to the health of women, especially older women. Rosenberg wanted an equally good name for the problem of muscle loss, which he saw as equally significant.

"I mean, the skeleton can't get around very well without muscle to push it around," he later said.

Of the two names he suggested, *sarcopenia*—meaning "lack of flesh"—was the one that stuck. The global medical establishment would not officially recognize sarcopenia as a disease for another thirty years. But right away, the name started doing its job. That day in Albuquerque, gerontologists began to talk about sarcopenia, and within a few years, the National Institutes of Health convened a symposium on the phenomenon, a first step toward funding research on how to address the problem.

"Once it had a name," Rosenberg tells me, "then it could become a disease."

Even in the short time of Fiatarone's eight-week study at Hebrew Rehab, when older people's muscles changed, their lives changed.

Dorothy Tishler was ninety-two. In her first one-rep max test, she lifted 17 pounds with each leg. Eight weeks later, her strength had more than tripled. She lifted 60 pounds with each leg. "I love it," she told the reporter from *The Jewish Advocate* who visited the gym. "I become younger. When I came here five years ago, I could hardly walk. Now I walk better than my daughter, who's only seventy-two."

Fiatarone streamlined this and all the others' transformations into academic understatement when she wrote up the study's results. "Two subjects no longer used canes to walk at the end of the study," she noted. "One of three subjects who could not initially rise from a chair without use of the arms became able to do so." She sent the paper to *The Journal of the American Medical Association*, and in June of 1990 it was published without a single edit, as she recalls—and with a title that, decades later, still sounds almost as exotic as near-future science fiction: "High-Intensity Strength Training in Nonagenarians."

Strength training is safe. This was the most important truth established by the Hebrew Rehab study. In structured regimens overseen by qualified trainers, high-intensity strength training involves very low risk, even for the frailest, oldest people. As Maria Fiatarone wrote, edging as near sarcasm as the idiom of medical literature allows, "The known hazards of immobility and falls seem to outweigh the potential risks of muscle strengthening interventions in this population."

Since then, research has consistently shown that resistance training injuries are rare, especially when people follow planned exercise programs under expert supervision; and this is true for people of all ages, at all levels of ability. In 2019, the National Strength and Conditioning Association's position statement on resistance training for older adults cited a systematic review of exercise programs designed to reduce the risk of falls among people between the ages of seventy and ninety-two, a review that found "only one case of shoulder pain with resistance training out of 20 studies and 2,544 subjects." Yet older people can and do sometimes get hurt while lifting weights, the position statement said, mainly owing to factors such as inappropriately "heavy and repetitive workload, unfavorable positioning or incorrect technique, and exercise selection." At the other end of the age spectrum, "data examining acute resistance training-related injuries in youth" aged eight to thirteen "reveal that approximately 77.2% of all injuries are accidental," according to a 2014 international consensus "Position Statement on Youth Resistance Training" published by the *British Journal of Sports Medicine*. Most such injuries, the document asserted, "are potentially avoidable with appropriate supervision, sensible training progression based on technical competency and a safe training environment."

The Hebrew Rehab lifters showed that strength training can be not only safe for practically everyone but also feasible: easy and convenient. Fiatarone would later say, "To me it was always the idea to keep exercise simple, keep focused on what was the most feasible thing to do." Strength training is feasible in nursing homes, she says, whereas aerobic training may not be.

To some that may sound counterintuitive, but it's the commonsense conclusion of her experience working with the oldest people. She continues: "When you go to a nursing home, it's quite clear that people can't do lots of kinds of aerobic exercise. Simple *walking* is really hard. Yet people *can* sit at a machine and lift heavy weights. It's very much the opposite of what people think. Where it seems like aerobic exercise is the easiest thing to do and it's much harder to lift weights, it's really the exact opposite. The frailer you are, the more opposite it is."

The Hebrew Rehab strength training study had extensive practical implications. Its philosophical implications were at least as far-reaching. What happened at Hebrew Rehab upended the traditional story of aging. The cliché of inevitable decline—as age increases, function wanes—turned out to be false. What happened in Dorothy Tishler's legs suggested nothing less than a new way of considering the course of human life. The cliché that it's never too late turned out to be true. Even into oldest age, even in dire situations, every person has some power to change how time changes the body. Especially if you have help, and knowledge.

Still, the Hebrew Rehab lifters had not conquered time. Their example offered hope to others, but demanding hope. "All subjects resumed their sedentary lifestyle" after the study ended, Fiatarone found in follow-up visits with them. They resumed their sedentary lifestyle because the nursing home had no system or program in place to help them continue training. One month after the nine stopped lifting, they had lost, on average, a third of what they had gained during three months of workouts.

"Changes in muscle function are not maintained in the absence of continued training," Fiatarone wrote in the published study. Ongoing improvement would require "an ongoing program of muscle reconditioning."

Changes in muscle are not maintained for long periods of time, in the absence of continued training—for anyone, of any age. Even during puberty, when growth spurts might be expected to make up for detraining, "adaptations to exercise training are transient and will steadily decay once training ceases," as Neil Armstrong and Joanne Welsman write in the textbook *Young People and Physical Activity*. Up to the age of thirty-five or so, most younger people are, however, able to maintain muscular strength and mass with minimal frequency of exercise—as little as one workout per week, involving as little as one set per exercise. Likewise, most older people who have gotten stronger at the gym can maintain their increased ability to exert force with the same modest number of workouts, just one per week. Maintaining muscle mass for older people seems to require more exercise, though—at least two workouts per week, involving at least two or three sets per exercise.

Little to none of this information was known in 1990—researchers had not yet begun to contemplate the minimum effective dosages of weight training that could help older people maintain muscle mass and strength—but Maria Fiatarone did know that the residents of Hebrew Rehab would require follow-up training programs to help keep them lifting weights after the data collection for scientific studies had ended. She knew these programs would also need to be based on larger amounts of data than her pilot study had produced, and she wanted to collect some of that new data in a different kind of study.

Her first study of strength training at Hebrew Rehab had been a quasi-experimental, uncontrolled trial. In uncontrolled trials, subjects are measured; they are put through an intervention; and they are measured again to discover the intervention's effects. Such trials produce what scientists call poor quality evidence, because people tend to improve their behavior when they know that behavior is being studied. There was no way to say how much of the nonagenarians' increased strength could be attributed to the exercise alone, and how much came from their wish to please the charismatic young doctor who coached them or to experience camaraderie with other participants in the study.

To establish a cause-and-effect relationship between lifting and strength among the nonagenarians, Fiatarone needed to follow up her pilot study with a true experimental study, a randomized, placebo-controlled trial. Randomized controlled trials can influence public policy, medical practice, and economic choices better than other forms of evidence, because they are designed to minimize the influence of chance, bias, and confounding on the results. Randomized controlled trials involve much larger numbers of research subjects than uncontrolled trials, and they assign people to the control or intervention groups randomly.

Until the 1990s, there had been few placebo-controlled exercise trials. It's hard to control exercise trials because people who exercise almost always know that they are exercising. But it's easier to control an exercise trial using older subjects because exercise, as many older people experience it, is rarely robust enough really to count as exercise. A gentle exercise class for senior citizens is usually not much more effective than no exercise at all—"placebo exercise," Fiatarone believed. When she read about a gentle exercise program for older people in Minnesota—the program was called SMILE, which stood for So Much Improvement for So Little Exercise—she scoffed, "That's great—that's going to be the control group for my next study."

Gentle exercise does not build strength, except perhaps marginally or in the most severely deconditioned muscles, because it doesn't give muscle what it needs to thrive. To thrive, muscle needs to work hard, it needs rest, and it needs nutritious food, especially food rich in protein. The first study of weight training at Hebrew Rehab showed how muscle could respond to regimens of work and rest without an intentional change of diet. The new study would show how a nutritional supplement affected workouts, too.

Compared to the first cohort of lifters, the second cohort would be very frail, very sarcopenic, and fairly malnourished; and there would be ten times more of them: one hundred, instead of ten. Each person would be randomly assigned to one of four groups, receiving four different interventions. One did strength training. The second received a nutritional supplement. The third did strength training and received the supplement. The fourth was the control group: They did gentle exercise. To manage all these interventions, especially the training, Fiatarone needed help. She found Evelyn O'Neill.

Growing up on the south coast of Massachusetts in New Bedford, Evelyn O'Neill had been the middle child of nine. Her dad worked in a box factory, her mom taught elementary school, and the family went out to eat exactly once a year, for pizza. Summer vacation meant going to the other end of New Bedford, to the beach. O'Neill grew up to be resourceful, always working multiple jobs. She worked nights waiting tables at Franco's on Route 1 in Norwood, and during the days she organized activities at Hebrew Rehab. When she applied to help run the next strength training study, O'Neill was thirty years old, and she had never really exercised. She told Fiatarone that she had not had time or money for sports in high school, and had been busy studying and working in college—was that okay?

She remembers the doctor's answer: "You don't know the exercise piece, but I do, and I can teach you everything I know so that you can do a good job. I need someone who is compassionate and has a lot of caring for the residents."

The pair of them bonded over their shared Catholic faith, the basis of O'Neill's compassion and caring, and of Fiatarone's, too. But meek and mild these two were not. Instead, Fiatarone aspired to an ethic she named with the Greek word *agapē*, defined in some popular Christian writings as "selfless love that is passionately committed to the well-being of others." O'Neill brought a champion's spirit to the Hebrew Rehab physiotherapy room gym that was like a visitation from Nike, the goddess who bestowed the gift of victory on ancient Greek athletes. O'Neill's motivational talent, one of her former colleagues says, was "the dirty little secret" of Fiatarone's strength training research. "What really made the difference was the coach and the motivator."

O'Neill motivated residents by saying that strong muscles would help them do some of the things they most wanted to do, and she showed them that she had absolute confidence in their potential ability to do those things. "To put it in terms from *Game of Thrones*," O'Neill's former colleague says, "Evelyn could get those people to climb that frozen ice wall."

Helene Freundlich was an articulate, wheelchair-bound, ninety-six-year-old Holocaust survivor. She wore big glasses with big cat's-eye-shaped frames, and she was one of the first volunteers for the new strength training study.

Before training commenced, Evelyn O'Neill decided that Helene Freundlich was going to need a bra. O'Neill remembers telling the older woman, "We can't exercise you in a skirt, and with your bosoms down below—we need to get you fitted up for this new situation."

Getting fitted up was complicated, because bras are not generally designed to be worn by ninety-six-year-old bodies, and trying them on would be no small exertion for a tiny woman who was so weak that lifting up her arms took almost all the strength she had. To buy a bra, Freundlich would need someone to take her, in her wheelchair, to the ladies' store, and go into the dressing room with her; help take off her sweater and her shirt, and put the new bra on her, fasten the clasp, adjust the straps; and then, after all that was done, help her decide how it looked.

O'Neill was glad to do the favor. "I did anything for the residents. It didn't bother me," she says. "I took her to Lady Grace, which is a lingerie store for older women," and did some maneuvering in the dressing room. When that whole process was complete, O'Neill remembers, "She was so happy. I don't think she'd worn a bra for years."

The new study at Hebrew Rehab involved two exercises, a pair of lifts activating major muscles used in basic movements of standing up and walking. In addition to the knee extension, the residents now did a hip extension movement. Hip extension happened on a leg press machine. The machine had a high-backed seat reclined at a slight angle, facing a raised rectangular metal plate perpendicular to the ground. When the lifter sat and pushed the plate out with her feet, the machine enabled a kind of horizontal squat.

"It was hard to find a machine that could deliver the intensity we were looking for," recalls another of Maria Fiatarone's collaborators, Lewis A. Lipsitz, who ran Harvard's Geriatric Medicine Program and directed medical research at Hebrew Rehab. "The machines that one might want to use were set up for younger people," Lipsitz adds. "You don't start at a quarter of a pound."

One leg press machine could be dialed down to a quarter of a pound of resistance, though: the Keiser Leg Press, manufactured by Keiser Sports Health Equipment (now called the Keiser Corporation) in Fresno, California. Keiser resistance training machines operate by a pneumatic system, in which compressed air provides variable resistance that can be almost instantly adjusted on a range of weight from 0 to 1,200 pounds, calibrated to the tenth of a pound or kilogram. Many elite athletes train on Keiser machines, which are expensive—the 2023 model of the Keiser Leg Press retailed for $9,920. In the summer of 1990, though, after the machines' inventor, Dennis Keiser, read about the study of strength training in nonagenarians, he contacted Fiatarone to ask if she would like to have some equipment to use at Hebrew Rehab, and soon a shiny new Keiser machine showed up in Roslindale.

When training sessions for the new study began, Evelyn O'Neill would press her palms against her own legs and hips, to show the residents which muscles were involved in each lift. For motivation, she constantly referred to the big picture, helping residents see how it would help them to be strong. "I always kind of related strength to an activity that meant a lot to them," she tells me. "Like going out with their children for lunch. They don't want to be a burden. So being stronger meant not being a burden. Being independent. That made it real to them."

Sometimes she had to drag out all the reasons in the world to motivate a pair of sore old legs. After their first workouts, the residents felt intense discomfort in their muscles, as is typical for beginning lifters of any age. O'Neill had to figure out how to talk about these sensations without using bromides like "No pain, no gain." "We're talking about ninety-year-old people!" she exclaims. "They're already in a lot of pain. Why would they do something that would make them feel *more* pain?"

For the same reason that Eleanor and Thomas DeLorme engineered the inviting name of "progressive resistance exercises"—in order to make lifting more appealing to more people than the imposing name of "heavy-resistance exercises"—Evelyn O'Neill collected her thoughts in order to hammer out some diplomatic language that would help her in training the residents.

"Instead of talking about pain, we talked about soreness. Soreness that will go away, not soreness that will last forever," she tells me. To the residents, she explained, "This is DOMS—delayed-onset muscle soreness. It's normal." O'Neill says "DOMS" in a friendly tone, pronouncing it like the affectionate nickname for a person named Dominic—as in, *That's Dom's dumbbell.* To the residents, O'Neill emphasized that DOMS is a universal experience: "I get it. Olympic athletes get it. *Everybody,* if they're working in a right way and exercising to their limits, will get the same kind of soreness."

She reassured them, "That's okay, to feel some soreness. Exercise is physical *work.* Damage to the muscle, and repair."

Of all the words O'Neill used to help the residents divine what happened in their muscles, she found one that most dependably prompted insight. *Work* was the magic word—and when O'Neill invoked that term during training sessions with residents, she says, "They were like, 'Exercise is *work?*'"

Yes, she answered, "Exercise is physical work. And when you do that work is when you're going to see results. The change in the muscle. If you're not working hard enough, it's not going to change your strength."

When muscles are not accustomed to the kind of hard work these residents were doing in the weight room, delayed-onset muscle soreness sets in without much delay. Especially for beginning lifters, the day after a workout, muscles can feel tender or stiff or raw. People describe the sensation in a variety of ways, but the gist of it is, muscles feel sore when they move. Usually they're also weak. The discomfort and the fatigue can last for a few days, while fading, and then DOMS becomes another of life's problems that has solved itself. Soreness can return as workouts become more intense, but DOMS almost never makes most people as uncomfortable as it does at first.

Delayed-onset muscle soreness does not come from exercise in general, and it is not meaningless suffering. It comes from one particular type of movement, called eccentric contraction—which, in weight training, happens not while you lift the weight, but while lowering the weight back down. (The first syllable of *eccentric*, in this context, is pronounced as a long *e*—so *eccentric* rhymes with "me-centric.") When people move heavier weights than they are used to handling, eccentric contractions can cause damage to muscle fibers. This damage, in turn, triggers the process of the muscle's repair and regeneration, making it less liable to damage in the future.

The whole process is a complex neuromuscular phenomenon that scientists understand imperfectly, but some of the known details have an uncanny quality about them, seeming almost to suggest that paradox is built into the cellular structure and function of our muscles.

When you pick up a weight and do a biceps curl, the muscle contracts to its shortest length at the top of the movement. Then when you lower the weight slowly, under careful control, your biceps keeping up the effort to contract, the fibers are still trying to shorten, even while they're being lengthened by the force of gravity pulling down the weight. One physiologist describes eccentric movement as "extension while flexing." Another calls it "the active stretch of an activated muscle."

When higher-force eccentrics lengthen muscles more than they are used to being lengthened, muscle fibers are damaged. Only some of the fibers in the muscle are damaged, not all of them. But the damage is dramatic. Before eccentric exercise, muscle fibers are arranged in orderly fashion, like neat rows of two-lane highways:

After eccentric exercise, those structures look very different. They look blurry. As if a worker on the road crew went rogue and drove a bulldozer across those highways when the concrete surface was freshly poured:

The name *eccentric contraction* is a bit nonsensical. Muscles do not shorten during eccentric contraction, so it's strange to call the movement a contraction, but that's what it's called. Eccentric contraction triggers a chain of events that can seem a bit nonsensical, too. Muscle damage in eccentric contraction begins the process of its own repair, a process that another physiologist found so flummoxing, he remarked: "It is difficult to imagine why a structure would be injured when performing the very act for which it is designed!"

It is difficult to imagine if you assume that injury is always insult. But eccentrics refute that assumption. Muscle damage from eccentric exercise, and the repair and regeneration that follow, mark the nonlinear progress by which muscles grow stronger.

This process of repair begins with inflammation—which also continues to damage the muscle. Delayed-onset muscle soreness corresponds with inflammation, and a lot of people, including people much younger than ninety, don't enjoy it. In hopes of easing the soreness and getting back to workouts more quickly, many take non-steroidal anti-inflammatory drugs such as ibuprofen. With these drugs, pain may more rapidly subside. But experiments show that these drugs may also interfere with the process of repair and recovery, and may therefore inhibit strength and hypertrophy in the longer term. Whatever causes the soreness may also be necessary for the proper function of the muscles.

Boiled down, some of the physiological fundamentals of lifting weights can sound as strange as anything Alice heard in Wonderland:

Eccentric contractions, which are not contractions, damage muscle to rebuild it.

When muscles feel sore, it can mean that they are healing.

Muscles weaken to grow stronger.

Drawing these weird observations into one, the central and paradoxical principle of strength training made sense to Evelyn O'Neill, as it would to almost anyone who's ever worked two jobs: *No struggle, no prize.*

O'Neill had to stay vigilant about her vocabulary, to convey the imperative of effort to frail elders. The same way she talked to the residents about soreness instead of pain, the same way she made sure they understood that weight-lifting exercise was a type of work, she framed the appropriate degree of struggle as "Working at a level that feels challenging. Not hard. *Challenging.* Challenging for *you.* Everybody works their own challenge."

All the residents brought their own personal challenges to training sessions, and Evelyn O'Neill and Maria Fiatarone worked all those challenges with them. Esther Pepi, who had severe muscular dystrophy, could not walk. She had to use a wheelchair. She had just a few strands of muscle left in her hamstrings, which had been almost completely replaced by fat. Ed Rosenthal—this was the Ed who, when asked to bequeath his brain to science, said no because he didn't want to get teased in heaven for being brainless—had retired from teaching biology at Girls' High School of Boston. His peripheral neuropathy was so severe, he could barely feel his feet on the ground, and his legs were so weak he did not feel safe walking up stairs. Helene Freundlich, who wore her new bra when she came to the gym, could lift no more than the weight of her own two arms when she started working out, and so her two arms were what she lifted.

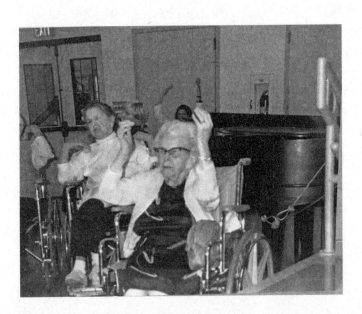

The New England Journal of Medicine published the study results in 1994, with Maria Fiatarone and Evelyn O'Neill as the lead authors. The residents in the group that only lifted weights increased their strength by 100 percent. The residents in the group that lifted weights and took the nutritional supplement increased their strength by 250 percent. Training increased walking speed, stair-climbing power, and spontaneous physical activity. Training and taking the nutritional supplement increased all these measures slightly more. The residents in the group that just took the supplement, on the other hand, gained no strength, and they experienced no improvement in function.

No residents recovered so much function that they were able to leave the nursing home, but no one had been expecting time would run backward. Fiatarone and O'Neill hoped that training would improve the residents' abilities to act upon the world and more independently navigate the place where they lived, and that is what happened. Four people in the strength training group who used walkers at the start of the study were afterwards able to walk with only the help of a cane.

Helene Freundlich, who "never missed a session," O'Neill says, regained so much strength, she was no longer dependent on her wheelchair. For O'Neill, seeing Freundlich "able to walk again with a walker and do things for herself was a joy," she remembers. "I think that is what the study gave to a lot of people," she adds. "Freedom to feel strong again."

According to the data gathered by this study, the most reliable predictor of people's strength gain was the starting level of muscle mass, and training was associated with increasing mass. The area of individual muscle fibers and the size of whole muscles increased by 10 to 15 percent in people who trained and took the supplement. Size decreased by about the same amount in the control group. Discussing these results, Fiatarone once said, "If you do nothing, your muscles are shrinking away before your very eyes."

Ed Rosenthal did not shrink, he grew. His legs grew strong enough that he could climb stairs again, though due to the neuropathy he still couldn't feel the stairs underfoot. He had always been slender, but lifting made the muscles in his chest and neck and arms grow so much that, when he went to buy a new suit to wear to the bar mitzvah of the nursing home director's son, the tailor did a double take at the customer's new measurements: *You have to wear a portly!*

Esther Pepi also grew. After coming to the gym for a while, she said, "The weight training is helping to keep some of my muscles alive." She became strong enough to transfer herself into and out of her chair more independently. She wrote a poem about the gym, her poem was posted in the gym, and before long, everyone at Hebrew Rehab called the gym by a new name, a phrase plucked from the poem's first line: "The Circle of Fitness is the place to be."

"Life is short, play hard," was the ubiquitous tagline of a Reebok advertising campaign in the early 1990s. When Maria Fiatarone saw that, she read it in light of what she saw happening in the Circle of Fitness, and she adapted the phrase, thinking to herself: *Life is long, play well.*

The residents of Hebrew Rehab were learning to play well because their muscles were learning to work well. Their muscles learned—their muscles adapted to the new demands of training—because doing so is quite literally the nature of each individual muscle cell.

Muscle cells adapt. They are always growing or shrinking, always remaking themselves in response to the demands that are, or are not, made of them. To better understand how that adaptive process happens—how muscle repair is managed—physiologists probe to the molecular level of muscle cells, where much depends on a special kind of nucleus.

Skeletal muscle cells are the largest and longest cells in the human body. They are also the most densely populated with the structure that middle school biology textbooks sometimes call "the brain of the cell," the nucleus. Each brain cell, liver cell, bone cell, and skin cell has one nucleus. Each skeletal muscle cell has multiple nuclei. Many have hundreds or thousands. These muscle cells need all those extra nuclei to coordinate simultaneous transmission of signals from the nervous system and the structural changes in muscle fibers these signals mandate. Different kinds of nuclei perform different kinds of tasks.

Among the many nuclei arrayed along muscle cells are a few called satellite cells. When muscle cell walls are damaged, satellite cells appear. Satellite cells attach to muscle cell walls as to scaffolding, while the immune system removes damaged cell components. Satellite cells fuse together, forming the proto-muscle fibers called myotubes, and myotubes produce replacement parts for the muscle. The parts include myofilaments, which are the stuff of contraction: Molecules of actin and myosin grab onto one another to make muscles contract after regeneration is complete.

Scientists long believed that humans were born with relatively large numbers of satellite cells in their skeletal muscles—perhaps almost one-third of the total number of muscle cells—but that, by oldest age, all the satellite cells in a person's muscles, or all but a very few, were dead or defunct.

That belief was called into question when Walter Frontera's and Maria Fiatarone's studies showed older people's muscles getting bigger and stronger. When those older muscles grew, it made the doctors wonder: Was the whole process of regeneration, down to the molecular level, also still taking place?

Before and after the second Hebrew Rehab training study, Fiatarone took biopsies from the quadriceps muscles of residents in each intervention group. Slices from those chunks of muscle were prepared and stained, then photographed through an electron microscope. The photographs suggested that every aspect of regeneration could indeed still happen in the muscles of the oldest people—but these things happened only in the muscles of the people who lifted weights.

In the muscle biopsies taken from people who did strength training, Fiatarone found a hormone, insulin-like growth factor 1, that mediates the effects of growth hormone to stimulate the body to make muscle. She found an increased volume and density of mitochondria, the cellular components ultimately responsible for muscle power. She saw tiny buds of myotubes branching out from satellite cells, reaching toward buds that branch from other satellite cells—the branching that starts the process that becomes their fusion into muscle fibers.

When she stained the sample, the shapes lit up green, purple, blue—different kinds of molecules taking on different colors. Green was embryonic myosin. Purple, neonatal myosin. Until that point, those molecules had been seen only in the muscles of embryos and fetuses, and during recovery from neurological injuries. No one had any idea that adult muscles could generate these molecules as a result of weight training, much less that it could happen in the tenth decade of life.

These findings mattered to Fiatarone as scientific discoveries, but they mattered more to her for what they symbolized about human potential, for the individuals from whom the cells had been collected.

Almost twenty-five years after publishing her findings, she shows me photographs from the electron microscope, and then she shows me a portrait of Ed Rosenthal. The portrait shows him looking at his hand,

gripping a dumbbell, near the top of the range of motion of a biceps curl. "He's not looking at his sarcomeres," she says. "He's looking at himself as a whole person."

Both aspects of existence and experience do connect, she believes. They connect in "the idea that you can have regeneration at ninety-five. The body isn't just static and declining. It's actually capable of growth and development and improvement."

Back in 1987, in the same week that Walter Frontera's milestone study of strength conditioning for older men was accepted for publication, a new pair of characters appeared on the season premiere of *Saturday Night Live*: two sneering, buffoonish Austrian bodybuilders who worshipped Arnold Schwarzenegger and denigrated everyone else, in a desperate sales pitch for their $9.95 workout video, "Pumping You Up with Hans & Franz."

Thirty Rockefeller Center and Irwin Rosenberg's lab at Tufts were very different worlds. Yet Hans and Franz and Fiatarone and Frontera had something in common. They all lived and breathed muscle. The Austrian fools and the Boston doctors presented polar-opposite views of muscle's importance in human life, though, and these views were in combat in society at large. In this contest, the actors playing the Austrians had an edge, because they fed off and carried on a tradition much older than the nonagenarians: the tradition of skepticism, ambivalence, anxiety, and fear about muscle and weight training. Those feelings were as widespread and intense as ever. Laughter in the studio audience proved it when Hans and Franz made their taunting sales pitch: "*Maybe you don't undestahnd* ENGLISH. *Maybe the only time you undestahnd is when you are* PUMMELED *and* BEATEN *with fists full of* PUMPED-UP MUSCLES!"

In the background of the Tufts lab's breakthrough studies on aging muscle, a string of events repeatedly demonstrated muscle's penchant for making friends like Hans and Franz.

A few months before Maria Fiatarone started teaching residents of Hebrew Rehab to lift weights, a psychiatrist at Harvard diagnosed psychotic symptoms among forty-one bodybuilders and football players who had used anabolic steroids. After the report on those symptoms was published, *'roid rage* became a byword for big men's violent, aggressive outbursts.

During the summer when the first Hebrew Rehab strength training study commenced, so many weightlifters tested positive for banned anabolic drugs at the Olympic Games in Seoul that an official of the International Olympic Committee later recommended dropping weightlifting from the games. Olympic weightlifting was so inherently compromised by athletes' drug use, some thought the only possible solution was to get rid of the sport.

While some of the first Hebrew Rehab lifters made their amazing initial gains of strength and muscle, a survey that would be published in *The Journal of the American Medical Association* found that, among high school senior boys in the United States, one in fifteen used or had used steroids. Forty-seven percent of the boys said they used the drugs mainly "to improve athletic performance," and about 27 percent said they took steroids mainly for the sake of "Appearance."

Not long after Irwin Rosenberg proposed *sarcopenia* as a name for age-related muscle wasting, legislators in the British House of Commons and the U.S. Congress proposed criminalizing the use of anabolic steroids without prescription.

As Hebrew Rehab's second strength training study was coming to a close, the judges of the Mr. Olympia contest, the world's most prestigious bodybuilding competition, selected Dorian Yates as the winner. He was the first Mr. Olympia to weigh more than 250 pounds. And the same psychiatrist who had associated steroid use with 'roid rage psychosis reported on a compulsive drive for muscularity among 108 bodybuilders he interviewed. He suggested naming the compulsion *reverse anorexia nervosa*. Four years later, he suggested a different name, *muscle dysmorphia*.

All these intersections illustrate a dismaying development. During the same period when medical science showed that muscle, built and strengthened by means of weight training, was integral to the well-being of all people—including and especially some of the people least likely to consider themselves as muscular beings—muscle was repeatedly marked, in the popular imagination, as a troublemaker. Even when the medical research was recognized and celebrated, it was often presented with a shiver of anxiety.

"*What she learned*," said Diane Sawyer, introducing a TV news segment about Fiatarone's research at Hebrew Rehab, "*defies the common logic.*"

By the common logic—expressed through popular culture, psychiatry, sport, and politics—lifting weights to build muscle and strength was excessive, basically ridiculous, and often dangerous. By the common logic, muscle's best buddy was steroids. Muscle was a sign of vanity and greed, associated with cheating, crime, and self-destruction—but still muscle was ogled and glorified. Except when it wasn't. Muscle was an instrument of the male aggressions, violence, and sexism now commonly referred to as toxic masculinity. Muscle might look like power, but really it was misery.

Psychologists did not spend much time studying muscularity until the early 1990s, according to the American Psychological Association's 2007 handbook of research on the topic, *The Muscular Ideal: Psychological, Social and Medical Perspectives*. The book's editors, J. Kevin Thompson and Guy Cafri, call attention to "the deleterious effects of an extreme focus on physical perfection," which seems to become more prevalent alongside "the trend toward greater body dissatisfaction," a trend that grew dramatically between 1972, when 25 percent of women and 15 percent of men reported dissatisfaction with their bodies, and 1997, when 56 percent of women and 43 percent of men were dissatisfied. By some estimations, the trend has continued. In 2008, one report found that almost 72 percent of women and almost 61 percent of men were less than very satisfied with their bodies.

The Muscular Ideal focuses mainly on men's and boys' concerns about muscularity. "Perceiving one's body as muscular has been shown to be important to men's psychological health," Cafri and Thompson write. The reasons go back to boyhood. By the age of six or seven, boys begin to show a preference for large, muscular male bodies, and "being underdeveloped is the most common body concern among boys."

The Muscular Ideal also summarizes research showing that, for men and boys, regular exercise reduces anxiety and improves levels of body satisfaction—effects, they clarify, that may be intensified by one kind of exercise: lifting weights. "The more involved men are in weight training activities, the more satisfied they are with their bodies, irrespective of their actual muscle mass," research shows. One of the earliest studies in this area, in 1988, "demonstrated that bodybuilders had a more positive body image than runners" and sedentary men.

When weight training changes men's body image for the better, how does that work? What's the mechanism for change? A systematic review of literature on resistance training and body image, published in *The Journal of Strength and Conditioning Research* in 2017, cites evidence that "subjective improvements in perceived body composition" may explain the change in men—meaning that lifting weights can help men see themselves as being more muscular and having less body fat.

This review, by Nicholas J. SantaBarbara and colleagues, finds that most research on resistance training and body image has been of poor quality, and there have been few randomized controlled trials. It also points to evidence suggesting that some groups of people may have especially strong potential for improving body image by means of resistance training. Studies show that body dissatisfaction and poor body image are especially high among some gay men who "place a greater emphasis on physical attractiveness, and more specifically, on levels of muscularity, making them an ideal target for a resistance training intervention." The authors add that female-to-male transgender individuals "could also benefit from resistance training, as objective improvements in muscle strength and subjective improvements in muscle size resulting from resistance training could compliment their existing habits aimed at achieving a more masculine body."

Unlike the earlier studies cited in *The Muscular Ideal*, more recent research on resistance training and body image has mainly focused on women and has found more various mechanisms of change in women than in men. Women's body image improves as their subjective perceptions of body composition improve, as is true of men, but in addition, "objective improvements in muscular strength and endurance are particularly important to women," SantaBarbara and colleagues write. With respect to this gender difference, they add, "Resistance training programs structured to support and promote increases in muscle size may be most effective to improve body image in men, whereas training targeting physical function may be the most effective to improve body image in women."

This contrast sounds a bit like the difference between the young Arnold Schwarzenegger's reasons for lifting and Jan Todd's—the difference between a bodybuilder's reasons for lifting and a powerlifter's reasons. Lifting weights can improve men's body image mainly because of how it makes them look. Lifting weights can improve women's body image, too, and not just because of how it makes them look, but because of what it helps make them able to do.

For a small portion of the small segment of the population that lifts weights, muscle dysmorphia is a grave problem, associated with risky and destructive behaviors including drug abuse and suicide attempts. Drawing attention to this disorder was a goal of *The Muscular Ideal*, and attention has been paid. In 2013, the condition was recognized by the American Psychiatric Association as a valid disorder and included in the *Diagnostic and Statistical Manual of Mental Disorders, Fifth Edition*. The DSM-5 lists muscle dysmorphia as a type of body dysmorphic disorder, which involves "preoccupation with one or more perceived defects or flaws in physical appearance that are not observable or appear slight to others," and this in turn is a form of obsessive-compulsive disorder.

Some psychiatrists and psychologists have questioned the validity of muscle dysmorphia as a disorder, however, and questioned whether scientific evidence of muscle dysmorphia is sufficient to avoid pathologizing disciplined regimens of athletic training. Federal University of São Paulo psychiatrist Celso Alves dos Santos Filho and colleagues found, in a 2015 systematic review, that reliable research about muscle dysmorphia was so rare, they concluded, "the existing data are yet insufficient to cement it as a valid construct." Many gaps in knowledge and shortcomings in studies of muscle dysmorphia were listed in a 2016 review by the sport psychologist David Tod and colleagues of Liverpool John Moores University, who found no reliable information about how common the disorder might be and little information about how to treat it—there have been no randomized controlled trials.

Despite any possible shortcomings as a psychological construct, muscle dysmorphia has been an abundantly successful media phenomenon, enhanced by its links to anabolic steroid use. Harrison G. Pope Jr., the psychiatrist who described and named muscle dysmorphia, cowrote a book published in 2000, handsomely titled *The Adonis Complex: The Secret Crisis of Male Body Obsession*. Almost twenty-five years later, the book is still in print.

As a media phenomenon, muscle dysmorphia has been far more successful than sarcopenia. If coverage in *The New York Times* indicates the scale of cultural impact, Harrison Pope's research on steroid abuse would seem to be more significant than Maria Fiatarone's contemporaneous discovery that strength training can alleviate the functional impairments of sarcopenia. As of this writing in late 2023, the newspaper had mentioned him, in connection with his views on steroid use, twenty-two times—compared to six mentions of her.

To flesh out the comparison, consider a few more numbers. The *Times* has published sixteen stories mentioning muscle dysmorphia, "reverse anorexia," or "bigorexia," compared to eleven stories mentioning sarcopenia. In the ten years between 2013 and 2023 alone, the *Times* published hundreds of stories mentioning athletes using steroids. Numbers like these surely help explain why, when most of us hear the word *muscle*, we're more likely to think of big guys on steroids than to think of our grandmothers.

But so what? What's really at stake here? Is anyone truly harmed by these kinds of assumptions we make about muscle?

To help answer those questions, here is one last set of numbers, showing how much attention anabolic steroid abuse, muscle dysmorphia, and sarcopenia receive, compared to the population-wide prevalence of each—a comparison that highlights their relative importance as public health issues.

The first numbers concern anabolic steroid use. The American College of Sports Medicine's most recent pronouncement on that topic, published in 2021, says estimates of prevalence are unreliable, partly because "users are secretive," before citing a range of estimates. The highest of these estimates says 4 million American men may be on steroids. If true, that would be 1.2 percent of the country's population. Prevalence of anabolic steroid use among females is estimated to be much lower—perhaps between one-tenth and one-half as common as among men.

The second set of numbers concerns muscle dysmorphia, for which there are no reliable estimates of population-wide prevalence. But in Germany, Sweden, and the United States, the nationwide prevalence of all body dysmorphic disorders, of which muscle dysmorphia is one type, has been estimated to be between 1.7 percent and 2.9 percent.

The third set of numbers concerns sarcopenia. Adults over the age of sixty represent more than 23 percent of the U.S. population—that's almost 78 million people. Estimates of sarcopenia's prevalence in older adults vary, but one authoritative study says that low levels of muscle mass put 45 percent of the older American population—some 35 million people—at risk of physical disability. And of course the disease of sarcopenia is not just an American problem, it is a global problem. Seven hundred and seventy-one million people in the world are sixty-five years old or older, according to the United Nations. The global prevalence of sarcopenia in otherwise healthy older adults appears to be roughly 10 percent. A team led by Sousana K. Papadopoulou of International Hellenic University in Thessaloniki estimated in 2020 that almost 25 percent of the world's older men and women who are hospitalized have sarcopenia. The same scholars estimated that in the world's nursing homes and other residential aged care facilities, about half of the men have sarcopenia, and almost a third of women. But other experts believe these figures are gross underestimates.

Clearly, if media coverage had any connection to the actual scale of these problems, we would all be hearing a lot more about age-related muscle loss and sarcopenia, and a lot less about pathologies of bodybuilders and elite athletes.

Media coverage of sarcopenia is comparatively rare, though, probably in part because it can scare us about the prospect of our own aging and inevitable death, and it can make younger people feel guilty for not doing all they can to help older people realize their potential strength. And even if learning about sarcopenia does make younger people want to take constructive action to help older people they care about, or to get out ahead of this later-life problem that may develop in their own lives, they will pretty soon run up against challenges that are familiar to practically every older person who would like to build muscle or strength. The challenge of finding trainers who operate according to a long-term vision of lifting; the challenge of affording to pay for trainers' help; and the challenge of securing access to quality equipment that allows for incrementally progressive training all contribute to a great barrier of public access to the regimen. On the rare occasions when older people lifting weights *are* depicted in the media, they tend to be trotted out as kooky, aspirational oddballs instead of being treated as what they are: normal people doing what's best for their own good and for the good of their communities.

As public health issues, age-related muscle loss and the disease of sarcopenia are not comparable to the problem of anabolic steroid abuse and the disorder of muscle dysmorphia. As public health issues, muscle loss and sarcopenia are much more important. And our general tendency to pay much more attention to the much rarer problems of a much smaller number of younger people is in fact actively harmful to the grandmothers and grandfathers of the world, because our misallocation of attention compounds our forgetfulness and negligence of the needs and potential of older people, especially those who spend the last years of their lives in chronic pain, with limited mobility, and without the knowledge and resources to help themselves.

There is still no good answer to the question about muscle in older age that Irwin Rosenberg asked his colleagues so long ago in Albuquerque: *Why have we not given it more attention?*

Walter Frontera's and Maria Fiatarone's discoveries that older people could build muscle, strength, and functional abilities by lifting heavy weights were reproduced by many other researchers around the world. One of them was Jan Lexell, a Swedish professor of rehabilitation medicine at Lund University and physician who directs neurological rehabilitation at Ängelholm Hospital. Around when Walter Frontera published his study of strength conditioning in older men, Lexell and a few colleagues did "a very small study on five elderly women who were training, doing biceps curls" at 80 percent of their one-repetition maximum strength, he tells me. The women were in their seventies, and Lexell recalls that "over a period of six weeks, they doubled the weight, went from two kilos to four kilos, which is 100 percent improvement. Which is phenomenal if you look at it in relative terms. But in absolute terms, it's just: two to four. I can remember being asked, *How do you dare to do it? Where is your rescue team? Where is your defibrillator team?* Then I asked, *Have you ever tried this?*"—a rhetorical question: The answer was no. In medical research on strength training, Lexell says, "There was a sudden dramatic change simply because a few people challenged the common belief" that older people should avoid training their muscles at high intensity.

Nevertheless, in most of the medical profession, as in global popular culture, ignorance about older people's potential to build muscle and strength persists, as part of a general failure to recognize the centrality of muscle in human life and health for everyone, throughout the lifespan.

But it never crossed Maria Fiatarone's mind that Hans and Franz or anabolic steroid scandals or muscle dysmorphia or any news coverage of such things had anything to do with her work, she says.

On her own terms, she went on investigating muscle.

By 1992, four years after the first small group of residents at the Hebrew Rehabilitation Center began lifting weights, workouts had become a fixture of the culture of this nursing home. Before the end of the decade, Maria Fiatarone created a nonprofit organization, Fit for Your Life, to teach strength training to residents of other aged care facilities and to older individuals living independently. Evelyn O'Neill quit her second job waiting tables at Franco's, and she took on a different second job, much more to her liking—O'Neill became the coordinator of Fit for Your Life. Fiatarone and O'Neill enlisted Ed Rosenthal to act as a roving ambassador for weight training among New England's older population. He visited other nursing homes and invited his peers to participate in new strength training studies organized by the lab at Tufts; and he passed out brochures illustrated with a portrait of himself flexing his biceps, to help show what was possible. The brochures were also emblazoned with the motto Fiatarone had adapted from Reebok's ad campaign: "Life is long, play well."

At Hebrew Rehab, about half the residents had adopted some kind of strength training regimen by the late 1990s. O'Neill and Fiatarone adapted high-intensity progressive resistance exercises for anyone who wanted to do them, including people who could not lift weights. Even people who were unable to move their joints through a full range of motion—due to severe arthritic inflammation of the hips, for instance, or extreme degeneration of rotator cuff muscles—could participate in strength training, from the starting point of isometric contraction. *Isometric* means "same length"—*iso* for "same," *metric* for "measure" or "length" or "size"—and in isometric contraction, muscles tighten but joints do not move, or barely move. When range of motion contracts or totally collapses, with pain flaring to enforce the lockdown, isometrics can preserve and build strength until the pain subsides. As structures surrounding the joint grow stronger, range of motion naturally increases.

Strength training in all its forms at Hebrew Rehab was the subject of another article in *The Jewish Advocate* in 1999, marking ten years since the newspaper covered Fiatarone's first strength training study. "Is the program merely the outgrowth of a perverse and impossible forever-young culture," asked an intrepid reporter, "or is there meaning behind the madness? What is it?"

"It's a godsend, that's what it is," Ben Engleman said. He was ninety-three years old. He had been a shoe salesman in Swampscott, Massachusetts. When he arrived at the nursing home, he could not comb his hair or tie his tie. Four years later, because he followed a strength training regimen, he was able to make himself look sharp again, without needing a hand from anyone else.

To help people persist in training, Evelyn O'Neill says, she learned to emphasize that strength is cyclical, not linear. The cyclical nature of strength has long structured athletic training regimens, as described in writings by authors from Philostratus to Pavel Tsatsouline; and these regimens have proven their effectiveness by the record-setting performances of many athletes at many levels of competition, from Soviet Olympians to Jan Todd to Charles Stocking.

Strength falls and rises, O'Neill assured the residents of Hebrew Rehab. It ebbs and flows with all the unexpected turns of life. The residents "go in the hospital. They get pneumonia. Their cats get sick. And then they don't sleep. So *they* get sick," O'Neill tells me. Through all this, she was there to remind them, "Okay! So we back off. But then I know you're going to get your strength back. I know you're going to feel better physically. The illness will subside. And then we can start again."

Helene Freundlich started again, many times, and never gave up. When Evelyn O'Neill took her shopping for a bra at Lady Grace, the lingerie store for older women, Freundlich could not have imagined anything so outlandish as celebrating her one hundredth birthday with a birthday cake *in a gym*. But that happened, and she kept lifting until she died at the age of a hundred and three.

Ed Rosenthal, too, kept coming to the gym. Even when his vision faded and he lost the ability to walk, he was in the gym for thirty minutes, three times a week, always looking forward to his favorite machine, the chest press. He always kept on with his other activities at Hebrew Rehab, too. He wheeled himself out to the garden in his wheelchair and dug in the dirt, he played percussion with the klezmer group, and one day when he turned on the TV, the movie *Oklahoma!* was playing, and the musical number for "The Surrey with the Fringe on Top" brought back a boyhood memory, almost as old as the century, of the morning when relatives rolled up to his family's house in the factory town of Malden, Massachusetts—they rolled up in a real-life horse-drawn surrey with an actual fringe on top, and they took him for a ride so long he saw towns he'd never seen before—that ride must have covered ten whole miles—and now as an old man he wrote a story about it all, about the song and the ride and the remembering, and he published his story in the nursing home's own newspaper.

"He kept up with the weight-lifting machines to the end," Ed Rosenthal's son tells me. The son, David Rosenthal, is an old man now himself. A few years ago he moved into an assisted-living facility not far from the nursing home where his dad lived. Today at Hebrew SeniorLife, the son works out on the same kind of weight-lifting machines that his dad used at Hebrew Rehab in the 1990s. "I daresay, when I do it, I think of my father doing it," says the son. "I just envision him right beside me, doing the same thing. I try to do everything he does."

CHAPTER 8

Push and Pull

Imagine a patient walking into a doctor's office.

The patient says, "I'm sick."

"Take a drug," the doctor says.

There is a long pause. The patient, confused, stares at the doctor.

What kind of drug? the patient wonders. *How much should I take? When should I take it? Should I take it with food or on an empty stomach? Should I take it in the morning or at bedtime? Should I take it for a day or for a week? Or a month? Or a year? Do I have to keep taking it for the rest of my life?*

The doctor repeats himself. "Take a drug."

Just imagine that.

The scenario above conveys a problem that Maria Fiatarone began trying to solve in the 1980s. When most doctors talk to most patients about exercise, they rarely say much more than *Take a walk*—though telling most patients to take a walk is as useless, and to frail older patients it can be as dangerous, as it would be to hand them prescriptions that said only, *Take a drug*. Still today, she and some of her like-minded colleagues relate versions of this anecdote about the indeterminate prescription when they suggest a solution to the problem. The solution, they contend, is for doctors to prescribe exercise the same way they prescribe drugs.

To bolster this argument, she quotes the definition of a drug, as established by the U.S. Food and Drug Administration. The FDA defines a drug as an article that is "intended for use in the diagnosis, cure, mitigation, treatment or prevention of disease" or "intended to affect the structure or any function of the body" or "intended for use as a component of any articles specified" in those clauses. By that definition, she says, "Exercise is clearly a drug."

The committed minority of doctors who agree with her derive their views from a broad base of evidence. Two expansive papers summarize the research. The first, "Lack of Exercise Is a Major Cause of Chronic Diseases," by Frank W. Booth, Christian K. Roberts, and Matthew J. Laye, published in 2012, reviews evidence that physical activity and exercise serve as "primary prevention" and can serve as "rehabilitative treatment" for thirty-five chronic conditions. The second, "Exercise as Medicine—Evidence for Prescribing Exercise as Therapy in 26 Different Chronic Diseases," by Bente Klarlund Pedersen and Bengt Saltin, published in 2015, focuses on evidence of exercise as a medical first-line treatment. In 2018, a much more extensive analysis of evidence that physical activity and exercise can prevent and treat chronic diseases was put forth by about forty leading physicians, physiologists, and public health experts from the United States, Canada, and the United Kingdom. The document, which runs to 779 pages, was published by the U.S. Department of Health and Human Services as the *2018 Physical Activity Guidelines Advisory Committee Scientific Report*.

Most of the evidence about physical activity's effects on chronic disease concerns aerobic exercise, but Maria Fiatarone's early studies at the Hebrew Rehabilitation Center established foundational evidence that progressive resistance training is a healthy practice for older people. In the following decades, she designed and carried out numerous international research collaborations involving colleagues in Australia, Canada, Finland, France, Indonesia, Iran, Israel, Norway, and Spain. She is an author of more than four hundred peer-reviewed scientific papers. Her longitudinal studies and randomized controlled trials, and her reviews and analyses of research by other doctors and scientists, have investigated and assessed how progressive resistance training can prevent or treat a range of conditions affecting people through many stages of life, including chronic diseases such as depression and diabetes, and afflictions more common in later life, such as renal failure, heart failure, cancer, osteoporosis, and osteoarthritis.

"Almost every chronic disease that you can think of now has randomized controlled trial evidence," suggesting answers to the question of "whether or not you can treat disease with exercise," she says. The answers to that question, while complex, are on the whole affirmative.

Physical activity, she emphasizes, is "*one* of the things that contributes to health and well-being. It's not everything. Obviously, there are genetic factors that influence how people age and what diseases they get. There are ecological factors—things like where you live, what toxins you're exposed to." All those factors modify the others. People even modify their genes by actions they take. "Those epigenetic modifications, which can change over the lifespan, are probably in some cases even more important than the actual genes themselves," she says. "Because the epigenetic modification tells your DNA whether or not to express itself in proteins that then go on to, maybe cause, or prevent, disease. So you have a lot of control," even if you're "not genetically blessed."

You can wield that control through exercise, she says. She names three ways that exercise can interact with traditional medicine. "In some cases it's complementary—for example, in the care of diabetes. In other cases, it's an alternative—like depression. In other cases, we have no other treatment—like dementia or frailty. There is no drug to treat those things. But there's more and more evidence that exercise is good for those conditions. So depending on what condition you're talking about, it might be a stand-alone treatment, it might be alternative treatment, or it might be adjunctive treatment."

In 1993, a young doctor from Australia who met Fiatarone was impressed by her research. Nalin Singh had come to Harvard for training in geriatric medicine. Fiatarone was one of his instructors. Fresh off her Hebrew Rehab studies, Fiatarone envisioned next steps in exercise research with such persuasive clarity that, to Singh, it seemed the whole field of geriatrics was about to change. "Aging had been around forever," he thought when he learned about her studies, "but the core tenet of it all—frailty, the progression of frailty over time—had never been tampered with."

The main difference between old people and younger people, as Singh understood it, was that the old are much less robust. "They couldn't withstand small perturbations. But I'd never really thought about, *Could that be tampered with?* I just thought it was the genetic process of aging," he told me.

Nalin Singh pored over Maria Fiatarone's research, which respected basic facts about how exercise works. Different kinds of exercise change people in different ways. Different amounts of each kind of exercise done at different times and with different degrees of effort make different kinds of change. Different amounts and qualities of rest and food can affect the changes, too.

In line with those facts, Singh observed, Fiatarone's early research had prescribed strength training as medicine for frailty. In her trials, exercise was carefully selected—to build stronger, bigger muscles and improve people's capacity to function in everyday life. Exercise was precisely measured—starting with tests of one-repetition maximum strength. Exercise was regularly administered—in dosages of three sets of eight reps, three days a week. Exercise intensity was adjusted over time—based on follow-up strength tests, administered at regular intervals. And exercise was supported, to help make it more effective, with proper nutrition.

Considering the results of Fiatarone's exercise prescriptions, Nalin Singh recalled, he came to see weight training as "the first thing that showed some evidence that frailty *could* be tampered with."

Frailty is a syndrome involving muscle loss, strength loss, slowness, and fatigue. Frailty is an economic burden on health care systems. Data from Germany suggest that medical expenses for people with frailty may be about five times higher than expenses for people who are not frail. Being female, being poor, and living in a poorer country increase the risk of frailty, but much of the world's older population is frail or at risk of frailty. Doctors have some differences of opinion about how to define frailty; but in 2021, a review of 240 large studies conducted in sixty-two countries found that 24 percent of people aged fifty and older are frail, as defined by one reputable frailty index, established by Kenneth Rockwood of Dalhousie University in Halifax,

Nova Scotia. The same review, by Rónán O'Caoimh of National University of Ireland Galway, found almost half the world's older people—49 percent—to be at substantial risk of the condition: They are "pre-frail."

Frailty is often described as a function of three factors: aging, disuse, and disease. The first factor, aging, affects everyone. Eventually, we will all lose some functional capacity. Highly trained athletes in many sports are able to maintain their physical condition well past the age of sixty, but even they usually experience significant declines in physical performance by the time they are in their seventies, according to a study of record-setting performances in masters sports by the University of Sydney anesthesiologists A. Barry Baker and Yong Q. Tang. The same study found that older athletes in "strength-dependent sports" such as Olympic weightlifting "show the earliest declines" in performance, beginning after the age of fifty.

The second factor, disuse, suppresses and deadens organ systems that need stimulus to thrive. Disuse is a clinical term for the "lose it" part of "use it or lose it." Data from a longitudinal study of the health of thousands of Harvard alumni shows the truth of that old saw. The Harvard alumni study found that people who played sports in college but then stopped engaging in vigorous physical activity after graduation were about as likely to develop coronary heart disease as their classmates who had not been varsity athletes. "However," the study's principal investigator wrote, "college laggards who became physically active as alumni achieved the same benefits as their classmates who had been active all along." Maria Fiatarone's studies of strength training at Hebrew Rehab had found even more certain proof that exercise counteracts disuse.

When Nalin Singh met Maria Fiatarone, she was turning to investigate how lifting weights affects the third factor, disease. She believed that if strength training could help treat other major chronic diseases as it treated frailty, while at the same time compensating for aging and disuse—especially if it could provide benefits similar to drugs, without risk of harmful side effects—then it could become, in her words, "mainstream medicine, and not just an ancillary sort of treatment that you would refer people to in addition to your medical care. It would substitute for less effective or more harmful versions of care, for the same problem. That was the thought."

In mainstream medicine, in the treatment of frailty, her thought is reshaping reality. That is shown by "The Asia-Pacific Clinical Practice Guidelines for the Management of Frailty," published in 2017 by the *Journal of the American Medical Directors Association*. Clinical practice guidelines, as the paper noted, "are evidence-based recommendations systematically developed by expert panels who have a working clinical knowledge of respective medical conditions." Doctors look to clinical guidelines for instruction and counsel on how to treat patients with various conditions. Doctors who consult this clinical guidelines document find the main suggested treatment for frailty named, in a headline, as "a progressive, individualized physical activity program that contains a resistance training component" as the primary mode of exercise. As secondary treatments, the guidelines recommend balance training and walking, based on individual abilities. The 2017 guidelines designate these modes of exercise in the order of priority that Maria Fiatarone's publications began to suggest in the 1990s.

Speaking about the frailty guidelines in 2021, she said, "Think about somebody trying to get out of a chair: What do you need to stand up? You need strength. What do you need to stay up? You need balance. And what do you need to walk across the room or the nursing home? That's where your aerobic capacity comes in. But trying to do it in a different order, trying to stand up and walk without having the strength to get out of a chair, is actually inappropriate. Flexibility or gentle exercise, stretching—none of that really has any role to play" in the project of getting an older person to stand up from a chair. "Yet when you look at nursing home so-called exercise programs, primarily they are stretching programs, and they really don't attend to this statement or any consensus guidelines about the treatment of frailty and sarcopenia. Once somebody is able to get up and stay up without falling over, that's when walking can be added. And if you add it in a different order, if you let somebody walk who doesn't have sufficient strength, that's been shown to actually increase the risk of falls and fractures. So the order in which you get out of a chair is the order in which these different modalities of exercise are needed."

Between 1987, when the U.S. Food and Drug Administration approved Prozac to treat depression, and 1993, when Nalin Singh traveled to Boston to study, American doctors wrote more than 6 million prescriptions for those green and beige capsules. *Listening to Prozac*, a bestselling book by Peter D. Kramer about the antidepressant phenomenon, was published during Singh's first months in the United States.

No one knew the long-term consequences of taking new classes of antidepressant medications, but research showed the short-term consequences, for some, included side effects that could be especially noxious in older age: dizziness, fatigue, weight loss, nausea. As treatment for depression, especially for older people, Maria Fiatarone wondered if exercise might have a more favorable risk-benefit ratio than drugs.

Yet in the mid-1990s, research on exercise and depression was meager. By 1997, only eleven trials had tested exercise as a means of helping to manage mild depression. Only two of those eleven trials had involved strength training. The other nine focused on aerobics. All the trials had found exercise to be "significantly better than no treatment" for depression, and about as effective as psychotherapy, according to a review of the scientific literature that Fiatarone and Singh completed together.

The studies had shown no clear advantage for aerobic or strength training as ways of helping to manage mild depression. And informed comparisons of aerobic and strength training were impossible to make, since reports of the two "strength training" trials included no details of what was meant by strength training. Published reports of those trials suffered from common weaknesses of much strength training research. They did not specify the intensity of the exercise—they did not say how heavy the weights were that people lifted. Nor did they describe the effectiveness of the exercise—they did not say how much strength people gained.

Nevertheless, in early published trials of exercise as treatment for depression, one contrast between strength and aerobic training was clear-cut. Subjects in the strength training studies tended to do better at sticking to their programs. The strength training studies had higher compliance and lower dropout rates than the aerobic studies.

The early trials of exercise as treatment for depression, Fiatarone and Singh noticed, had another weakness, too. Those trials involved only younger people. Older people were not included in early randomized trials of exercise and depression, not even in the aerobic exercise trials, except for one in Montreal in 1991. A six-week walking program for moderately depressed older people reduced self-reported measures of depression by about a third—which was significant, but not much greater than a placebo effect. Running or jogging had antidepressant effects, too, according to the research on younger subjects. Telling most ninety-year-olds to go running, though, would not do them much good.

Running was not always feasible for large and growing numbers of younger people, either, due to the increasing prevalence of obesity—and obesity and depression are reciprocally linked. Over time, obesity increases the risk of developing depression by 55 percent, and depression increases the risk of developing obesity by almost 60 percent.

In the 1990s, thinking mainly about older people, Fiatarone speculated that strength training "might be more tolerable, have the same benefit for depression, and perhaps a spectrum of other benefits that aerobic exercise just doesn't address." Unpleasant and potentially hazardous side effects of antidepressant medications, for instance, were known to be improved by exercise. Could strength training be used in combination with drugs to counteract these effects? Or was it possible that exercise and drugs might turn out to be equally effective for treating depression?

Nalin Singh gravitated toward such questions, with his affinity for complex patterns of causation. When I asked him about his background, he answered by describing his maternal grandparents, born in India, who purchased their freedom after working as indentured laborers in the cane fields of Fiji; and his paternal grandparents, who lived in a house without electricity and encouraged their four children, all boys, to become doctors. He told me that the eldest son, Harold—Nalin's father—became the first Indian from Fiji to emigrate to Australia to study medicine.

Harold Singh was among the small number of nonwhite emigrants allowed into the country when the "White Australia" immigration policy was beginning to be dismantled in 1950. When he settled in Sydney, violent racism was still so common in the city that he never knew when a white stranger might throw a rock at him. In alternate years through most of that decade, Harold Singh studied, then worked to earn money to fund his brothers' emigrations from Fiji. Harold's medical practice was the setting of some of his son Nalin's earliest memories. In those memories, Harold's patients expressed their heartfelt gratitude for his personalized care.

Following his father's example, Nalin became a general practitioner, and he chose to specialize in geriatrics because, he told me, "I just loved the people." He was fascinated by the stories of older people's lives. He loved the breadth of skills their care demanded of him, too. Geriatricians care for every part of the whole person, from legs and lungs to hearts and brains. Many older people who suffer from depression initially seek treatment not from specialists, but from general practitioners in primary care.

Within months of arriving at Harvard and meeting Maria Fiatarone, Nalin Singh began collaborating with her to design a randomized controlled trial of high-intensity progressive resistance training as treatment for depression in later life.

The trial involved thirty-two people, aged sixty to eighty-four, who suffered from some form of significant depression. About half had minor depression, and about half had major depression, as measured by a reliable complex of diagnostic criteria for the disease. These criteria included the Beck Depression Inventory, rated and reported by patients themselves; the Hamilton Rating Scale for Depression, rated by therapists; the Geriatric Depression Scale, geared to measure distinctive aspects of depression in older people; and patient interviews conducted by physicians, following guidelines in the *Diagnostic and*

Statistical Manual of Mental Disorders. Few trials of exercise for depression prior to this one by Singh and Fiatarone had so stringently defined and measured depression by validated clinical standards.

Singh and Fiatarone designed the trial's intervention, a ten-week regimen of weight training for several major muscle groups, including muscles of the legs, back, and chest. In 45-minute sessions, three days a week, working at 80 percent of their one-repetition maximum strength, people completed three sets of eight repetitions of exercises such as the chest press, leg press, and knee extension.

Singh personally administered the study's initial strength tests and oversaw practically all its training sessions. Having been captain of his high school cricket and rugby teams in Sydney, he knew his way around a gym; and recreationally, he still lifted weights. Singh's longtime training had given him a good eye for effort, which enabled him to adjust the loads for each lifter during every session, as appropriate. After the study's first four weeks, everyone took another maximal strength test, to reset the baselines of their regimens.

During the same period when Singh ran these resistance training sessions, he also met with the trial's control group, teaching an interactive health education program on a variety of topics, from nutrition to first aid. The control group did not exercise during those ten weeks, but they did take one-rep max strength tests at the beginning and end of the period, to set points of comparison with the strength training group.

After ten weeks, almost 60 percent of the group that lifted weights had reduced their symptoms of depression by more than 50 percent. This change met the standard threshold of clinically meaningful response to treatment that was used in psychopharmacology trials. When everyone's scores on the depression scales were tabulated, Singh found that fourteen of the sixteen people who lifted weights no longer met diagnostic criteria for the disease.

Progressive resistance training worked in this trial as treatment for the disease of depression in multiple forms, from persistent minor depressive disorder to major depressive disease. While lifting weights, participants in this trial had also grown considerably stronger. According to their final strength tests, they were on average 33 percent stronger.

About a quarter of the control group also found relief from depression, perhaps merely from participating in the health education class—a fairly typical placebo effect. At the same time, because they were not exercising, those in the control group lost muscular strength, by an average of 2 percent.

The trial then proceeded to a second phase, in which almost all the people who lifted weights continued to train, but without supervision, for ten more weeks. In that period, the antidepressant effects of the exercise persisted. Evaluations showed that people who had been most depressed at the beginning of the trial experienced the greatest antidepressant benefits from weight training.

For a majority of those who took part in this and later trials by Singh and Fiatarone, high-intensity strength training treated depression as well as the best antidepressant drugs. Training did not relieve all cases of depression in these studies. About 25 percent of the people received no relief, or no lasting relief, from the exercise. But in the majority for whom it worked, it worked regardless of whether people believed that it would work. Low-intensity weight training was also effective, but much less so, for fewer people, and in a different way. It worked like a placebo. Low-intensity lifting worked if people believed it would work, and it didn't work if people believed it wouldn't.

But at high intensity, strength training took effect about as quickly as drugs; and it remained effective over longer periods, too. What exactly caused these changes remained unknown. The mechanism did not seem to be endorphins, aerobic capacity, or social interaction. "It is likely that exercise works by multiple mechanisms," Singh and Fiatarone speculated, "and individual responses may depend more or less heavily on a given mechanism depending on the characteristics of the individual, their life circumstances," and the form of their depressive illness.

Dumbbells were not just a diversion. Heavy training protocols were not placebos. This was the main finding of Singh and Fiatarone's research on strength training as an antidepressant: Lifting weights affected the brain like a drug. And because exercise had no adverse side effects for most people, it had a more favorable risk-benefit ratio than drugs.

Evidence that exercise treats depression has accumulated since Nalin Singh and Maria Fiatarone's first collaboration. Few other researchers have focused on resistance exercise, though, especially high-intensity resistance exercise, despite evidence of its effectiveness. In 2016, a review of literature on exercise and depression, published in the *Journal of Psychiatric Research*, identified just three randomized controlled trials focused on resistance exercise. Two of the three were published by Singh and Fiatarone. The same review identified nineteen trials focused on aerobic training.

"Exercise has a large and significant antidepressant effect in people with depression," and evidence shows that exercise most effectively treats major depressive disease, especially when supervised by professionals, the review concluded. Yet as the reviewers noted, no one had ever conducted a trial of resistance exercise exclusively as treatment for major depression. The *2018 Physical Activity Guidelines Advisory Committee Scientific Report* also noted that most trials of exercise as treatment for depression were still focused only on aerobics. The report recommended that more research on resistance training should be done.

Much of the most reliable knowledge about general physical activity and depression, as summarized by the 2018 report, was foreshadowed by Singh and Fiatarone's 1997 discoveries about resistance exercise and depression. Randomized controlled trials "definitely demonstrate that physical activity is an effective treatment" for reducing symptoms of depression, according to the report, which found the antidepressant effects of activity to work "across the adult lifespan." The report corroborated that physical activity tends to have stronger treatment effects for individuals with major depression. "When physical activity is compared to either cognitive behavioral therapy or anti-depressant pharmaceutical treatments," moreover, "the groups show no significant differences, indicating that physical activity is as effective for treating depression as these other common approaches for treatment."

Likewise for preventing depression, "greater amounts of physical activity are strongly associated with a reduced risk." This, too, holds true for adults of all ages. The committee also found evidence that physical activity has a dose-response effect on preventing depression. "Even low amounts of activity (less than 150 minutes per week) were associated with significantly reduced risk of depression, although more activity was associated with larger effects. Engaging in more than 30 minutes per day of activity reduced the odds of experiencing depression by 48 percent."

After reviewing such evidence, the World Health Organization concluded that exercise reduces risk of depression and "may be comparable to psychological and pharmaceutical therapies in reducing symptoms." Another international medical organization formally recommended that doctors prescribe exercise for its antidepressant effects. In Australia and New Zealand, clinical practice guidelines for treatment of mood disorders suggest that psychiatrists should begin treatment of depression with lifestyle interventions. The guidelines name exercise as a primary intervention and unequivocally describe resistance training as "beneficial in the management of depressive symptoms," while noting that "a combination of aerobic and resistance exercises are probably optimal." The speculation that combined exercise is superior is only speculation, however. The guidelines cite no experimental evidence that combined exercise manages depressive symptoms better than resistance training alone.

One powerful medical organization that appears to be more guarded about the prospects of exercise as an antidepressant is the American Psychiatric Association. The APA's *Practice Guideline for the Treatment of Patients with Major Depressive Disorder*, third edition, which was published in 2010, states that "Data generally support at least a modest improvement in mood symptoms for patients with major depressive disorder who engage in aerobic exercise or resistance training." The document cites six studies of aerobic exercise, and just two studies of resistance training—and Nalin Singh is the lead author of both. Despite these data, the guideline presents exercise as a treatment option in a tone that sounds almost grudging: "If a patient with mild depression wishes to try exercise alone for several weeks as a first intervention, there is little to argue against it." (To be fair, the association's website does say the 2010 guideline "can no longer be assumed to be current.")

Yet another mental health care organization has remained skeptical of exercise as an antidepressant. The American Psychological Association's most recent clinical practice guidelines for treatment of depression, issued in 2019, makes only a "conditional recommendation for use" of exercise in treating the disease, and only for adults "for whom psychotherapy or pharmacotherapy is either ineffective or unacceptable." Although the document was published in 2019, it cites no scientific literature on exercise published after 2013.

Through the late 1990s, Maria Fiatarone's research on weight training and chronic disease branched out in more collaborations. In one, lifting weights improved sleep and served as treatment for insomnia, she found, working again with Nalin Singh—who, by the time he returned home to Australia in 1999, was her husband. Now she was Maria Fiatarone Singh.

At the University of Sydney Medical School, she became the John Sutton Chair of Exercise and Sport Science, a post she has held for more than twenty-five years; and Nalin Singh, too, served on the university faculty for more than two decades.

In the early 2000s, the Singhs also began to design, run, and oversee investigations of progressive resistance training as a complementary treatment for type 2 diabetes, and for some of the many other conditions that often accompany it. Their trials, as well as their illuminating interpretations, reviews, and analyses of other people's work, anticipated and helped to stimulate wide-ranging international research efforts on exercise and diabetes.

Type 2 diabetes is a disorder of metabolism, the body's ability to turn food into energy. Most of the food we eat gets broken down into glucose, or sugar. When glucose enters the bloodstream, it triggers the pancreas to release insulin, the hormone that helps dispose of blood sugar by moving it into the body's cells, to be turned into energy. With chronic overeating and poor nutrition, though, people's metabolism can be deranged. Blood sugar can build up and overwhelm our capacity to handle it. We can lose the ability to make enough insulin. We can also become insensitive to insulin.

The condition of insulin resistance can be exacerbated by shifts in body composition such that people carry too much fat (especially visceral fat, the fat around the waist) and too little muscle. Because skeletal muscle is an organ of metabolism—because muscle is the biggest sink for glucose disposal in the human body—people who have proportionally less muscle are unable to metabolize glucose as well as people who have more muscle.

Dramatic evidence of the difference comes from Helsinki University Hospital in Finland, where researchers measured the insulin sensitivity of a group of distance runners, a group of people who lifted weights, and a control group of people who did no regular exercise—all subjects were men. The measurements showed that "sensitivity to insulin is directly related to muscle mass and inversely proportional to adiposity." Lifters had the highest levels of glucose metabolism, an advantage owing to the bulk of their muscle. People who lifted weights had glucose disposal rates that were 40 to 50 percent higher than people who did no regular exercise.

In 2021, about half a billion people—more than 10 percent of the world's population—had type 2 diabetes. Almost half again as many were predicted to develop the disease by 2045. Prevalence of diabetes grows along with obesity. According to the World Health Organization, the number of people with type 2 diabetes almost quadrupled between 1980 and 2014, while the number of adults who are overweight or obese almost tripled between 1975 and 2016. Other major risk factors include inactivity and poor physical condition. In the Swedish military, measurements of aerobic fitness and muscular strength of more than 1.5 million male soldiers at the times of their conscription and retirement showed "the combination of low aerobic fitness and low muscular strength was associated with a 3-fold risk" of developing type 2 diabetes later in life. These associations were independent of their body mass index, a common indicator of body composition. The associations were also independent of family history of diabetes.

The risk of developing type 2 diabetes increases with age. Globally, the disease afflicts almost 25 percent of people aged seventy-five to seventy-nine. But large numbers of young people also are susceptible. Between the years of 2001 and 2017, prevalence of type 2 diabetes doubled among children and adolescents in the United States. In roughly the same period, the prevalence of prediabetes, the disorder of chronically elevated blood sugar indicating a high risk of becoming diabetic, almost tripled among American youth aged twelve to nineteen. Almost 30 percent of the country's young people had prediabetes by 2018.

Over time, diabetes can take a devastating toll on individuals, and diabetes threatens also to be extremely damaging to societies. Type 2 diabetes is the main cause of kidney failure, adult-onset blindness, and lower limb amputations; and it increases risk of osteoarthritis and of cardiovascular diseases, including high blood pressure, stroke, coronary heart disease, and heart failure. Drugs and surgeries can alleviate some suffering caused by type 2 diabetes, but these therapies are expensive. Average annual cost of medical care for diabetics is at least twice the average cost for people who do not have diabetes; and with complications such as those just mentioned, the cost can multiply by 500 percent or more, according to data from a range of countries, from Italy to the United Arab Emirates. For health care systems around the world, the burden of diabetes threatens to become unmanageable. By 2045, the number of adults with diabetes in China, India, and Pakistan is predicted to be larger than the total population of the whole of Western Europe.

But research shows that type 2 diabetes is preventable for a great many people. A number of randomized controlled trials—such as large trials in China, Finland, India, Japan, and the United States—have shown lifestyle changes, involving combinations of diet and exercise, "usually are very effective for preventing the onset of diabetes," even in "very, very high-risk individuals," says Maria Fiatarone Singh.

One of the largest such trials, the Diabetes Prevention Program, which involved about 3,200 individuals, found that, for younger people, taking the drug metformin worked as well as structured lifestyle changes—diet and exercise regimens—for controlling blood sugar. But in older people, she says, "metformin was no better than just advice to take a walk and eat better."

Doctors who prescribe exercise for diabetes prevention the same way they would prescribe drugs will tailor their recommendations for individual patients depending on many factors, but Fiatarone Singh says that "certainly age is one of the things that dictates how you prescribe this drug. So would you prescribe metformin to prevent diabetes in an older adult based on this data? You wouldn't. And I think that's the important thing, looking at the intricacies of the data, not just the overall blanket message."

In older adults, type 2 diabetes accelerates age-related losses of strength and muscle, too, and abilities to function in daily life. The acceleration can be extreme. One large study tracked the body composition of almost 2,700 older adults for six years and found that women with diabetes lost twice as much muscle in their thighs as women who did not have diabetes. Those who lost the most muscle were men and women with undiagnosed diabetes, "suggesting that the effect of type 2 diabetes on skeletal muscle mass seems to be manifested in the early stages of the disease." Building on that observation, the study's authors, led by Seok Won Park of CHA University in Seongnam, South Korea, noted another dire downstream effect of diabetes: "If older adults with undiagnosed type 2 diabetes were left untreated, they would be at higher risk for developing sarcopenia." People with undiagnosed diabetes are also at higher risk of vascular complications, including reduced blood supply to the muscles.

Because diabetes can be so detrimental to muscle, and because muscle is integral to fending off many of the ravages of diabetes, the American Diabetes Association recommends two or three sessions per week of resistance training, on nonconsecutive days, for people with type 2 diabetes. The group's position statement on physical activity and exercise for diabetics says that resistance training can produce "improvements in glycemic control, insulin resistance, fat mass, blood pressure, strength, and lean body mass." The document specifies that "heavier resistance training with free weights and weight machines may improve glycemic control and strength more"—and on that point, cites Maria Fiatarone Singh.

Yet in the clinical treatment of diabetes, as in clinical treatment of depression, patients rarely benefit from research that shows progressive resistance training can help to manage their disease. Only 12 percent of people with type 2 diabetes lift weights or do any "activities that would increase muscular strength," according to one of the few large assessments of physical activity among diabetics that specifically asked about weight training—a survey of almost 1,200 adults in the Canadian province of Alberta.

After people are diagnosed with diabetes, prescriptions follow a fairly standard course in routine medical care, as Fiatarone Singh describes it. Doctors usually prescribe weight-loss diets; but research shows that most people are unable to follow these diets—and even when people do follow the diets, she points out, they "lose significant amounts of muscle and bone in doing that" and also decrease their metabolic rate. "So diet alone is not the answer," she says.

"What about adding aerobic exercise? Well, that does work, and it's good if you can do it," she says, but for overweight or obese people, especially those who suffer from related afflictions such as knee arthritis and cardiovascular diseases, regular or sustained aerobic exercise may not be feasible.

"So because neither of these things work very well," she continues, most doctors "very quickly go to drugs." One of the drugs most often prescribed for diabetes is insulin, and insulin "causes you to gain about five to ten kilos in the first year of using it. So although it does lower your blood sugar, it actually makes the underlying problem, which is the visceral obesity, worse than it was before." And many doctors observe significant losses of muscle and bone density in patients who take the weight-loss drug semaglutide, better known by brand names such as Wegovy and Ozempic.

Doctors use four standard indicators to monitor the progress of diabetes—blood pressure, control of blood sugar, lipids, and body mass index—and the *2018 Physical Activity Guidelines Advisory Committee*

Scientific Report cites "strong evidence" that regular aerobic or resistance exercise, or both in combination, can reduce all four, "though the strength of evidence varied somewhat by risk factor." Resistance training may be especially helpful in controlling blood pressure and blood sugar, according to evidence cited by the report, and this evidence has been corroborated and in some cases strengthened by later publications.

Most people with diabetes also have high blood pressure, or hypertension, a titanic threat to individual and public health in its own right. More than 30 percent of the world's population has hypertension, which is the main risk factor for cardiovascular diseases. In addition to being the world's number one cause of death overall, cardiovascular diseases account for half of all deaths of older adults with type 2 diabetes. In the United States, doctors diagnose high blood pressure more often than any other disease; and for Americans over the age of fifty, doctors write more prescriptions for high blood pressure medications than for any other drugs.

Considerable research and analysis since 2018 have shown, however, that "exercise is equivalent to medication as antihypertensive lifestyle therapy," according to a 2022 review by Alberto J. Alves of the University of Maia in Portugal and colleagues, in *Current Sports Medicine Reports*, a journal published by the American College of Sports Medicine.

In 2019, when a group led by Huseyin Naci of the London School of Economics and Political Science analyzed almost four hundred randomized controlled trials involving almost forty thousand people—trials of exercise and of pharmaceuticals as stand-alone treatments for hypertension—they found that aerobic, resistance, and combinations of those two modes of exercise were "as equally effective as most antihypertensive medications" for people with high blood pressure. Doing both aerobic and resistance exercise in combination, their analysis suggested, may be about 50 percent more effective than doing either one by itself.

Reflecting on this and similar findings by other researchers, a group of exercise physiologists and medical doctors led by Pedro L. Valenzuela of the University of Alcalá in Madrid wrote in 2021, in the journal *Nature Reviews Cardiology*, that standard medical treatment of hypertension does not apply the full range of effective therapies at its disposal.

"Although endurance exercise is probably the most commonly prescribed exercise modality for patients with hypertension," the authors observed, "the benefits of resistance exercise (such as weight lifting), or the combination of endurance and resistance exercise remain largely unknown in the clinical setting, despite eliciting similar (or even greater) reductions" of blood pressure "to those achieved with endurance exercise in patients with hypertension."

Other exercise regimens, including yoga, tai chi, and qigong, may relieve hypertension at least as much as the conventional Western exercise categories of endurance, resistance, and combined training, according to the 2022 review by Alberto Alves and his colleagues.

Those authors also named a problem with most of the trials that have directly compared effects of exercise and drugs on high blood pressure: Many people who took part in the exercise trials were also on blood pressure medications, "so that the reported [blood pressure] reductions were not due to exercise *per se*." That does not invalidate the findings of the exercise trials, but it does raise questions about how exactly exercise and medications interact as treatments for high blood pressure. Does exercise treat high blood pressure mainly by boosting the effects of pills?

Linda S. Pescatello, a professor of kinesiology at the University of Connecticut, Storrs, sought to answer that question as the lead author of a strict analysis of the best quality randomized controlled trials "that isolated and then directly compared exercise and medication combined to exercise and/or medication alone." Her team's findings challenged conventional beliefs. Treatment effects of exercise and medications for high blood pressure, the analysis concluded, "are not additive or synergistic." Instead, exercise by itself may reduce high blood pressure more than medication by itself; and this appears to be true not only of aerobic, resistance, and combined training: It may also and to a greater extent be true of yoga, tai chi, and qigong.

These findings raise more questions about standard medical care for people with hypertension. In 2020, the surgeon general of the United States published a "Call to Action to Control Hypertension," articulating one most basic principle of usual care for this disease—the principle is that drugs are primary, and exercise is secondary. "Although antihypertensive medications tend to reduce a person's systolic blood pressure more than a structured exercise routine, all types of activity (endurance, resistance, or a combination of both) are effective" for lowering high blood pressure, the document stated.

In light of Linda Pescatello's analysis, though, conventional treatment of high blood pressure may put the cart before the horse. If Pescatello is correct, exercise does not boost the effects of drugs for treating high blood pressure. Instead, drugs boost the effects of exercise.

Another indicator of type 2 diabetes disease progression that can be improved by regular exercise is control of blood sugar. Doctors test glycemic control by measuring the amount of a protein in red blood cells that bonds with sugar. The protein is called hemoglobin A1C, often abbreviated as HbA1c, or A1C. The more A1C you have in your blood, the higher your blood sugar. The less A1C you have, the better your glycemic control. If you are diagnosed with diabetes and are able quickly to reduce the levels of this protein in your blood, it can yield big benefits. Reduce your A1C by 1 percent, and you may reduce your risk of heart attack or death from diabetes complications ten years later by nearly 20 percent.

Aerobic exercise alone reduces A1C a bit more than resistance training alone, according to the 2018 physical activity guidelines scientific advisory committee, and combining both kinds of training—doing some resistance and some aerobic exercise—improves the control of blood sugar more than either form of exercise by itself. For people who cannot or will not do aerobic exercise, though, a 2019 review of twenty-four trials involving nine hundred and sixty-two people with diabetes found "profound benefits" of resistance training on its own for reducing A1C, and also insulin.

When people with diabetes lift weights, "intensity should be the primary concern to accomplish the maximum benefits," according to the authors, led by Yubo Liu of Zhejiang Normal University in Jinhua, China. Resistance exercise at "both high- and low-to-moderate intensities substantially reduced" A1C, they found, but training at high intensity reduced A1C by almost twice as much as training at lower intensities. High-intensity resistance training also reduced insulin, "while low-to-moderate intensity had no effect."

Findings like these did not come as news to Nalin and Maria Fiatarone Singh. With a team led by one of Maria's former doctoral students, the exercise physiologist Yorgi Mavros, they have demonstrated the importance of progressive resistance training intensity for type 2 diabetics in a series of publications based on a randomized controlled trial called Graded Resistance Exercise and Type 2 Diabetes in Older Adults—abbreviated as GREAT2DO.

In the trial, more than 100 people aged sixty-five and older, all overweight or obese, and all with long-standing diabetes, were randomly assigned to one of two groups, to begin a yearlong exercise regimen in addition to their usual care for diabetes (the medications and diets prescribed by their doctors). One group followed a regimen of high-intensity, high-velocity progressive resistance training. Three times a week, they did full-body workouts involving seven exercises—three sets of eight repetitions, in most cases, lifting loads equal to 80 percent of their one-repetition maximums, adjusted monthly with individual strength tests. The other group followed a regimen of low-intensity, low-velocity resistance training. Three times a week, they did the same exercises and the same number of sets and repetitions—but they lifted very light weights, they did not take strength tests, and they never progressed to lifting heavier weights.

After working out for a full year, people in the group that had exercised at low intensity had lost more weight, on average, than people in the group that trained at high intensity. Body composition tests showed that fat loss, by contrast, was about the same in both groups. When each person's body weight and body composition were considered alongside their blood test results, the Singhs and their colleagues found some remarkable associations.

Glycemic control, insulin resistance, and systemic inflammation had improved only for people who lost weight in this one way: by means of high-intensity progressive resistance training. "It seems to matter not only how your body composition changes," Maria Fiatarone Singh says, it seems also to matter "how you got to those changes. And just losing weight was not in fact beneficial" for glycemic control, insulin resistance, and systemic inflammation, for the people who took part in the GREAT2DO study.

"It is time that the insufficient goal of 'weight loss' be replaced with 'muscle gain and fat loss'" as a standard medical recommendation "if optimal metabolic health is to be achieved" for people with diabetes and for anyone who is overweight or obese, the Singhs and their colleagues concluded.

Type 2 diabetes also creates high risk for developing chronic kidney disease. In the United States, about 40 percent of adults diagnosed with diabetes have chronic kidney disease, too. Treatment of chronic kidney disease may involve prescription of a low-protein diet; and protein deficiency will lead to muscle wasting. In turn, functional abilities and quality of life can decline on steep trajectories, to end stages involving dialysis or kidney transplant surgery. But high-intensity progressive resistance training can protect against loss of muscle strength and function while substantially counteracting a range of other side effects, with results including improved quality of life: These are a few of the findings of randomized controlled trials that Maria Fiatarone Singh has run with colleagues, especially the physician Carmen Castaneda and the exercise physiologist Bobby Cheema.

People with end-stage kidney disease lifted weights *during* their dialysis sessions, in a study that Fiatarone Singh ran with Cheema. Before being hooked up to the dialysis machines, people would train the arm that was about to hook in. Then while the machines did their work, people did full-body workouts for 45 minutes.

When kidney disease advances to the point where transplant is necessary, resistance training again can be critical to treatment. To control inflammation and to protect against rejection of the new organ after kidney transplant surgery, or indeed any major organ transplant surgery, doctors often prescribe corticosteroids such as prednisone. That cure creates problems, too. Taking prednisone for three months can wipe out as much muscle as a full decade of aging, according to research on heart transplant recipients by Randy W. Braith of the

University of Florida in Gainesville. Yet strength training at medium intensity, prescribed in combination with corticosteroids, not only can stop the muscle loss but also can produce significant gains in both size and strength, according to a randomized controlled trial that Braith published.

"All organ recipients require anabolic exercise"—exercise that builds muscle—"as well as aerobic exercise," in Maria Fiatarone Singh's view. "This is *not* part of routine care" at the present time, she adds.

Most of the best proof that progressive resistance exercise can help to prevent and treat chronic conditions is based on measurements of movements that individuals have made. Those measurements add up to a compelling body of evidence, but they do not show how a person learns to lift weights and how the skill of strength training can transform a person's whole life. I learned about these things from a woman named Ramanee, whom I met at the gym in Maria Fiatarone Singh's University of Sydney laboratory. Ramanee took a break from her workout one day to tell me how she came to be there, to describe the help she found there, and to recount the methodical yet often unpredictable process by which progressive resistance training freed her from debilitating pain and restored her power to act upon the world.

From the first moment of every day, Ramanee was consumed by a problem. Waking up in bed, shifting carefully to the edge of the bed, she would touch her feet to the floor and think to herself, *You can't get out.*

Moving slowly, slowly, and fighting through the pain, Ramanee could take half an hour to straighten her legs and stand up. And even after this, when she had managed to stand, everything that lay ahead of her, every minute of the day, was a problem. Walking to the bathroom. Walking to the kitchen to start the morning coffee. Ramanee states the problem: "When you're walking, you bend your knee." Every single step, she had to think, *bend the knee*.

Ramanee speaks softly. Her gaze is direct. English is her second language. Sinhala is her first. In the 1960s, growing up in the Sri Lankan capital city of Colombo, she was never very sporty, she says. But she did like to play netball.

A variant of basketball developed in the nineteenth century as a game for women, netball has rules limiting players' movement on the court. The rules are based in part on nineteenth-century medical beliefs that women's exercise capacity was more limited than men's. Netball's fairly modest cardiorespiratory demands suited Ramanee because she was asthmatic. When her family left Sri Lanka for New Zealand, when Ramanee was fifteen years old, the popular sports at her new school were much more vigorous—softball, tennis, gymnastics—so she took every excuse not to play.

The next year, she was diagnosed with scoliosis; and then until she was eighteen years old, she wore a brace that enclosed her from hips to armpits. At first she wore the brace most of the time, every day and night, except when she bathed and when she did physical therapy to strengthen her back. Sometimes, though, she would take off the brace and borrow her brother's bicycle and go pedaling on paths near the hospital where she worked after school helping the nurses. The year after Ramanee graduated from high school, her doctor said she no longer needed the brace. So she left it in his office and she went home, feeling free.

Ramanee went to university, where she loved science courses, but difficulty with language led her to drop out. She married. Her parents moved to Australia, and she and her husband moved there, too. She got a job at a mental hospital. She worked as a trust clerk. When people were admitted to the hospital as patients, trust clerks took care of their valuables, recording and packing each item for storage in a safe. Trust clerks also processed the patients' income, such as pension checks, and dispersed spending money so patients could buy things at the hospital's cafeteria and shop. Ramanee had a baby, and she left her job, and she had another baby, and she raised the children, "and so," she says with calm good cheer, "time passed."

In the passing time, she tells me, "here and there walking, here and there cycling, here and there some strength exercises. And that was my life to the age of fifty. But at the age of fifty, all of a sudden, my knees started to hurt. I went to the doctor, X-rays were done, and that's when you discovered your knees are worn out."

Referring to herself, Ramanee frequently, easily shifts between first and second person—as when she recalls her diagnosis: *I went to the doctor . . . and that's when you discovered . . .* She had osteoarthritis.

In a person's joints, slippery coatings of cartilage cover the ends of bones. "Cartilage serves the purpose of a grease for the joints," wrote Galen, almost two thousand years ago, and that's a good way to picture it. Cartilage keeps joints smoothly gliding. Every time you take a step, turn a faucet, type a word, or nod your head, and those movements are easy and pain-free, be thankful for your cartilage. Over time, and especially after trauma to the joints, cartilage can degenerate. When cartilage degenerates, joints can become stiff, with movements difficult and painful. Degeneration of joint cartilage is a disease called osteoarthritis.

Though any joint can be affected, osteoarthritis is most burdensome in the knee. Globally, almost 25 percent of the world's population aged forty or older has osteoarthritis of the knee. In some countries, even larger numbers of people live with knee arthritis: in Korea, more than 35 percent of the total adult population; in Japan, almost 40 percent; in Thailand, more than 45 percent. Knee arthritis is more prevalent among women than among men, and prevalence grows with age. Half the people in the world who are eighty years old and older have knee arthritis. Globally, the number of people with knee arthritis more than doubled between 1990 and 2019.

Pain is the main symptom. Morning stiffness is also common, as are crackling sounds or sensations in the joint. But the pain itself and the limitations that pain imposes on people are what make osteoarthritis of the knee the most common cause of disability among older adults.

Excess body weight is the main risk factor. People who are overweight are more than twice as likely to develop knee arthritis as people of normal weight. Mutually reinforcing biomechanical and biochemical effects exacerbate the disease: Excess weight puts knees under increasing load, while causing inflammation throughout the body. Other main risk factors include muscle weakness and a history of knee injury. Having weak legs—not pathologically weak, just on the weak end of a normal range of strength—increases the risk of knee arthritis by 65 percent. And risk almost triples for athletes who have had sports injuries, though higher risks are associated with some kinds of injuries for some kinds of athletes—such as the minimum fourfold increase in risk of knee arthritis among female athletes who suffer ACL tears while in their twenties, as mentioned earlier.

Osteoarthritis is another disease that Maria Fiatarone Singh has learned to treat with exercise. One September afternoon, she talks about this to a class of exercise physiology grad students, in a small lecture room on the Cumberland Campus of the University of Sydney, across the street from the largest cemetery in the southern hemisphere. As always, she is impeccably dressed for work—today in a bright red wraparound dress and a necklace studded with the birthstones of her seven children. While lecturing, occasionally she shifts her weight fully to one leg, then to the other.

Because knee arthritis is related to obesity, she says, most people who have it also have other conditions, such as diabetes, heart disease, peripheral vascular disease, and high blood pressure; and so treatments for knee arthritis should also account for an individual's other conditions. She gives these future exercise physiologists practical advice about how to care for their future patients: "Think about them as a person. Don't think about them as a knee OA." (Osteoarthritis is often named with the abbreviation OA.) "These people may have more than arthritis. They're more than just their joints."

Erosion of cartilage is a problem that feeds on itself. Once it starts, it tends to worsen. When the knee is properly aligned, ground reaction force goes straight up through the center of the joint. When cartilage erodes, the angle of the ground reaction force shifts. Force travels upward, typically through the proximal side—the inside—of the knee. To compensate, people may begin to walk bowlegged.

"The joint is meant to be held in place by strong muscles and tendons. So if it's meant to go like this"—Fiatarone Singh makes two fists and presses them together, stacked vertically so her knuckles fit like jigsaw-puzzle pieces—"and instead it goes like this"—she repositions her fists, knuckles misaligned, akimbo—"the cartilage degenerates."

In severe cases, the cartilage can be completely worn away, damaging the overall structure of the joint. Knees lose the gliding quality of the normal walking gait. Instead they grind, bone on bone.

During the first couple of years after Ramanee was diagnosed with knee arthritis, her pain grew worse. She went to see an orthopedic surgeon about her knee. The surgeon said that eventually Ramanee would need surgery. He said that when her pain worsened, she should come back to see him again. He recommended that in the meantime she should see a physical therapist and build strength in her legs. She went to appointments with the physical therapist for six weeks, and then for a while she did exercises on her own. Then she stopped doing the exercises because they made her knees hurt.

More years passed. Ramanee's pain kept getting worse. Bending her knees was so difficult and required such extreme concentration, she stopped doing much of any walking outside the house, aside from what was strictly necessary.

Grocery shopping was an ordeal. Just walking from her car to the store was an ordeal. Once inside the store, she managed to walk down the aisles by leaning on the handle of a shopping cart, relieving some of the pressure on her knees. The worst part of the errand was the end. When she returned home after shopping, it took all the energy she had to carry the groceries from the car to the front door. By that time, she says, "you *had it*. Now your muscles are tight, you're tired, and then you have to get the groceries inside the house and to the fridge. Especially the perishables. You can leave the other stuff for later."

I ask her, What was the pain like?

"Every step," she quickly answers. "Every step when you're walking and you put weight to the knee, it hurts." She never had a break from this. "Every bend," she says. "Every time, it's harder to walk." From one step to the next, there was no predicting how her legs would work: "And when you're walking, the knee used to lock up. In a sense, you can't move."

On a scale of 1 to 10, she rates the pain almost a 9. It was "really, really hard," she says, "because the thing is, you had to keep moving. You had to do the things." The demands of daily life did not stop for her arthritis.

Something had to change. She says, "I was looking for answers. To keep on going. Because what was in front of me was surgery, which I didn't want to do."

Then she gestures to the door of the gym where she is now a regular, and she says, "So that's how I ended up here."

Osteoarthritis has no known cure. No treatment has proven to reverse degeneration of cartilage. Slowing the damage and managing the symptoms are the best that can be done. Most people with knee arthritis want two main things. They want the pain to go away. They want also to function independently for as long as possible. To be able to walk and to go up and down stairs without help.

Hope for relief from intense pain caused by arthritis has inspired a plethora of therapies. These include arthroscopic surgeries and various types of injections: stem cells, platelet-rich plasma, collagen, viscosupplementation, and corticosteroids. Physiotherapy techniques for arthritis include ultrasound, cold therapy, kinesiology tape, acupuncture, special shoes, and knee braces. Over-the-counter supplements for arthritis include glucosamine chondroitin, vitamin D, turmeric, and pine bark extract. Maria Fiatarone Singh says there is little or no peer-reviewed evidence to show that any of those therapies work. With extreme circumspection, she says, "Lots of people do lots of things that are non-evidence-based."

Evidence shows that other therapies do relieve the pain of knee arthritis. Clinical guidelines for treatment emphasize lifestyle interventions, mainly diet and exercise. "This is the one condition where doctors seem to agree, exercise trumps drugs for treatment," she says, adding that there is "probably no other condition where that's the case, in all of the clinical guidelines."

Some of the most popular exercise therapies for osteoarthritis are the gentlest forms of exercise—stretching, walking, and aqua aerobics. Yet they are not very effective for relieving pain or improving physical function, according to extensive reviews of medical literature on the topic.

The Cochrane Library's 2015 review, "Exercise for Osteoarthritis of the Knee," analyzed the findings of forty-four randomized controlled trials of land-based exercise as treatments for the condition. The review found that exercise programs with the largest effect sizes, indicating the greatest reductions of pain or improvements of physical function, involved strength training focused on the quadriceps muscles. The effect size of these quadriceps strengthening programs was only moderate, but it was about 25 percent greater than walking programs for reducing pain, and about 50 percent greater than walking programs for improving physical function. Based on these numbers, Fiatarone Singh tells her students, "it does look like strength training probably is at least a primary recommendation, if not the only one, for arthritis."

Another systematic review, published in the *British Medical Journal* in 2013, analyzed more studies: sixty randomized controlled trials of flexibility, aerobic, and strength training programs, both land-based and aquatic, for osteoarthritis of the lower limbs. For reducing knee arthritis pain, Fiatarone Singh says, among land-based exercise programs, "the strongest effect was in programs that used strengthening exercise." Among aquatic exercise programs, she continues, "the highest level of effect was with the combination of flexibility and strengthening." And "flexibility and aerobics in the water"—like an aqua-aerobics class—"which is what most people do," was found to have "almost no effect whatsoever."

For improving physical function, too, this review found the most effective therapies involved strength training, in combination with flexibility or aerobic training—or all three together. "It's a very multifactorial condition, so approaching it from one angle is probably never going to be ideal," Fiatarone Singh says. In that way, "it's like diabetes: Exercise works; diet works; in combination they probably work better."

For preventing knee arthritis or slowing its progression, she adds, diet is at least as important as exercise because functional independence stems largely from the interaction of two factors, force capacity and fat mass. Existential physics, according to Fiatarone Singh: "If you've got not enough muscle and too much fat, you have a hard time dragging your body across space."

Most clinical practice guidelines for treatment of knee arthritis recommend weight management for patients, but this doctor reminds her students of a problem with dietary weight loss: "If you lose weight by diet, you will lose muscle and bone." When people lose weight by diet alone, or by diet with a predominantly aerobic program of exercise, most of the weight lost is fat, but much of it—around 30 percent, or more, according to some studies—is muscle and bone. She says, "In order to prevent that, you need anabolic exercise to offset the lean body mass losses."

The medical literature on knee arthritis symptoms, progression, and treatment suggested to Fiatarone Singh that comprehensive long-term strategies for managing the disease would need to intervene in multiple interrelated processes simultaneously: losing fat, gaining muscle, improving metabolism, reducing joint strain, and enhancing function.

She was especially inspired by what she calls "the definitive study of diet and exercise" in the treatment of osteoarthritis, the Intensive Diet and Exercise for Arthritis trial designed by Stephen Messier of Wake Forest University and colleagues. The IDEA study found that diet in combination with aerobic and strength training, along with stretching, did more to reduce pain over the long term than diet or exercise as stand-alone interventions. In the IDEA study's published results, however, Fiatarone Singh noticed especially large losses of lean mass in all groups who dieted, no matter what kind of exercise they did, and she believed she understood why this happened. Even the people who did strength training lost lean mass, she thought, because "the strength training in IDEA was not very robust."

She wondered what might happen if a somewhat similar trial included a more robust strength training program. So she worked with one of her doctoral students, Yareni Guerrero, to design a randomized controlled trial involving four types of intervention: diet, strength training, gait training, and a combination of all three. "If you had this weight-loss diet and weight lifting, opposed to the diet by itself," Fiatarone Singh hypothesized, "you would have the best shift: increased muscle, decreased fat, improved metabolism, improved function. Our theory was: That would be the best way of attacking everything at once." The strength training program focused on certain muscles, the hip abductors, that could reduce pressure in the joint.

Yareni Guerrero, who is a former member of the Mexican national women's wrestling team, set up the trial's strength training program, guided by Fiatarone Singh's questions: "You need to look mechanically at, where is the weakness? And what can you do to get the joint as aligned as possible?"

A member of Ramanee's family heard about the University of Sydney arthritis study and told her about it. Ramanee's husband drove her to the campus for initial evaluations. She met with Fiatarone Singh, who took her full medical history.

The conversation was not what she expected. "Professor Maria was prepared to listen to you," Ramanee remembers. "Not all doctors do. She was prepared to listen to your side of the story." Ramanee told the doctor everything. She talked about her asthma, her high blood pressure, the awful pain in the mornings. She talked about her weight and the yo-yo effect of her dieting: how she would repeatedly lose a few pounds, only to gain them back again. "Every time the weekend comes, you go out for dinners and things like that, and then you're back to where you were before," she says.

Then Ramanee went to the gym, where Yareni Guerrero administered one-repetition maximum strength tests. After Ramanee did her one-rep maxes, she was so tired that she fell asleep on the car ride home.

Randomly assigned to the diet group, Ramanee met with a dietician whose demeanor and technique surprised her as much as the doctor had surprised her. Instead of imposing a diet or meal plan, the dietician started by asking questions—not just *What do you eat?* but also *What is the recipe for this dish? Do you fry the fish or broil it? Oil or butter?*—and then suggested changes of ingredients and cooking techniques. Conversations built a one-year plan. The goal was for Ramanee to lose 10 percent of her body weight without losing any muscle or bone mass. "The diet was a high-protein, low glycemic index diet, rather than a traditional low-fat diet," Fiatarone Singh tells me. "This was designed to maximize retention of muscle, lower inflammation, and be more sustainable than low-calorie, low-fat diets usually are."

The dietician, Yian Noble, suggested such slight changes that Ramanee's husband never realized the recipes had changed. Noble also gave Ramanee strategies for avoiding the yo-yo effect of weekend restaurant meals, such as "the idea to make your own food and take it with you when you're going out," Ramanee says. "Like, have extra vegetables with you, and then you decide, looking at the menu, whether you're going to eat what's there, or have your own meal, or mix it up." By drinking less soda and alcohol, Ramanee eliminated more calories from her diet. When Ramanee cut down on the beverages, she realized she cared so little for them, she didn't mind giving them up.

Full-body scans by dual X-ray absorptiometry, or DXA (pronounced "DECKS-uh"), measured Ramanee's body composition at the beginning of the study, after six months, and at the end of the year. By the midpoint of the study, her waist was slimmer, her weight had come down, and she had not lost any muscle or bone. All those things were still true at the end of the year; and by that time, she had met her goal of losing 10 percent of her body weight, without losing muscle or bone. Her asthma was noticeably better. She didn't need an inhaler anymore. The pain in her knees was still awful, but some days she thought that maybe it was fading a little bit.

"So luckily for me, I was interested in cake decorating," Ramanee says when she tells me about another aspect of the arthritis study. At the beginning, Fiatarone Singh referred Ramanee to a physical therapist who prescribed a simple program of body-weight exercises. The physiotherapist suggested that Ramanee should do the exercises for five minutes a day and then gradually build up to fifteen minutes. Exercising in front of a mirror, the physio said, could help Ramanee keep an eye on her form—and when the physio mentioned mirrors, Ramanee thought of cakes.

Ramanee had extra mirrors at her house because cake decorating was one of her favorite hobbies. While decorating cakes, Ramanee would prop a mirror on her kitchen counter behind the cake or off to one side. That way, wherever she was standing, she could see how the decorations looked from another angle, without having to walk around to see the whole situation.

Now, as she began her new physiotherapy program, she hung one of the cake mirrors at the end of a hallway in her house, so she could watch her form as she moved along the corridor doing her exercises.

When she exercised, that is. She tried to do her exercises in the early mornings, but then sometimes when she was in a hurry, she would skip them. When she was exercising, or even when she was thinking about exercising, five minutes could seem like forever. "Then I realized, if you don't put exercise on the top of your do-list on the day, you're not going to do it," she says.

Those exercises, at the beginning, were not satisfying for Ramanee. They were frustrating. She thought her balance was not good. She thought her movements were too slow. But frustration fueled resolve. She made two decisions about her training. Some of the people closest to her thought her choices looked like cheating, but these choices were pivotal in her moving from frustration to success.

She decided to get a walking stick. Sometimes she used it at home, but mainly she carried it when she left the house. With the stick, she could walk farther and for longer periods of time. Most important, it helped her to go up and down stairs.

Knee arthritis makes stairs a special plight. Up is difficult, and down is torture. Descending a staircase inflicts some of the highest forces on the knees that most people experience in daily life. Ramanee is hard-pressed to describe the challenge. "Going down doesn't happen," she says. "You don't *win*."

Then she looks mischievous, like she's telling a secret, when she adds, "But if you go *backwards*—you can do it easily. I discovered that it can be done." Walking downstairs backwards decreases the force on the knee, and it shifts the burden of the center of body mass from the knee and ankle to the hips, which can better absorb the force.

Ramanee also decided to expand her walking-stick experiment. She decided to use the walking stick all the time. Even at home and even while doing her exercises. Ramanee's husband, her physical therapist, and everyone else with an opinion said they thought it was a bad idea. "But for me," she says crisply, "I think that's the best thing I started doing."

She also changed the metric of performance that her physical therapist had suggested. She stopped trying to exercise for the prescribed duration of five minutes, failing to meet that mark, and feeling frustrated by falling short. Instead, she kept count of how many repetitions of the exercise she was able to do, and little by little, she began to increase that number. Changing the metric from duration to volume gave her a new sense of accomplishment. "I started counting instead of timing," she says. "Instead of doing five minutes, I did five *steps*. And getting better became: *ten* steps. From the kitchen to the bathroom, I counted the steps."

To her, changing the rules was not cheating. It was adapting. She adapted to the shifting circumstances she encountered while navigating her environment. She adapted the system she learned from the physical therapist to make it suit her life.

She trusted herself. At the same time, she kept a skeptical eye on herself. She tested the effectiveness of her choices by keeping records of her exercises. "I started writing it down," following the habit she had developed during the yearlong arthritis study, when she had filled out lots of questionnaires for the researchers, documenting the ups and downs of many aspects of her life: her psychological state, the pains she felt, and her daily activities. This experience of observing herself while also being observed—of having company in paying attention to all the choices and habits that make up the whole shape of her experience—coincided with significant changes in her behavior, she noticed. "That's when I started doing regular everyday exercises," she says, with half a smile.

When the study ended, Ramanee had another meeting with Maria

Fiatarone Singh. Ramanee asked the doctor, "What's next?"—and the doctor sent Ramanee to M-block.

The main site of strength training for Maria Fiatrone Singh's clinical trials was then a gym in a plain gray concrete building, a brutalist rectangle known as M-block on the University of Sydney's Cumberland campus, in the city's western suburb of Lidcombe. (Parcels of land on the campus are named with letters of the alphabet. This building happens to stand on the parcel of land named with the capital letter M.)

The manager and head trainer in the gym at M-block at that time was the exercise physiologist Guy Wilson, who had held a variety of jobs—sometimes tending bar, sometimes laboring on construction sites—until his mid-thirties, when he started working toward his doctorate with Fiatarone Singh as his adviser.

When Fiatarone Singh sent Ramanee to M-block, Guy Wilson suggested that, in addition to the exercises she did at home in the mornings, Ramanee should start a program of progressive resistance training at the gym two or three times a week.

Ramanee replied that she would come to the gym one day a week.

"Because you are in pain and you need to go home," she reasoned. "You have to do what's there at home, and you should be able to do that. If you don't do what's to be done at home, this is going to fail. Because my husband won't be very happy, and I won't be very happy," she tells me. Then she raises her eyebrows.

Ramanee started going to the gym. She lifted weights once a week, on Mondays. For the first few workouts, Guy Wilson trained her. He adjusted the weight machines to fit the proportions of Ramanee's limbs, and he taught her proper technique for the exercises. Each workout was a full body workout, alternating upper-body and lower-body exercises, one after the other, through all the major muscle groups of the legs, back, chest, shoulders, and arms. For upper-body lifts, Wilson taught her to do high-intensity power training at high velocity, pushing as fast as she could. For lower-body lifts requiring knee flexion, Wilson taught her to do high-intensity strength training at low velocity, to protect her joints from damage.

Listening to him, Ramanee learned some basic principles of

strength training program design. "Don't do the same exercise every day in the same order. Your body gets used to it," she says.

When she started, every aspect of lifting weights felt new to Ramanee: all the tiny adjustments of effort and balance required to move the weights up and down through space. When she did the dumbbell shoulder press, she began by raising dumbbells that weighed 1 kilogram. "It's hard," she remembers, "but then you get used to it. Then when you're comfortable, you go for the next one"—the 2-kilogram dumbbells.

When she started, also, her ankles hurt after every workout, which made her worry. Was this normal? Was she injured? Telling the difference between good pain and bad pain can be difficult for anyone who's learning to lift weights. It can be especially challenging for someone who has a condition defined by pain.

Ramanee always told Guy Wilson what she felt, and he always asked her questions: *Is it new pain? Is it sharp?* Wilson said, "If it's a sharp pain, that's an injury. If it's a dull pain, then it's just the arthritis pain or exercise pain."

Ramanee was never absolutely certain she was on the right path. She always had some doubts, because she was doing something so far removed from her prior experience. "But the thing is, you kept going," she says.

"I stuck to the once-a-week" regimen of lifting weights at the gym, she says. She stuck to it for a year.

At the end of that year, Guy Wilson asked Ramanee if she would think about coming to the gym more often—maybe twice or three times a week.

Ramanee told him, "It's not practical."

She had to do the housework. She had to make dinner. She liked having some hobbies, and she barely had time for those now. Her garden was suffering, because sometimes there was not enough time in the morning to water the plants, after she had cooked breakfast and done her exercises.

She kept to her routine of lifting once a week.

After a while, she took a vacation from the gym when she traveled to Sri Lanka to visit relatives. From day one of that trip, she was in trouble. She was in pain. Pain in her knees, pain in her muscles. Back home in Sydney, she talked to Fiatarone Singh about the pain, and then she went to see a specialist, and the specialist told Ramanee that only two things could make her feel better: surgery or prescription painkillers.

Ramanee disagreed. Surgery was out of the question. She would not risk the complications. Painkillers were also out of the question. If she used pills to mask the pain, how would she know when her body truly needed rest?

Ramanee decided to take a different course of action. When she felt tired, as she often did after lunch, she sat down and rested. There was one slight problem with her strategy. Sometimes when she rested, she fell asleep; and when that happened, she had to make more time for housework in the afternoon, which in turn pushed her bedtime later into the evening. "But it doesn't really matter," she says, "as long as I get things done. You can balance. And if you take medication, that is not there—you go further than your limit."

She did not go back to see the specialist again. She went on doing her exercises at home and lifting weights at the gym one day a week. "The thing is, I was getting slight improvements. Tiny improvements," she says.

Even tiny improvements could have big effects. When her daughter was married, "I was fit enough to do the cake," she says. Her voice blooms. "My weight was down. I looked good."

At the end of her second year of weekly strength training, Ramanee decided to take Guy Wilson's advice to lift weights more often. She started coming to the gym twice a week, which in some ways was difficult. "That took more time. I was getting time-short. I had to drop certain things. Definitely dropped the garden."

For two more years, then, she lifted weights at the gym twice a week. Very gradually, she found, she built more capacity to move without pain. As she felt more comfortable, she also felt she had more spare time. "Bit by bit, the fitter I got, I was able to fit in more," she says.

Every so often Guy Wilson told Ramanee that more frequent workouts—three times a week—would probably bring more improvements, and when he said this, Ramanee would tell Guy Wilson that she was able to come to the gym two times a week, no more.

Compromising, Guy Wilson gave Ramanee a third day's worth of exercises to do at home; and Ramanee did the exercises, not all in one day, but spread across the rest of the week. "And that's how my improvements are a lot more this year," she says to me—in her fifth year of lifting—and her narration rounds with detail, rising to a triumph.

She says, "I can bend the knee when I'm walking."

She can walk longer distances now, she says, and her knee pain is almost gone: "When you're walking and you put weight on it, it used to hurt a lot. But now it's almost disappearing. But you could still feel there's discomfort in there. But the direct pain that you used to have is disappearing."

She continues: "That means you're faster. You have a sort of flowing feeling. Earlier, you really had to think and try harder to keep moving. But this year, all of a sudden, I can bend it."

Now that she can walk faster, she says, the flowing feeling lasts longer: "Flowing feeling is the sense that you are not trying hard to walk. It's—you're moving quicker. Flow is *quicker*. The leg movement. The tightness of your muscles is reduced. In the morning it's the same thing." When she wakes up in the morning, she no longer has to struggle for half an hour, pushing through pain, just to get out of bed and take her first steps into the day. "When you get up, you can straighten your leg, go quicker," she says. "The first ten steps, you're not *trying*. It's getting easier to walk."

Her knees feel so much better now that, when she drives to campus to lift weights, occasionally she will park her car some distance from the gym, because she wants to enjoy taking a walk as a warm-up to her workout. She does not often use her walking stick anymore, but she still keeps it around because, she says, "Sometimes I need a kick-start." When Ramanee feels uncertain on her feet, as a kick-start she will sometimes use the stick for the first five steps. Then, she says, "You lift it up and you don't use it for the rest of the walking."

In the gym, she is lifting big weights now, compared to where she started. For the shoulder press, she has worked her way up from 1 kilogram to 6 kilograms. The dumbbells that weigh 6 kilograms are coated in yellow rubber, the color of daffodils. Ramanee thinks 6 kilograms may be the limit of her ability. "I don't think I will be able to do seven kilograms," she says, "because you're not able to control the weight"—her arms wobble when she tries to lift the sevens. "Maybe next year, I might. I don't know. I might be able to."

Progression is surprising, she finds, and not really predictable. At one time, Ramanee thought that she would never lift the sixes, but then she did it. Back in the time when she was lifting fives, a trainer persuaded her to try six—and she got the weight up, but barely. So she stuck with the fives until the start of this year, when she felt strong enough to give six another try, and "I was able to do it. So I dropped the five-kilogram and started doing the six," she says. "So on this schedule, you start with the lower weight, so you are able to do it. Then you have the confidence to do more."

Confidence in the gym carries over. When Ramanee goes shopping now, she has no problem carrying the groceries from her car to the front door of her home. "You can do it! It's—it's *good*," she tells me. "And when you go out on trips, now you can walk longer. Which means you look around. And you are better at spending that time."

The long-term effectiveness of exercise as treatment for osteoarthritis is unknown. No randomized controlled trial has tested the effects of an exercise program on the symptoms of arthritis over as many years as Ramanee has been training. No one has established the optimal intensity, frequency, volume, and modes of exercise for people with arthritis. No one is entirely certain, either, of exactly *how* exercise for this condition affects the person who does it. Doctors and physiotherapists like to believe the pathway to disability follows a logical course. They like to believe that physiological decline causes functional impairment, and the result is disability. Fixing the problem, by this logic, means reversing the process.

Maria Fiatarone Singh does the math, which is not really math: "You think, if I get rid of the muscle weakness, get rid of the physiological problem, that's *why* it will be better. But in the arthritis studies, even the ones about strength training, when people get better in terms of their symptoms, they often don't get stronger. Or there is no correlation between how much stronger they got and how much better their pain was. So we *think* that's what we're doing, we *think* that's how we're doing it, but in fact, so far, the trials have not shown that direct correlation."

Arthritis studies, unlike studies of depression, diabetes, and some other conditions, have not shown a strong dose-response relationship between intensity of exercise and beneficial outcomes. For arthritis, strength training works at both low intensity and high intensity. For controlling pain, Maria Fiatarone Singh's research has shown that sham exercise—lifting very light weights, with no progression—works just as well as lifting weights at 80 percent of one-repetition maximum. But for people who can work at high intensity, she thinks high intensity is best because of its other benefits, including higher bone density, less depression, and decreased inflammation and insulin resistance.

With these goals in mind, Guy Wilson and other trainers at the M-block gym went on helping Ramanee remember to keep the intensity of her workouts as high as safely possible. Helping her to lift weights that were heavy for her, while taking every precaution to minimize risk of injury.

After she had spent some time following her new, personalized, three-days'-worth-of-workouts-spread-across-each-week progressive resistance training regimen, Ramanee boarded an airplane and flew to London to visit family and do some sightseeing.

The last time she had gone to London, every day ended at noon, because her knees hurt and she needed to rest. On this trip, Ramanee tells me, she stayed out late, "to about four o'clock in the evening."

More important than her performance on the clock, though, was the fact that on this trip, Ramanee had the ability, the strength, and the power to go everywhere she wanted to go. She went to Buckingham Palace. She went to the British Museum to see the Egyptian galleries and the Asian galleries. But those grand things could almost be called forgettable, compared to what happened one day in the London Underground—the Tube.

Making their way around the city, Ramanee and her husband arrived at a Tube station, and this particular Tube station, they discovered, had no working elevator or escalator. The only way to reach the underground train was by descending a stairway. A stairway that, if you were in your twenties with knees made out of rubber bands and magic, you might blithely tromp right down. But if you were in your sixties, with knee pain a central fact of your life, that same stairway might make you feel a flash of fear or terror, almost as if it were the passage to the underworld.

Unless you were Ramanee.

Ramanee stood at the top of that stairway. She looked down. And she knew what to do.

She turned her back to those stairs, and then, one step at a time, she made her backwards descent.

And at the bottom of the stairway, on the train platform, she felt a whoosh of air—the Tube arrived—red doors retracting—and she stepped into the train, and she was on her way.

On a deep breath, a sigh—contentment—she says, "The thing is: you finished the day."

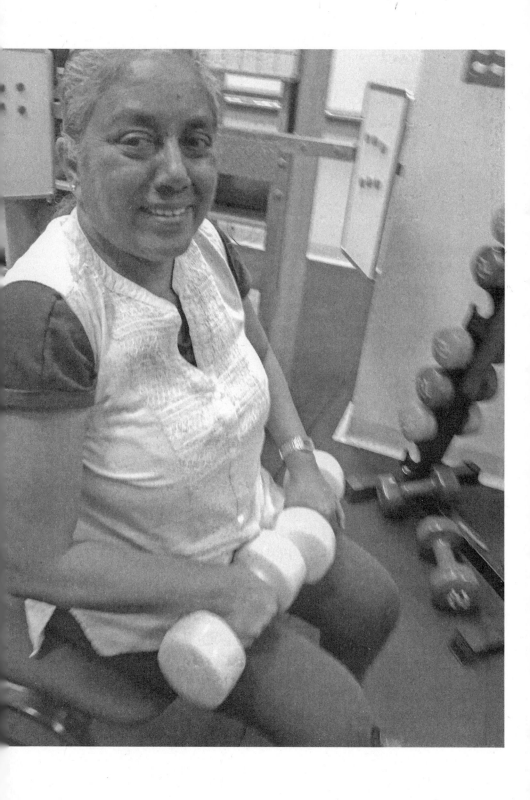

CHAPTER 9

Fall and Rise

Consider how hard Ramanee worked to manage her pain and overcome the limits that pain put on her life. Consider the bravery it took for her to change her everyday habits—cooking, eating, and drinking; socializing and scheduling—and to make these changes in her late fifties and early sixties, while at the same time laboring through long years of learning to exercise in ways that would improve her ability to function. Consider how much help she had from doctors, dietitians, and trainers—all the assessments, instructions, and guidance that she trusted, tested, and adapted to her own purposes. Consider how lucky she was to find this help, and to gain access to facilities with equipment that could aid her transformations. Then consider those aspects of Ramanee's experience in light of more data.

These data come from the World Health Organization's *World Report on Ageing and Health*, published in 2015, based on reviews of experimental, epidemiological, public health, and economic research. The title of one review prepared for the WHO, by the Australian physiotherapist Andrew M. Briggs and colleagues, sounds an alarm: "Musculoskeletal Health Conditions Represent a Global Threat to Healthy Aging."

By 2050, the population aged sixty-five and older will double, growing from 771 million to 1.6 billion, according to the United Nations. This group will grow from 10 to 16 percent of the planet's population—from one in ten people to one in six. As the older population grows, so, too, will the number of people who, like Ramanee, suffer from the most common musculoskeletal conditions—which in addition to osteoarthritis include low back pain, osteoporosis, fractures, and falls.

"The prevalence and impact of musculoskeletal conditions increase with aging," Briggs and colleagues note, and "this burden far exceeds service capacity." The burden is exacerbated by compound afflictions that can develop when musculoskeletal problems limit physical activity. Global health care systems emphasize "maintaining an active lifestyle to reduce the impacts of obesity, cardiovascular conditions, cancer, osteoporosis, and diabetes in older people. Painful musculoskeletal conditions, however, profoundly limit the ability of people to make these lifestyle changes."

What can help? "Multilevel strategies and approaches to care that adopt a whole person approach are needed" to engage the complexity and sequelae of musculoskeletal impairments, Briggs and colleagues write. These strategies must be evidence-based and feasible, they must be "person-centered," defined as "care where the unique impairments profile of the individual is considered in identifying the appropriate components of care and their sequencing." Effective strategies will also involve "co-care," defined as combining "both care delivery by health professionals and an active self-management role" by patients.

In every detail, this model of effective health care sounds like what Ramanee experienced at the University of Sydney, when changes in diet and exercise were prescribed as medicine for her knee arthritis, prescriptions that Ramanee actively managed and creatively adapted in collaboration with expert caretakers.

Part of Maria Fiatarone Singh's vocation is to expand access to that kind of health care so that experiences such as Ramanee's will cease to be so exceptional. The doctor once told a group of her colleagues, "I think that our mission really is not to make masters athletes of anyone or everyone, but"—quoting one of her mentors, the former Stanford School of Medicine professor Walter Bortz II—"the mission of medicine is the assertion and the assurance of the human potential."

Beginning in 1999, many of the findings of Maria Fiatarone Singh's clinical trials of progressive resistance training, including her collaborations with her husband Nalin Singh, were implemented as part of usual medical care at a clinic that Nalin created. In his practice, Nalin Singh started prescribing weight training as medicine to older people, to prevent and help treat chronic diseases and to preserve and improve abilities to function in daily life.

Over time, Maria's research and Nalin's treatments evolved, with early work on building muscular strength leading to later work on training muscle power. In addition to their research on resistance training and chronic diseases, they found that lifting weights could be a central pillar of a comprehensive treatment program for reducing older people's risk of the worst outcomes of hip fracture and falls. Maria's research especially has encompassed exercise throughout the lifespan—and she has shown weight training to be a feasible and healthy exercise for some demographics of people who statistically are highly likely to be physically inactive, such as overweight and obese children and adolescents.

Epidemiologists are finding at the population level evidence of some of the same associations of resistance training and health that the Singhs and others have shown in research and clinical practice. A growing body of epidemiological research suggests that progressive resistance exercise and the physiological results it can produce, including increased muscle mass and strength, may reduce the risk of several chronic diseases and reduce mortality risk. This evidence raises the possibility that, for many people, lifting weights may be nothing less than a matter of life and death.

As a geriatrician, Maria Fiatarone Singh is highly sensitive to matters of life and death. Even her emails carry intimations of mortality. The signature at the foot of her messages reads, "Love extravagantly. Life is not a dress rehearsal."

To start with, it was one particular life and death that taught her about exercise and determined her vocation. Her maternal grandmother, Jeanne Marie Céline Torre Saint-Gaudens, had the joyful, embracing force of life that can be a grandma's prerogative. This made for a good balance to the young Maria's opinionated, somewhat contrarian childhood temperament.

By the time she was six years old, in 1960, Maria was a person of firm preferences. She loved horses, and she hated being modern. Loved sports, hated being a girl. Why couldn't she wear the same clothes and do the same things boys did? She was at least as good at any sport as any boy.

Something else she loved was visiting her grandmother, who lived in a big Victorian house in San Francisco. Jeanne Marie was born in 1895 on the street called Rue de Napoléon in the city of Bastia, on the faraway French island of Corsica. She immigrated to California with her parents in 1896. Now that she was a grandmother, she liked to tell her granddaughter incredible family stories from old times. Like the one about the San Francisco earthquake of 1906, when Jeanne Marie was ten years old, and her mother—Maria's great-grandmother—was on the toilet when the ground began to shake, and she got *stuck* on the toilet, and it took *three people* to pull her out! Picturing that was so funny, it always made Maria laugh.

Jeanne Marie grew up to stand four feet nine inches tall, and she weighed 90 pounds for her whole adult life. She had thick, dark hair, set off by a complexion that drew lots of compliments (her secret, she told everyone, was Oil of Olay); and she cleaned house with gusto—apron, rubber gloves, bustling around, a roll of Scott Paper Towels tucked under one arm. But then she always took off her apron, her granddaughter remembers, "when we were doing Jack LaLanne."

The Jack LaLanne Show had been broadcasting from a local Bay Area TV station since a few years before Maria was born. By the time Jeanne Marie and Maria started watching, the program—one of the earliest TV shows that taught people to exercise—was a sensation. People even watched it on the other side of the world, in Australia.

On the boxy console TV in Jeanne Marie's living room, Jack LaLanne would appear, a handsome, happy husband with big round muscles; and often, so would Elaine LaLanne, a pretty, cheerful wife. Jack would tell the program's viewers, "I'm here for one reason and one reason only: to show you how to feel better, and look better, so you can live longer." Then he and Elaine, often wearing coordinated workout clothes, would stretch, high-step, march, and do jumping jacks, moving their bodies in unison while the young Maria and her grandmother tried to mirror all the moves; and when everything aligned, all four were doing exactly the same thing at the same time—they were doing Jack LaLanne.

Today, Maria Fiatarone Singh thinks Jack's and Elaine's and Jeanne Marie's examples helped to teach her, as a girl, that exercise was for everyone, regardless of age or gender. "It wasn't like, 'Women do this, and men do that.' It was like, '*People* do this.'"

As the grandmother and her granddaughter grew older, they did different things with their bodies. When Maria was in high school, she found the sport of fencing; and Jeanne Marie went on dueling with the dust.

Then Jeanne Marie turned seventy-five, her husband died, and in grief, depression overtook her. Frequently disoriented, she would become paralyzed with indecision at the grocery store, looking at the canned vegetables, unsure what to buy. After depression came delirium, and treatment in a psychiatric hospital.

The next year, at age seventy-six, Jeanne Marie fell and broke her hip. The fracture brought complications. Confined to bed, she became dehydrated and malnourished. She developed a bedsore. Then sepsis. She had a stroke. Various doctors treated her for various conditions with various drugs, including antipsychotics. Her weight dropped to 70 pounds.

When Maria and her family visited Jeanne Marie in the hospital, they found her tied down in restraints. Maria ripped off the restraints and took them outside and threw them in a dumpster, and "that was when I decided," she recalls, "*that* was going to be the place I needed to go, with medicine." The place she needed to go, as a doctor, was old age.

Her goal was to help others avoid the most painful parts of her grandmother's fate. She dreamed of building a geriatric sports medicine clinic and preventive health care center. She would combine the skills of a sports medicine doctor, diagnosing and managing disorders and injuries related to exercise, with the skills of an exercise medicine doctor, diagnosing and managing physical activity to prevent and treat disease. Her clinic would treat the rich and serve the poor, with payments from affluent clients subsidizing treatment for those with less money. Old people at risk of falls would come to the clinic, learn to exercise, and get stronger so they wouldn't fall.

Yet future ambitions, no matter how worthy or passionate, did not solve Maria's present problems. Maria could not simultaneously hold in mind the whole careening mass of her grandmother's multiple, mutually reinforcing dysfunctions, diseases, and syndromes, and the many drugs that doctors had prescribed to treat all these problems. Maria and the rest of Jeanne Marie's family, like most families, believed they had no choice but to accept the doctors' best advice.

For a time, Jeanne Marie managed to walk again, but with another downturn, she was bedridden for her last year of life, at home with her daughter Marie-Louise, Maria's mother.

Jeanne Marie died a few months before Maria graduated from medical school in 1981.

Maria Fiatarone Singh's research on exercise is a tribute to her grandmother. Like the grief it grew from, the research is both fathomless and sharp. From her earliest exercise trials at Tufts, her training protocols and techniques have been precisely structured, targeted, and validated, always following the principle of *specificity*, that physiological term for the way organ systems, including the neuromuscular, musculoskeletal, and cardiovascular systems, adapt in specific ways, developing specific attributes, in response to specific demands made of them.

In Boston, back in the late 1980s and early 1990s, like most of her colleagues studying resistance exercise, she mainly focused on one attribute of muscle: maximal force, or strength. Then the field shifted. Early in the new century, more researchers focused on another attribute: fast force, or power.

Walter Frontera, whose study of strength conditioning for older men made way for the first studies of high-intensity strength training for nonagenarians, says that in breaking a fall, "the issue is not maximal force generation, it's generating a certain level of force quickly." The difference between being able to lift 60 pounds or 80 pounds on a leg extension machine matters less, if you lose your balance, than having muscles crossing the joints of your hips, knees, and ankles that can exert force fast enough to stop the pull of gravity on your whole self.

One of Frontera's most distinguished former students, Jonathan Bean, now a professor at Harvard Medical School, confirmed the primacy of muscle power for movement in older age by using data from an Italian longitudinal study, InCHIANTI—measurements of the strength, power, and functional abilities of more than 800 people in the Chianti region of Tuscany. More than strength, Bean found, power determines people's abilities to walk and climb stairs safely and speedily in older age.

Muscle power, more than muscle strength, helps people handle sudden changes of pace and unpredictable scenarios. As people grow older, "we lose power faster than strength—and the difference between these two is really *time*," Walter Frontera says. "Power is force times distance, divided by time. Or: force times velocity. So the main issue, if you think about it, is time."

By the eighth decade of life, many people have lost up to 75 percent of the muscle power they had in their twenties, along with about 50 percent of their strength. These magnitudes of decline are not inevitable. But walking and general physical activity, and even maintaining a regimen of higher-intensity aerobic training, as mentioned earlier, are not enough to make up for these losses.

One early study suggesting how different types of exercise might preserve some youthful qualities of muscle into older age took place in Denmark and was published in 1990. The study compared a group of healthy but sedentary men in their late twenties to four sets of men in their late sixties and early seventies—an inactive group who did no exercise; a group of committed, recreational runners; a group of similarly dedicated swimmers; and a group that lifted weights. The older men were not elite athletes, but they had been regularly training in their various ways about three times a week for twelve to seventeen years.

Researchers found that muscle fibers associated with strength and power in the legs of the inactive older men were about half the size of those in the young men's legs. These fibers, type II fibers, were not significantly better preserved in older aerobic athletes. Despite years of running and swimming, their type II fibers were about the same size as the corresponding fibers in the muscles of their inactive age peers. But in the leg muscles of older men who lifted weights, researchers found that type II fiber size was "extraordinarily large"—and in fact "identical" to the muscles of the young men. In measurements of strength, power, and the cross-sectional area of whole muscles, the study found similar differences between the young men and three groups of older men—the runners, the swimmers, and the sedentary. But as Maria Fiatarone Singh summarizes the finding, the muscles of seventy-year-olds who lifted weights were of "similar size, function, and contractile properties as the muscles of men who were fifty years younger."

After Nalin and Maria Fiatarone Singh settled in Australia in 1999, research on muscle power inspired a new phase of their progressive resistance training studies. Maria remembers thinking, "What could I do that would be the very best way to deliver weight lifting to older adults? If I want to maximize their function. Maximize their bone. Stimulate the type II fibers the most. Maximize everything I could. Power training seemed to be the way to do it." Power training, she thought, might turn out to be a cornerstone of lifting as preventive medicine.

The technical difference between strength training and power training, tracking the difference between strength and power, is mainly about time—although, in still another strange and complex quirk of the neuromuscular system, effective power training for older people may depend on a lifter's psychological *experience* of time, not only on the durative, empirically measurable time of physics.

The Singhs' early weight training studies, like Walter Frontera's, had followed a common strength training practice of moving weight slowly: two or three seconds up, hold for one second, two or three seconds down. In the Singhs' strength training studies, the trainer cued the lifter to push "as hard as you can." In their power training studies, by contrast, the trainer cued the lifter to push "as fast as you can." (In both strength and power training, the lifter lowered the weight slowly.) But power training comes with a lot of provisos, especially where speed is concerned.

Fiatarone Singh emphasizes, "You have to be careful." Power training for athletes and young people typically involves fast movement of light loads, she says; but for older people—especially for people who have, or might have, any degeneration of the joints—fast movement of light loads can be dangerous. Power training at higher intensity, lifting heavier loads, is much safer for older people, even if that means they move the weight more slowly. She says, "for some joints, particularly the shoulder or the knee, you have to do it at slow velocity," to avoid the possibility of tearing muscles or tendons. But since "the *intent* is what stimulates the type II fiber recruitment"—intent being the signal traveling from central nervous system to alpha motor neurons to muscle fibers—it's important to make the *effort* to move the weight fast.

With the exercise physiologist Nathan J. de Vos, who was then one of Maria's doctoral students, the Singhs tested the effects of power training at various intensities and velocities; and to Maria, this was a main lesson of that research: "The good thing about high-force/high-velocity training is that you're slowed down by the force. And so even though you're trying to lift it fast, you can't. Which is good. Because you still have the intent. Which is what your brain wants. But you don't actually have the ballistic movement. Which is what your tendons *don't* want. So it's the safest way to do power training." She says, "Low-intensity/high-velocity training is potentially dangerous in older adults. Athletes can do it. But these guys can't. So we never do that."

From the time they met, the Singhs had been professional collaborators, but Maria was always more focused on research and Nalin more on clinical care. When they moved to Australia, Nalin seized his chance to start a new kind of doctor's office, realizing a dream the couple had long shared: the dream of integrating exercise prescriptions, mainly progressive resistance training programs, with mainstream medicine.

Almost twenty years after Nalin started this practice, I spent a few days with him in 2019 at the Centre for Strong Medicine, next door to a piano school in Pymble, a suburb in northern Sydney. In the clinic's main room, a gym that looked to be almost 2,000 square feet, I counted thirty-three resistance training machines, each labeled with names of lifts the doctor prescribed: squat, standing hip abductor/adductor, seated hip abductor/adductor, back extension, seated leg extension, seated leg curl, leg press, seated calf raise, upright row, triceps pushdown, chest press, lat pulldown, pec deck, biceps curl, lateral shoulder raise . . . The only mirror in this gym looked like a castoff from a Victorian bedroom—full-length oval dark wood frame tilting between fancy finials—and it stood in front of the biceps machine.

Singh told me that, over the years, he had prescribed resistance training for about 5,000 patients with a mean age of seventy-three and an average of four diseases, treated by five drugs. Typically, he had found, 50 percent of strength gains happened in a patient's first three months of lifting. Measures of function—walking speed, chair stand, and balance tests—usually improved between 20 and 50 percent. He also emphasized that people's responsiveness to training varies widely, and there is no way of predicting how any given individual will respond.

For each patient, Singh selected a program of exercises based on a functional and physiological evaluation. Training began with one-rep max tests, to set appropriate weights. The individualized exercise programs approximated Thomas DeLorme's original prescription, and patients learned to lift weights at about 80 percent of 1-RM. In every workout, patients consulted with their trainers in order to find a level of exertion that was challenging but achievable, following DeLorme's basic principle of progression. Every three months, after a new set of strength tests, each person's lifting routine was adjusted as needed, with each prescribed movement tailored to the individual patient's needs.

For example: Singh prescribed the triceps pushdown for a patient named John.

When I see John working out in the clinic's gym, his snowdrift of short hair is parted to one side, and three buttons of his plaid dress shirt are unbuttoned for ventilation. (Few of Singh's patients wear special workout clothes to the gym—most of them lift weights in the same outfits they might wear to a restaurant: trousers or skirts and dress shirts or sweaters.) Behind John, the floor-to-ceiling windows frame a view of tall trees where white cockatoos flap their wings and hop from branch to branch, but he is too busy lifting weights to do any bird-watching.

Seated on the triceps pushdown machine, he is focused on retracting his elbows—retracting and raising them—so that his upper arms extend behind him and his forearms are perpendicular to the floor. His palms press against a pair of handles that are parallel to the floor. Then he pushes the handles, contracting the complex of triceps muscles on the back and sides of his upper arms, extending his arms straight down, in parallel to his sides.

When he finishes a set, I ask him, Why do you need to do this lift? When you do the triceps pushdown in the gym, what specifically does it help you to do out in the world?

He glares at me as if the question struck a nerve. Then he answers, "Get off the bloody lavatory!"

We hold eye contact, and while I wait for him to say more, he takes a few shallow breaths. Finally he offers, "See, you've got to *push* to get *up*."

I ask if that's what he thinks about while he's doing this exercise, and he raises his voice, correcting me: "*No!*" Then there is another pause, in which he seems to decide, *This guy needs a lesson*—and he starts to tell me about a challenge he experiences almost everywhere he goes. The challenge of standing up from a chair.

"See, that's a very difficult thing," he says. "If the chair's too low, I always push from the back, not from the front"—he demonstrates with his palms pushing the arms of an imaginary chair. His elbows are retracted and raised, upper arms pulled back behind his trunk, as in the starting position on the triceps pushdown machine. It's a trick of leverage. Pushing down from behind gives the body forward-moving momentum, to help him rise from the chair and propel his first steps away from it. The trick won't work, though, if the chair sits too low to the ground.

Standing up is complicated. Standing up coordinates movements of the whole body. Brace the upper back and shoulders as you contract the triceps, brace the trunk as you contract the quadriceps. If any muscles in this chain don't work, others will have to compensate. If the other muscles can't compensate, the whole job fails.

John does a lot of compensating. "I've got no quads at all," he says. "That's this one"—he rubs the front of his upper legs—"I've got none at all. You see, I can't—I can't put my foot forward." He chuffs, "It doesn't work! When I'm walking, I walk with my knees locked. Do you know what I'm talking about?" he asks, fixing me with a look. More subdued then, he says, "You probably don't."

He is saying that without strong quads, even triceps can't help you if your chair sits too low to the ground—and toilet seats tend to be low. Fortunately, "You can get extensions to your lavatory. That's a very important thing of living," he says. "*You* wouldn't have noticed it, you see. You would think it doesn't matter."

He cocks his chin, considers, and genuinely confides: "The thing that terrifies me the most is going out and having to go to the lavatory." He is scared of opening the door to a public lavatory, as in a restaurant, and seeing no safety bar that he will be able to grab onto when he needs to stand up afterwards. Unless he is able to find a handicapped-accessible facility, he says, the likely outcome of his entering there is wretched. "Can't get off the pot! You've got to call somebody to pull you up. It's frightening," he says. "That's a nuisance. That doesn't suit me. The way I operate in life, that doesn't suit me."

Helplessness does not suit John because he has spent his whole life being a person who helps others, often by literally putting roofs over people's heads. John is an architect. In a career that lasted half a century, he tells me, he built about four hundred houses in Australia's capital city of Canberra. "Good-quality middle-class houses. I wanted to try to lift the standard of what was available," he says. Behind his own house, John has a workshop. "Tool Heaven," his daughter called it. She imagined the tools were so fulfilled there, he says, "because I can do anything with my hands that's needed at all," and he laughs. It's the laugh of someone who is used to being liked, a laugh that says, *Let's have one more.*

John's life changed after he found out he had high cholesterol. To help lower his cholesterol, his doctor prescribed the statin Lipitor, one of the bestselling drugs of all time. In 2022, doctors wrote more than 114 million prescriptions for Lipitor's generic version in the United States alone. For John, the drug triggered debilitating neuromuscular side effects. Suddenly he struggled to walk. He could not even walk down the block. The doctor took him off Lipitor and prescribed another statin. Then his problems got worse. He says, "My neurologist thinks that probably triggered a latent problem I had," an inflammatory muscular disease called inclusion body myositis, abbreviated IBM, causing gradual and severe degeneration, resembling late-life muscular dystrophy.

All his life, John had been strong. "When I was young, I could lift the engine out of my car!" he says. "I only had a little car, *but . . .*" —he shakes his head, with a dark chuckle—"I can hardly pick up a loaf of bloody bread now!" Five years ago, IBM took away his strength. "It's affected *every*where. My hands are very weak," he says, now serious. After three years struggling with the disease, John heard about the Centre for Strong Medicine and came to see Nalin Singh, asking him to "help me to strengthen what I've got left."

Singh's prescriptions have worked for John. For two years, John says, "I've kept roughly the same." Holding steady, for someone who has an aggressive degenerative condition as John does, is actually improvement, Singh observes.

When John started training, his wife Therese started training, too. Therese has hair as white as John's. In the gym today, she wears a bright, flowing knee-length scarf from the Picasso museum gift shop in Paris. As she moves from one weight machine to another, she sometimes pauses to adjust the Picasso scarf, to keep it out of the way of the weights. Therese is not just keeping John company here. She has her own reasons for lifting. Therese has had two strokes.

"And she's gone *up*," John brags, meaning that her strength has increased. Turning to her, he marvels that today Therese pushed twice as much weight on the triceps pushdown machine as he did.

"With difficulty," she answers, being generous; and John's eyes soften, as if Therese had slipped her arm through his.

Courage courses through this place: the courage of the patients who had not lifted weights before they met Nalin Singh. Like most others here, John had never even been to a gym, because as he puts it, "I'm not a gym person." The doctor is courageous, too, because he meets his patients as complex individuals, recognizing that none of them has just one health problem. They all have multiple overlapping difficulties and diseases, chronic and acute, which he discerns from vital signs and test results, and also from the details of their stories. He considers their symptoms in light of the whole of those stories, and he chooses therapies that can help relieve the largest number of symptoms and preserve and build their functional abilities to live the lives that suit them.

In one of the clinic's examination rooms, Nalin Singh sits across from Millicent, a wisp in spring sweater set and headband. "You've fallen over," the doctor says, concerned for her. "What happened?"

"I'd been swimming in the pool," Millicent answers, "and as I came out, when I went to pick up my shoes, or something, I caught my toe and tumbled and hit my knee and jammed my left elbow." She is in her eighties. Even her voice is thin. She cups her elbow in her right palm, and she says she wonders if a bone might be broken. "When I fell about twelve months ago, I had a really bad fall, broke my ribs, and you feel *sick*. You sort of *know* that something's—"

Singh: "Do you? I haven't broken enough things to know."

Millicent: (with a kindly laugh) "You're *lucky.*"

This time when she fell, Millicent was lucky, too, in a sense. Lucky she fell this way instead of that, and lucky she landed on her elbow instead of her hip. Among older Australians, falls are the most common cause of injury requiring hospitalization. During the year when Millicent fell, half of older Australians who fell and ended up in hospitals were found to have broken bones. More than 30 percent of them had broken a hip. Most older people who suffer hip fractures will fall again during the following year, Nalin and Maria Fiatarone Singh's research has shown. Many of them will fall twice or more.

According to the World Health Organization, about 30 percent of people over the age of sixty-five experience a fall each year. The risk for people living in long-term care facilities such as nursing homes is even greater—more than half of them will fall. In residential aged care, according to some studies, falls are the leading preventable cause of death.

The abject pain and suffering caused by falls also entails enormous financial cost. The total cost of hospital care for fall-related injuries in the Australian state of New South Wales, estimated to be $268.2 million in 2008, was predicted in 2011 to increase over the next four decades by tenfold—to $2.85 billion dollars, not accounting for inflation—if hospitalization rates follow historical trends. If these trends do not change, by 2051 it would not be possible to care for the hundreds of thousands of older Australians who would be injured in falls every year without adding tens of thousands of beds to hospitals and residential aged care facilities, according to projections calculated by Wendy L. Watson of the University of New South Wales and her colleagues. The situation, these epidemiologists predict, "would be catastrophic for health provision in this state."

Why do falls happen? People like to point to something in the room—the electric cord, a wrinkle in the carpet—but most other people, coming across the same situation, would not fall. "The reason for the fall is not really that there's a rug in the way, it's that the person does not have the intrinsic resources to overcome the particular challenge," Maria Fiatarone Singh says in a lecture on falls and hip fracture midway through a graduate seminar she teaches. The course title is "Exercise Throughout the Lifespan."

Her lecture on falls is a tour de force, a synthesis of data and wisdom gathered during the forty-five years since her grandmother fell. Most of the information about falls in this chapter comes from her lecture.

One of the largest reviews of research on falls showed that the most effective single mode of exercise for reducing risk of falls is tai chi. According to the same review, says Fiatarone Singh, "Tai chi has only been found to be effective for those at lower risk of falls, not those in the nursing home, probably because of the kind of tai chi that's been done in the nursing home setting—it's very watered down."

She continues: "Other than tai chi, single exercise interventions have generally not been effective. Just strength training by itself, just aerobic training by itself, even just balance training by itself is often not effective for falls." In her analysis, "the combination of strength and balance is, clearly from the literature, the most effective way to approach the problem."

The common belief that consistent leisurely walking is sufficient exercise to reduce the risk of falls and fractures is not supported by evidence. According to the 2018 United States physical activity guidelines scientific advisory committee, research "does not support the use of low-intensity walking as a primary mode of physical activity to reduce the risk of fall-related injuries and fractures among older adults."

Fiatarone Singh mentions two studies in which "walking by itself has *caused* people to fall," on her way to making this larger point: "If somebody is at high fall risk, the last thing you want to tell them to do is to walk as their preventive strategy for falls reduction. You want to give them the resources they need to prevent falling. Which are generally a combination of strength and balance training, and then whatever else they need that's in the non-exercise realm. Whether that's getting their cataracts taken out" or, for instance, reducing or changing medications that may contribute to balance problems. "But from an exercise perspective, aerobic exercise is the least important thing to focus on," she says, "and could be hazardous to somebody who's at very high fall- and fracture-risk, if that's all that you do."

Back at the Centre for Strong Medicine, in Nalin Singh's examination room, I look at Millicent and think about more reasons that she is lucky. She is lucky that Nalin is Maria's husband and Maria is Jeanne Marie's granddaughter—because this pattern of relationship, though unknown to Millicent, makes her the beneficiary of a stupendous body of knowledge about falls prevention and care that has been built up by Maria's resolve to protect people from suffering what her grandmother suffered.

This body of knowledge is organized by one clear insight: Falls involve "a mismatch between demand and resources," as Maria Fiatarone Singh tells her class. "Extrinsic demands are high. Intrinsic resources are not high enough. And so you fall over."

Regarding health in later life, "people sometimes think of falls as the biggest problem of all," Fiatarone Singh says, "but I look at it as one aspect of frailty. I think frailty is bigger than falls. If you address falls only, which you can do by low-intensity strength and a bit of progressive balance training, you're not really addressing frailty. I also want to get to depression. I also want to get to bone density. I also want to get to frailty and sarcopenia. And so I am more inclined to use a high-intensity version" of weight training "that will reduce falls but will get everything else as well."

The convergent problems of frailty, falls, and hip fracture inspired a highlight of Nalin and Maria Fiatarone Singh's research collaborations. They conducted a clinical rehabilitation study of hip-fracture patients, with progressive resistance training a primary pillar of a thirteen-part intervention that lasted a year for each patient. "We weren't trying to be efficient," Maria said in a lecture at the annual meeting of the American College of Sports Medicine in 2014. "We were trying to treat everything that we thought they needed."

Unlike conventional treatment of hip fracture, which focuses on the broken bone, the Singhs also treated the underlying conditions that caused patients to fall. Before the rehabilitation study commenced, they did five years of diligence to identify what Maria called "all the things that were wrong with these people that could potentially be fixed." With a group of their colleagues, the couple investigated potentially treatable risk factors for falls, and they produced a list of thirteen: low muscle mass, low muscle strength, poor balance, malnutrition and weight loss, cognitive impairment, depression, osteoporosis, vitamin D deficiency, low self-efficacy, fear of falling, polypharmacy, physical inactivity, and inadequate social support.

The rehabilitation trial took place in Sydney at two public hospitals, Balmain and Royal Prince Alfred. Between 2003 and 2007, 124 people who were admitted to the hospital for surgical repair of hip fractures were recruited to participate in the study. All of them received twelve months of whole-body progressive resistance weight training, along with balance training as well as nutritional support. Anyone in the group who needed care for any other condition on the list of thirteen risk factors received those treatments, too. Most patients needed help with ten of the thirteen. The care they received may not have been efficient, but it could hardly have been more effective. The Singhs' program of treatment reduced the worst possible outcomes of falls—nursing home residence or death—within the first year of injury by more than 80 percent.

"It's not easy to do this kind of a study, but it actually is possible. And it's also not rocket science," Fiatarone Singh told her colleagues at the sports medicine conference. Her tone of voice was a little warmer than impatience. It was the tone of someone who thinks the world of you and can't quite wrap her head around why you don't do what's best. She said, "It's exactly what my grandmother would have told me she needed, when she fell and broke her hip at the age of eighty-five, because those are all the things that were wrong with her."

On another occasion, Maria Fiatarone Singh speaks to the problem of hip fracture from a different angle. Instead of talking about how to care for older people who break their bones, she talks about the possibility of reducing risk of late-life fractures by encouraging the types of physical activity that best build bone density early in life, when it's easiest to strengthen bones.

The best time to build bone density is prepuberty, she notes, while speculating that "probably on a population level one of the best ways to prevent osteoporotic fracture would be to bring the bone density by age twenty up to a higher level than it currently is." The best type of activity for building bone density involves "high-velocity, high-force movements," she adds, and this type of activity, "which we're seeing less and less of as there's more and more screen time, is not contraindicated, in fact is quite healthful."

She continues: "Many of the games that children used to play, skipping and hopping and jumping, the things that are now prohibited in every school because somebody might break their arm, those are great for bone. So when you prevent people from jumping on and off monkey bars, you might prevent the Colles fracture at age ten"—Colles fracture is a broken wrist, one of the more common childhood fractures—"but you're not preventing the hip fracture at age eighty." She says, "I think it's very important to think about how we're protecting our children in ways that are not as healthful as they might be. Put a helmet on them, I guess, and let them jump off the monkey bars."

When she speaks of treating conditions that became her grandmother's fate—hip fracture, osteoporosis, frailty, sarcopenia—Maria Fiatarone Singh speaks as a physician and as a granddaughter. When she speaks of preventing these conditions and averting such fates, she speaks as a mother, too. "At my children's grammar school at one point, they forbade any running on the playground. It was asphalt, and they were afraid—somebody, I guess, sued the school because somebody broke their arm. So there was no running," she says. "I think we have to be a little bit careful with being a little bit too timid with children, and the same thing has been true for older adults and continues to be true for older adults."

In 2019, when I visited Nalin Singh at the Centre for Strong Medicine, he was in remission from cancer, a rare soft tissue sarcoma of the lower back. The next year, the cancer returned, with complications that eventually forced him to stop working; and in the summer of 2021, at the age of fifty-nine, he suffered a debilitating stroke. A few days before Christmas of that year, surrounded by his family and with Maria at his side, he passed away.

At Nalin Singh's memorial service, family members spoke about his life and work. A couple of them invoked a maxim that he often repeated, a paraphrase of Aristotle made by the historian Will Durant: "We are what we repeatedly do. Excellence, then, is not an act, but a habit."

These words can serve as powerful affirmation for people whose lives involve exercise regimens: *We are what we repeatedly do.* Certainly, the words affirmed Nalin Singh's work to help patients learn progressive resistance training as a steady habit of excellent health. Yet for him and for the many people he cared for, progressive resistance training was never only repetitive. It was also investigative, a process of experiment, changing as individual circumstances change and as culture and science evolve.

While collaborating with Maria to advance the field of resistance training research, Nalin kept abreast of research by others, too. His clinical practice changed with the times, respecting new findings about muscle, health, and function. One significant shift that took place during his years of seeing patients was a growing medical consensus that, in older age, training to make muscles stronger and more powerful may be more important than training to make them bigger.

To learn more about that shift, I speak with Luigi Ferrucci, the scientific director at the National Institute on Aging in Bethesda, Maryland. Ferrucci is also the director of the Baltimore Longitudinal Study on Aging, one of the world's largest and longest-running scientific studies of health, function, and aging, and he is the principal investigator of the InCHIANTI longitudinal study in Italy. He recalls the 1990s, when medicine started reckoning with sarcopenia, as a time when scientific dogma practically equated size and strength.

"Everybody would have told you that we lose strength because we lose muscle mass. Billions, trillions of dollars were spent on drugs to increase muscle mass," Ferrucci says, referring to a competition among pharmaceutical industry giants, a race to be the first to develop drugs that could make older people more muscular. The race, which lasted into the early 2000s, included experiments with performance-enhancing compounds favored by many elite athletes, including drugs that are now banned from the Olympics.

"Those drugs did not work" for their intended purpose of restoring functional abilities, Ferrucci says. "There was some increase in muscle mass, and also a little higher strength, but there was no real effect on functional capacity because the entire assumption was wrong. The decline in strength is over and beyond the decline in muscle mass. Which means that mostly what we lose is the muscle *quality*, the ability of muscle to generate strength." Muscle quality is the ratio of muscle strength to muscle mass. High muscle quality, therefore, is generation of relatively high force with relatively low mass. Per cubic centimeter, "the volume-specific amount of strength generated by skeletal muscle declines dramatically after the age of seventy," Ferrucci adds. "We don't really understand why, but recent data suggest that the amount of contractile proteins—those that generate strength—actually declines with aging, while other proteins that do not contribute to strength—collagen or amyloid proteins—tend to accumulate. A new emerging technique that selectively estimates contractile proteins confirms this hypothesis."

Since Irwin Rosenberg named the problem of muscle loss as sarcopenia, there has been a protracted struggle among scientists to define the nature of that loss. Some saw the loss of mass and strength as being so distinct as to need separate names. Again reverting to the Greek, one group proposed to call the loss of strength *dynapenia*, the first part of that word from *dunamis*, one of the terms for human strength used in the epics of Homer and the medical writings of Galen. Others countered with *kratopenia*, referring to another Homeric "strong" word, *kratos*, meaning the strength of winning a contest, a type of strength that was almost always given by the god Zeus. "However, sarcopenia is already a widely recognized term, so replacing it might lead to further confusion," decreed an influential group of gerontologists, the European Working Group on Sarcopenia in Older People, and the name remained unchanged.

Irwin Rosenberg's former Tufts colleague Walter Frontera, whose study of high-intensity strength conditioning for older men was the precursor to Maria Fiatarone Singh's first strength training studies at Hebrew Rehab, thinks debates about the nature of muscle loss have been almost like struggles to overcome a kind of blindness. He tells me, "It's almost as if we didn't know how to *see* muscle." For many scientists and doctors, clear views of muscle were obstructed, rather oddly, by the material of muscle itself.

Size does matter, where muscle is concerned. Size does tend to correlate with strength—but as discussed earlier, the links between these aspects of muscle are complex. In 2010, that complexity was acknowledged by the first authoritative international medical consensus statement "on definition and diagnosis" of sarcopenia. According to the European Working Group on Sarcopenia in Older People, "Muscle strength does not depend solely on muscle mass, and the relationship between strength and mass is not linear." The group also mentioned a confounding phenomenon of body composition that researchers at Tufts and elsewhere had grappled with from the 1990s. With age, dynamics of size and strength are skewed by changes in muscle tissue such as "'marbling,' or fat infiltration into muscle," the working group wrote. For such reasons, "defining sarcopenia only in terms of muscle mass is too narrow."

Nevertheless, low muscle mass remained the primary criterion for diagnosing sarcopenia. Low muscle function, assessed as "strength or performance," was the other criterion. While finding "fewer well-validated ways to measure muscle strength" than muscle mass, the group endorsed the use of dynamometers—handheld devices that can be squeezed, a bit like squeezing the handle of a gas pump, to measure the force of a person's grip. "Low handgrip strength is a clinical marker of poor mobility and a better predictor of clinical outcomes"—such as walking speed—"than low muscle mass," the group stated, citing evidence gathered in Luigi Ferrucci's InCHIANTI longitudinal study.

Over the next decade, more evidence showed significant associations between low grip strength and negative outcomes including longer hospitalizations, poor health-related quality of life, and death. Grip strength is a strong predictor of mortality, says Maria Fiatarone Singh, "not because strong hands keep you alive, but somehow because hand strength is related to other kinds of muscle strength and function" in the body overall, though no one knows exactly why this is the case.

In the same years, research showed that measuring muscle mass was even more difficult than it had seemed in 2010. At present, no very accurate technique for measuring muscle mass or muscle quality is feasible for broad use. The most precise ways to measure muscle mass, which also allow for the only precise measures of the strength-to-mass ratio called muscle quality, are scans by computed tomography and magnetic resonance imaging machines. Both types of scans, CTs and MRIs, are prohibitively expensive and inconvenient for most people in most places. Full-body DXA scans are not as pricey, but still expensive, and DXA does not show how much fat has infiltrated muscles, as CTs and MRIs do. Maria Fiatarone Singh says that DXA also "significantly overestimates the amount of actual muscle tissue present and is therefore not a good predictor of dysfunction." The cheapest, most common ways of measuring mass, calipers and bioelectrical impedance devices, are often prone to error.

In 2019, the European working group, led since its inception by Alfonso J. Cruz-Jentoft, chair of the geriatric department at the Hospital Universitario Ramón y Cajal in Madrid, published a new consensus statement on defining and diagnosing the disease. The 2019 document reversed the 2010 statement's priority of emphasis on size and strength in sarcopenia. Low muscle strength became the first criterion for diagnosis, not low muscle mass. The group made this change for two reasons. First, "strength is better than mass for predicting adverse outcomes," according to their review of evidence. Second, mass is "technically difficult to measure accurately." The third diagnostic criterion is low physical function—poor abilities related to locomotion, such as walking, balance, and standing up and siting down without assistance. Sarcopenia that impairs physical function is classified as severe.

The European working group's counterpart in the East, the Asian Working Group for Sarcopenia, basically agrees with the Western diagnostic criteria; but the Asian group has led the European group in at least two important ways. The Asian working group, chaired by Liang-Kung Chen of National Yang Ming Chiao Tung University in Taipei, has put the primary diagnostic emphasis on muscle strength and function—not on muscle mass—since its first published consensus statement in 2014, five years before the European group made its corresponding shift. In 2014, the Asian working group also specified diagnostic, gender-specific cutoff points for measurements of strength, function, and mass; and in 2019 the European group followed suit, because in Asia, the European scientists observed, those cutoff points had turned out to be "very useful for implementation of recommended sarcopenia care."

When Maria Fiatarone Singh teaches students about muscle and aging, she teaches them a certain way of seeing the tissue. It is a new version of the perspective she began to form in the late 1980s, working at the Tufts lab: the view that body composition causes most of the main problems of aging.

In her graduate seminar "Exercise Throughout the Lifespan," she raises a series of questions: "What is it that really predicts functional decline? What body composition change is the most important? Is it an increase in adiposity, is it a decrease in muscle mass, or is it something else?" To work through those questions, she draws on a 2013 epidemiological review by Laura A. Schaap and colleagues of the VU University Medical Center in Amsterdam. The scholars analyzed results of longitudinal studies of large cohorts of people who were at least fifty-five years old. These longitudinal studies had first measured people's functional abilities and at least one of several measures of body composition and muscle strength, and years later had followed up to see if those people became disabled or experienced functional decline.

Fiatarone Singh summarizes two of the overall findings: "If people had low muscle strength at the beginning, they had an 86 percent increase in the risk of functional decline" in later years. That exceptionally high risk of disability associated with low muscle strength was surpassed, though, by risk associated with sarcopenic obesity, the combination of low muscle strength and excessive fat accumulation that threatens health. In the Dutch team's analysis, sarcopenic obesity increased the risk of functional disability by 250 percent.

Fiatarone Singh wants her students to take a message from this comparison. For preserving and enhancing functional ability, she says, "it's important to not forget about the other side of the equation"—important not to forget about excessive body fat—"and think that it's all about muscle. It's really the combination." People whose legs have to "carry a body weight with a lot of fat, with very little muscle to do that," have "the most impairing condition," she adds. "As opposed to somebody who just has low muscle, but is very small, and they don't need much strength to carry their body weight through space. So clearly, having sarcopenic obesity is probably the worst scenario of all."

Even the worst scenarios of all, however, can almost always be helped: This may be the overarching lesson to be drawn from Maria Fiatarone Singh's research and from its application in Nalin Singh's clinical practice.

As the first study of high-intensity strength training for nonagenarians proved the cliché that it's never too late for older people to regain significant amounts of the strength and muscle that time seems to have taken from them, another of Maria Fiatarone Singh's investigations proved that it's never too early for adolescents to build strength and muscle.

No one has conducted a longitudinal study of the possible lifelong effects of weight training in childhood or adolescence—and given current ethical and practical conventions of scientific research, such a study would be so complex and expensive that it is unlikely to occur in the foreseeable future—but Fiatarone Singh makes a reasonable inference that lifting weights in adolescence could help reduce the risk of sarcopenic obesity in older age. The European Working Group on Sarcopenia wrote in 2019 that "the development of sarcopenia is now recognised to begin earlier in life." And some doctors and scientists think development of sarcopenia may begin in early childhood.

Research by Richard Dodds and Avan Aihie Sayer of Newcastle University marshals evidence for taking "a lifecourse approach" to sarcopenia. The lifecourse approach is based partly on research showing links between lower weight at birth and lower muscle strength in adulthood, and partly on research showing "that muscle mass and function in later life reflect not only the rate of loss but also the peak attained earlier in life." (Studies of large populations in several countries, including Finland and the United Kingdom, have observed these connections.) Seeing sarcopenia this way might allow for effective prevention: "A lifecourse approach significantly broadens the window for intervention," as Dodds and Sayer have written.

In that context—in a lecture on sarcopenia and how it might be prevented—Fiatarone Singh describes a randomized controlled trial of high-intensity progressive resistance training for ten- to fifteen-year-old children and adolescents. She collaborated with Amanda C. Benson and Margaret E. Torode on the study, published in 2008, which measured outcomes related to obesity, body composition, and metabolic health; but "so as not to stigmatize the children, we actually invited everyone in the school to join the program," she recalls. When the researchers made initial assessments, more than half the young people were found to be overweight or obese by clinical standards.

The trial took place in a rural part of Southland, New Zealand, the southernmost region of New Zealand's South Island. In the small town of Gore, maybe best known for hosting the annual New Zealand Country Music Awards ceremony, about forty local students from two schools lifted free weights under supervision, twice a week for eight weeks. Their workouts, which included eleven exercises involving all major muscle groups, took place at a local gym called Everybodies Health & Fitness.

At the same time, about forty other students who did no regular exercise served as the study's control group. "When you're doing a study in kids, obviously they're growing," and by the mere process of growth, says Fiatarone Singh, "everybody's getting bigger and stronger, so if you have no control group, you don't know anything." Students in the control group understood themselves to be on a waiting list to receive the same training as the others. After the first group of students finished their eight-week regimen, trainers at Everybodies Health & Fitness began to work with students in the control group, too.

Fiatarone Singh describes the trial's intervention as "a very typical weight-lifting program. Exactly what we do in eighty-year-olds, in fact. Three sets of eight repetitions at 80 percent of the 1-RM." In this study of young people, though, she and her colleagues did not measure maximal strength with traditional 1-RM tests. Instead, strength was measured with reference to individuals' own perceptions of their efforts.

Since the late 1990s, when she and her husband began doing research together, Maria Fiatarone Singh had collected data on trial participants' ratings of perceived exertion when they lifted weights. Her framework for data collection was the Borg Scale of Perceived Exertion, developed in Sweden in the 1960s by Gunnar Borg, a psychologist who worked in a physiology laboratory. Measuring effort with the Borg scale involves asking one question about acts of physical exertion: *How hard was it to do what you just did, on a scale of 6 to 20?* A rating of 6 on the Borg scale means the effort was easy, involving "no feeling of exertion." A rating of 20 means the effort was "very, very hard," demanding all the energy you had.

By the time she helped design the study of weight training for young people in New Zealand, Fiatarone Singh had gathered thousands of people's reported ratings of perceived exertion and thousands of measurements from the same people's one-repetition maximum strength tests. Comparing those unpublished data sets, she determined that 80 percent of 1-RM was consistent with a Borg scale rating of perceived exertion in the range of 15 to 18, working "hard to very hard." She says the data showed this to be true in many adult populations, including the oldest people, blind people, and people with cognitive impairment. It worked for children and adolescents, too.

More than ten years after completing the study in New Zealand, when Fiatarone Singh describes how much stronger the young people grew when they lifted weights, she speaks in the present tense. The findings are impressive enough almost to demand the present tense.

"The control group gets stronger, even over that two-month period," by the normal process of growth, she says. "But the weight-lifting group gets a lot stronger than their peers who are just doing normal sorts of kid activities," she adds, pointing to a bar graph. The tallest bar on the graph is fire-engine red, and it shows that children and adolescents who lifted weights gained almost four times more strength in their upper bodies than the control group gained. Compared to the control group, young people who lifted weights also gained 30 percent more strength in their lower bodies.

Next she says, "we looked at their body composition," and in the control group, "their waist circumference, even at eight weeks, was getting bigger. Their body mass index was getting larger. Their percent bodyfat had gone up by 1 or 2 percent." For the young people who lifted weights, by contrast, all those metrics of body composition either decreased or stayed the same.

Numerous other studies, she says, have found that "weight training is not dangerous for children, but beneficial in many ways, including bone health, metabolic health, self-esteem, and depression." Speaking briskly, she wraps up her remarks about weight training for young people with one last lesson from the New Zealand trial, and her unsentimental tone gives the lesson some real bite. "If you have children who are a little bit timid about doing aerobic exercise because of their body size, weight lifting is something they can be really good at, and not feel intimidated or embarrassed about," she says, "which is often the case with aerobic activity. So in that way it's often a good substitute."

Guidance for pediatricians from the American Academy of Pediatrics in 2020 cites other studies showing that resistance training can help "provide positive options to engage children and adolescents with overweight or obesity in physical activity." The guidance finds evidence "that participation in a resistance-training program helps increase daily levels of spontaneous activity in school-aged boys, which suggests that resistance training may be a good place to start when trying to get inactive kids to be more active." For greatest benefit, lifting weights should be accompanied by aerobic exercise when possible, the guidance states: "Progressing into a combined program of resistance and aerobic training may generate added benefit because combined programs have shown favorable effects on the reduction of total body fat in youth."

Maria Fiatarone Singh is now almost seventy years old. More than thirty years after her paradigm-shifting study of high-intensity strength training in nonagenarians, she is a kind of younger-elder stateswoman of medical exercise science, especially as it pertains to older people. When the World Health Organization prepared its *World Report on Ageing and Health*, the agency looked to five experts on physical activity in older adults for background on that topic, and she was one of them. She is the lead author of the "Physical Fitness and Exercise" chapter in her medical specialty's authoritative reference book, *Pathy's Principles and Practice of Geriatric Medicine*. She has also helped to define many national and international exercise guidelines. Among thirty-six researchers from sixteen countries on five continents who contributed to a comprehensive "Expert Consensus Guidelines" statement of "International Exercise Recommendations in Older Adults," released at the annual International Conference on Frailty and Sarcopenia in 2021—an esteemed gathering of experts on medical care for the world's vulnerable aging population—she was one of the leading contributors.

The new international exercise guidelines drew attention to little-known evidence that "high-intensity progressive resistance training in older adults also improves aerobic capacity to a similar extent as moderate-intensity aerobic training." Some of this evidence comes from Fiatarone Singh's own lab in Sydney. With Yorgi Mavros and other collaborators, including also Nalin Singh and Guy Wilson, in 2017 she published results of a six-month randomized controlled trial investigating how high-intensity progressive resistance training affected cognition in 100 adults, aged fifty-five or older, with mild cognitive impairment. Lifting weights increased the participants' aerobic fitness by an average of 8 percent. An 8 percent improvement in fitness from lifting weights is nearly equal to the 7.78 percent improvement reported by a highly influential yearlong trial of walking that was published in the *Proceedings of the National Academy of Sciences*.

Strength training, in other words, builds more than strength. For older adults, strength training may also improve aerobic capacity about as well as walking at moderate intensity. Consequently, the 2021 consensus guidelines continued, "when initiating exercise in stages it may be easier behaviourally to provide a single exercise prescription consisting of a resistance exercise, which would target two major age-related changes in exercise capacity, before adding other modalities. By contrast, aerobic exercise alone does not improve strength nor balance and thus is insufficient as a single modality for older adults."

This comparison of aerobic and resistance training may be jarring, for people whose doctors have told them that walking is all the exercise they need. But these exercise guidelines are not wayward or unprecedented. They are the next inductive step in an ongoing process of the most credible medical authorities weighing the evidence about exercise and health and asserting the central importance of progressive resistance training for older people.

The World Health Organization's *World Report on Ageing and Health* attested that everyone needs every kind of fitness—aerobic, strength, and balance. Yet since resistance training is the only form of exercise that can improve every aspect of fitness, the WHO stated that for older people, especially those with limited mobility, the priority is clear: "strength and balance training should precede aerobic exercise."

The WHO document prioritized strength training based on "new evidence showing that progressive resistance training has favourable effects not only on muscular strength, physical capacity and the risk of falls but also that its benefits extend to cardiovascular function, metabolism and coronary risk factors for those with or without cardiovascular disease. However, the benefits of aerobic physical activities, such as walking, which is the main mode of exercise among older adults, cannot be transferred to improving balance, have no effect on preventing falls, and no clear benefit in relation to strength."

———

It would be nonsense to claim that any one type of exercise is best for all people in all situations. And it would be unwise to ignore the evidence showing that, from childhood onward, progressive resistance training is a viable alternative when people cannot or do not want to do aerobic exercise; and evidence that throughout adult life, the benefits of progressive resistance training complement those of aerobic exercise, while offering numerous benefits that aerobic exercise does not provide; and evidence that in older age, progressive resistance training becomes more necessary for more reasons, for everyone.

Much of this evidence—including evidence of the highest quality, from double-blind, randomized controlled clinical trials such as the Singhs have run—has important limitations, however. Three main limitations deserve special note.

First, as Maria has said, "Effects are diluted as the technique moves out into the world." Even in a doctor's office, Nalin explained to me, lifting weights tends not to work as well as it does in controlled laboratory experiments. Effects are diluted in the world because conditions are more carefully controlled in the lab, and also because people enrolled in trials are different from other people. This difference is a second limitation of clinical exercise trials: People who participate in weight training experiments tend to be homogenous. Researchers often intentionally recruit homogenous groups of participants, to minimize confounding variables in their experiments, and homogeneity limits wide applicability of the trials' results. That point connects to a third limitation: Even relatively large clinical trials involve small numbers of people compared to the overall population. The most dramatic evidence of positive outcomes from weight training by relatively small groups of people cannot be generalized as likely outcomes for whole populations. It is simply wrong to assume that everyone would experience the same benefits from lifting weights that participants in Nalin and Maria Fiatarone Singh's trials have experienced.

If everybody could have a trainer like Evelyn O'Neill or Guy Wilson, though, there's really no telling what might be possible.

These limitations of clinical trial evidence only heighten the importance of another kind of evidence about resistance training and health—epidemiological evidence about strength training. Epidemiology is "the study in human populations of frequencies and distributions of disease in terms of time, place, and personal characteristics," in the definition of Ralph S. Paffenbarger Jr., who conducted some classic epidemiological research on reducing risk of chronic disease and death.

In 1951, Paffenbarger began a long-term study of more than 6,000 California longshoremen. He found that men who worked more strenuous jobs had a lower risk of heart attacks and death, over a period of twenty-two years, than men whose jobs involved only exertions of low to moderate intensity. The California longshoremen study—like the earlier study in London that found ambulatory ticket-takers on city buses had a lower risk of heart attack and premature death than sedentary London bus drivers, and like the finding that more active people are less likely to develop heart disease even if they smoke—was a major event in building the base of evidence on which the field of public health began promoting aerobic exercise for disease prevention, the exclusive message of most physical activity guidelines through most of the late twentieth century. Roughly since 2010, however, when the World Health Organization added a recommendation of strength training to its guidelines, epidemiologists have turned more attention to resistance exercise, too.

They have been finding significant facts. Doing any amount of resistance training, compared to doing no exercise, reduces the risk of all-cause mortality by 18 percent, reduces the risk of cardiovascular disease mortality by 18 percent, and reduces the risk of cancer mortality by 16 percent, according to one review.

These percentages, like all statistics about mortality risk, describe reductions in risk of dying within a certain period. The numbers come from "Resistance Training and Mortality Risk," published in 2022 by Prathiyankara Shailendra and colleagues at the University of South Australia in Adelaide, who analyzed ten longitudinal studies with follow-up times ranging from seven to seventeen years.

Aerobic exercise alone is more effective than resistance training alone for reducing risks of all-cause and cardiovascular mortality, the same review shows—aerobics reduces both risks by about 10 percent more than strength training. But for cancer mortality, risk advantage runs the other way. Aerobic exercise alone has no effect on cancer mortality, while resistance exercise alone reduces cancer mortality risk by 16 percent.

Across the board, it is the combination of resistance training and aerobic training—habitual engagement in both kinds of exercise—that most effectively reduces these three risks of death. Lifting weights in addition to doing aerobic exercise is associated with a 40 percent reduced risk of all-cause mortality, compared to 25 percent for aerobics alone. Lifting weights in addition to doing aerobics is associated with a 46 percent reduced risk of cardiovascular mortality, compared to 29 percent for aerobics alone. Lifting weights in addition to aerobics is associated with a 28 percent reduced risk of cancer mortality, compared to 0 percent for aerobics alone.

Combining resistance training with aerobic training may be associated with many other benefits, too, compared with aerobic training alone, according to data from large cohorts—ranging from about 10,000 people to about 1.7 million people—in several countries, including Germany, South Korea, and the United States. Jason Bennie of the University of Southern Queensland, a prolific analyst of epidemiological data on muscle-strengthening exercise, lists these benefits as including "reduced prevalence of cardiometabolic (hypertension, diabetes, cardiovascular disease) and general health conditions (arthritis, chronic obstructive pulmonary disease, asthma); depression/depressive symptom severity; obesity; and prevalence of hyperglycaemia and dyslipidaemia." Because these data are cross-sectional, gathered from large populations at single points in time, they do not prove causality,

Bennie notes, although the findings "are consistent with evidence from clinical studies" of combined aerobic and resistance training.

That word of caution calls for another. Like the evidence from clinical trials, epidemiological evidence about exercise has limitations. "One of the key limitations of the literature on resistance training and the risk of mortality is the measurement of resistance training," according to the above-cited review of resistance training and mortality risk. Jason Bennie adds that epidemiological data regarding resistance training is "exclusively assessed by self-report, typically assessing its frequency only"—meaning the number of sessions per week. Researchers have almost never gathered information on the duration of resistance exercise sessions, levels of exercise intensity, or even the type of exercise that people do when they say they do "muscle-strengthening exercise," a phrase that covers a mishmash of activities, from sit-ups to Pilates to barbell squats.

That blurring of detail aggravates the aforementioned main limitation of epidemiological evidence about exercise and health. It does not prove causality. Even the largest, best-quality longitudinal studies of resistance exercise and risk of chronic disease or mortality do not prove beyond doubt that as a general rule, lifting weights can save your life.

What they provide instead, considered alongside the findings of clinical exercise research and the facts of physiology, is a call to action to fortify our powers to act upon the world with the highest level of physical functioning that our genes and environments allow, so that we can enjoy our lives together as well as we can for as long as we can.

Progressive resistance training can help all of us develop those powers—but not if we don't get the help we need to learn how to do it and how to keep doing it.

The twenty-first century is a transitional time in the history of medicine and exercise. We are at a crossroads. Research has shown that the old medical prejudice against athletics, including the prejudice against weight training, was only prejudice. Yet new knowledge about the regimen and its importance to lifelong health and well-being has not much changed the clinical practice of medicine, except on the margins in a few extraordinary doctors' offices such as the Centre for Strong Medicine.

"It's frustrating in a way, because it's been so long now," Maria Fiatarone Singh tells me, in that tone of guarded mirth that seems to run deep in her. Smiling as if on principle, she glances at the rosary, the book by the Dalai Lama, and the open box of chocolates that sit between us on the meeting table in her office. "Thirty years of data accumulated from around the world showing that it's very safe and very effective," she says.

She thinks the main reason progressive resistance exercise has not been more widely adopted as a medical treatment is that "there isn't a single entity to make money off it, like a drug. So there's nobody championing the cause, essentially. And you have to do something quite different to deliver it than you would for other kinds of exercise. Physicians still aren't trained in it in medical school. It hasn't gotten the same kind of backing that an effective drug would. Which doesn't make a lot of sense. The data is as strong, the effect size of the benefits are as large. You'd think it would be right there in mainstream medicine—but it isn't."

Another day, she attacks the problem again, but harder. "Most physicians don't see resistance exercise as medicine, whereas for me as a geriatrician, it's the most powerful medicine we actually have. There isn't anything more powerful than weight-lifting exercise. Full stop. Nothing is more powerful," she says. "So why don't we pay for it? Why aren't there leg press machines in every nursing home? Why aren't we telling people how to do it at home? Why is there no Medicare rebate for an exercise physiologist to come teach you how to do it? It's because no one thinks of it as medicine. If we thought of it as medicine, we would pay for it."

The great barrier of public access to progressive resistance training as a type of medicine, she says, is "a policy problem. It's an institutional problem. It's a physician education problem," even more than it is a problem of the choices that individuals make.

The physician education problem concerns exercise in general, not just progressive resistance exercise. The problem was measured, around when Maria Fiatarone started medical school, by the survey that said only 16 percent of medical schools taught their students about exercise. Data from later surveys show that time has not fixed the problem. In 2001, a survey of deans and directors of medical schools found that practically all of them—92 percent—thought that graduates from their institutions should be able to write exercise prescriptions for patients, yet the same survey showed that only 6 percent of medical schools required students to take even one course that would teach them to prescribe exercise. In 2009, virtually all medical students—94 percent—said they should know more about exercise: They said it was "moderately important" or "important" that they should be able to prescribe physical activity. Yet in another survey, involving medical school graduates in the classes of 2013, 2014, and 2015, only about 15 percent of future doctors said they were adequately trained to prescribe exercise.

In 2015, some of these data were compiled by Edward Phillips of Harvard Medical School and colleagues, as part of their work to incorporate lifestyle medicine—counseling and treatments focused on behavioral factors affecting health—in medical education. Phillips runs an Institute of Lifestyle Medicine at Spaulding Rehabilitation Hospital Boston.

Pointedly, the authors also noted that medical board examinations did not include test items directly addressing topics such as exercise and physical activity. Some scholars say that changing board exam test questions would create the only real incentives for medical schools to teach students about exercise. Marcas Bamman, a professor emeritus of cell, developmental, and integrative biology at the University of Alabama at Birmingham, and a leading resistance training researcher, tells me, "Until exercise-focused questions are on the board exams, I don't think anyone will ever care."

After taking their board exams, medical students who have been taught little or nothing about exercise generally become doctors who give little thought to exercise, which helps perpetuate another aspect of the problem Maria Fiatarone Singh has named. This is the institutional problem: Professional associations of doctors produce clinical practice guidelines to define standards of care that almost always recommend drugs and surgeries over exercise.

Even in the rare instances when clinical guidelines designate exercise as part of a first-line treatment, as for osteoarthritis, many doctors ignore the guidance to prescribe exercise and only prescribe drugs. A survey of general practitioners in France found that only 48.7 percent prescribed exercise for people with hip and knee arthritis, and 95.8 percent prescribed the drug paracetamol (also called acetaminophen).

Robert Sallis, a sports medicine and family physician who is a former president of the American College of Sports Medicine, raises a central question about exercise and standards of care. In an editorial he asks, "Why has the medical community neglected exercise as a standard treatment?" Answers to that question are complicated, he knows. He also thinks "it's just easier for most physicians to prescribe a pill" than to prescribe exercise to prevent or treat chronic diseases.

Since 2007, Sallis has led a global initiative called "Exercise is Medicine," a joint effort of the American College of Sports Medicine and the American Medical Association that aims "to make physical activity assessment and promotion a standard in clinical care." One goal of the effort is to persuade physicians that "physical activity should be regarded as a vital sign," equally with traditional vital signs such as blood pressure and temperature. Sallis argues, too, that health insurance coverage should be expanded to include reimbursements for exercise training. If "insurance will pay for bariatric surgery to help an obese patient lose weight," he writes, it would stand to reason that insurance should "pay for a fitness professional to help a sedentary patient become more active," given the mass of evidence that inactivity trumps obesity as a threat to health.

To be effective, exercise prescriptions must be precise, Maria Fiatarone Singh reiterates, again citing the trials of lifestyle interventions for diabetes prevention and treatment. "What does not work," she says, is "activity advice. If you just tell somebody, I think you should be more active—that's the control group of all these clinical trials—that actually doesn't work. You actually have to give somebody structured, supervised exercise, and in some cases a diet, and then it will work like a drug. So we have the solution in our back pocket. But unfortunately we don't use it. We don't pay for it the way that we pay for metformin, insulin, et cetera," and that's why general practitioners prefer to prescribe pills.

In her vision of health care, everyone in the system, from physicians to pharmacists, would account for exercise as part of every aspect of treatment. "If they write a prescription for a corticosteroid," she says, "they should be writing a prescription for weight lifting on the back of it." She is thinking back to that study of heart transplant patients, in which lifting weights according to a prescription counteracted a corticosteroid's side effect of rapid, significant muscle loss.

Yet even among the minority of doctors who promote exercise as medicine, resistance training is typically "an afterthought," says Marcas Bamman. When I see that the Exercise is Medicine advisory board includes no resistance training experts, I mention it to Robert Sallis. He replies, "I hadn't thought of it, but it is a kind of glaring omission."

That conversation took place in 2016. Seven years later, in 2023, the Exercise is Medicine advisory board still lacked a single doctor or scientist who specialized in resistance training research.

When doctors do *not* prescribe regimens of physical activity and exercise to their patients—especially when doctors do not prescribe robust progressive resistance training for older patients who are frail—those omissions should cause a scandal, Maria Fiatarone Singh believes. Prescribing exercise is in fact an ethical imperative for doctors, a requirement for clinicians who claim to follow the Hippocratic oath, she suggests—because doctors who fail to prescribe exercise to patients who need it, she says, will actively cause harm to those patients, in many cases.

She has developed this argument in several publications, cowritten with her colleague Mikel Izquierdo, professor and director of the

Health Sciences Department at the Public University of Navarra and Biomedical Research Center in Pamplona, Spain. Denouncing health care industries for widespread "failure to translate existing evidence" of how exercise prevents and treats chronic diseases, they call for action by drawing a stark parallel: "Withholding pharmaceutical agents with similar proven benefit, such as anticoagulants for stroke prevention in atrial fibrillation, would never be considered."

Until people mobilize strong movements to address the policy problems, the institutional problems, and the physician education problems that form barriers to access to progressive resistance training, answers to the question of how most of us can learn the skill of strength will continue to depend largely on luck.

People who are lucky enough to live in Australia and within easy traveling distance of the Centre for Strong Medicine, the clinic Nalin Singh founded in northern Sydney, now may have the chance to be treated by Maria Fiatarone Singh, who in addition to her teaching and research at the University of Sydney Medical School began running the clinic after Nalin died. She sees patients two days per week. When she prescribes progressive resistance training regimens for her patients, one of the people who implements those prescriptions is Guy Wilson—the lead exercise physiologist at her university laboratory gym, the one who guided the years of training that helped free Ramanee from so many of the restrictions that knee arthritis pain had put on her life. Four other exercise physiologists complete the clinic's team.

We can all hope to be lucky enough to meet doctors, physiologists, coaches, and trainers with the knowledge and the energy to help us. Even if we are lucky that way, though, our fates will always depend on our individual choices, too.

To learn the skill of strength at the higher levels honed by progressive resistance training, a person has to decide to learn and has to commit to practicing a regimen. "It requires a real shift in your brain," says Fiatarone Singh. "Encouraging people to walk more is much easier. You can just *do* that." But habits of daily life discourage the kinds of muscular exertion that build and preserve strength and power, she notes, because the general drift of culture since the Industrial Revolution has been to make life easier. Modern cultures have removed the necessity of exerting force to get things done. "That's the bottom line. Every single thing you think about doing has been designed to be easier. So you've got to take a step back and say, Okay, well, let me see if I can make that harder. How would I make that harder?"

Some years ago, with a team of University of Sydney colleagues led by Lindy Clemson, Fiatarone Singh collaborated on a study that integrated balance and strength training into activities of daily life. "We set up rules that people had to follow about how to move around their environment," she says. The rules included: "Never bend over, always squat. Always sit down slowly. Everything you could do with two hands, do with one hand. Use less muscle rather than more muscle to do the same activity so that you're forcing that muscle to work harder." Strictly following those rules does make people stronger, she says, but it also requires people to think constantly about the rules. A lot of people do not want to be so vigilant, she points out. They'd rather have twenty minutes of dedicated exercise time.

She circles back to the inevitable comparison. Strength training "isn't as simple to prescribe as walking. I mean, face it, it just isn't. It's harder. You can do it with body weight, but to do it in a robust way, it often, if not always, requires equipment, and requires a dedicated time of day in which to do it. So it's just not as easy," she says.

Even many twenty-year-olds, when rising from a chair, tend to use their hands and arms to push themselves up, she notes. They don't have to, but they do. "And so shifting that is really hard. It's not natural in any way. That's a barrier that you have to overcome. To go out of your way to make something harder to do. Where do you do that in life? You don't, right? Everything you do is to make life easier. So that's difficult: thinking about opportunities to make things harder. That's what you have to do. That's just not what we do."

But you do, I say to her.

"Yeah. I try to."

How?

"I try to stand on one leg when I'm lecturing. Use just my legs rather than arms and legs when I stand up from chairs or get out of the car. I don't take escalators or elevators unless there absolutely are no stairs." She stops, revises: "I *try*."

And to stay strong, Maria Fiatarone Singh lifts weights almost every morning—at 80 percent of 1-RM, or working at a rate of perceived exertion in the range of 15 to 18, on a scale of 6 to 20—doing upper body and lower body on alternate days. That kind of regimen—based on dedicated time for training—is what it takes, for most people, to overcome the brain's inclination to choose the easy thing.

In stroke rehabilitation, she says, therapists help people learn to use their weaker arm by tying the stronger limb behind them. "It's very effective, but it's not intuitive. You actually have to tie somebody's arm behind them, because otherwise they will definitely go for the stronger arm. Trying to go against the natural impulse to make things easier for yourself: That's hard."

Conclusion

Each act of voluntary movement—lifting a calf, turning a page, or walking backwards down a stairway to an underground train platform—can be described by a simple story. The one story that can account for all those moves we choose to make is extremely short, so short it only takes a moment to tell.

The story of movement, driven by conflict with gravity, is resolved by means of muscle. When movement is easy, when we feel comfortable in motion, the story has a strangely happy ending.

In the happy ending, muscle seems to disappear.

Muscle does not actually go away; but you feel as though it's gone. As a team doctor for the New York Knicks once said to me, healthy muscle "is supposed to feel *like nothing*."

In other words, healthy muscle draws practically no attention to itself, produces almost no discernible sensation. Unless it's working hard. Yet to enjoy the easy freedom of having healthy, pain-free muscles and joints, people have to be mindful of muscles. To take care of them by giving them the challenges they need.

That is the ultimate paradox of training. Muscle needs attention so that it can disappear.

What kind of attention does muscle need?

Thomas DeLorme found some practical, technical answers to that question in the 1930s when he was a young man, sick in bed, reading the muscle magazine that gave advice about how people could test their own strength and how they could use those tests to structure weight training programs that help muscles grow bigger and stronger. DeLorme carried those lessons into his adult work as a doctor in the 1940s, when he refined his knowledge of weight training through formal clinical experiments, transforming the tacit knowledge he found in muscle magazines by systematizing it in scientific terms. Ever since, DeLorme's published research has given athletes, trainers, and doctors better chances to find trustworthy answers when it was their turn to ask, *What kind of attention does muscle need?*

DeLorme's answers have also been challenged, not for being incorrect but for being incomplete. Researchers have shown that for some people in some circumstances, there are ways of lifting lighter weights that can be about as effective for building bigger and stronger muscles as lifting the heavier weights DeLorme prescribed. The goal of much of this newer research has been to reduce the joint stress of resistance exercise, to make strength training more accessible to more people.

To survey some of this work, the neurophysiologist Jacques Duchateau of the Université Libre de Bruxelles, in Brussels, collaborated with colleagues in writing the paper "Strength Training: In Search of Optimal Strategies to Maximize Neuromuscular Performance," published in 2021. "Over the last two decades, our concepts and practice of strength training methods have been strongly shaken," they wrote, and went on to argue that "uncertainty should be used as opportunity to advance the field" of research on how to make muscles grow bigger, stronger, and more powerful by resistance training.

The Belgian team cited analysis by Brad Schoenfeld, the expert on muscle hypertrophy, showing that lifting relatively lighter weights—as light as 30 percent of the lifter's one-repetition maximum—can build as much muscle, but not as much strength, as lifting much heavier weights, "provided that sets are performed to failure."

The Belgians also cited a string of experiments, beginning with the work of Minoru Shinohara, professor of biology at Georgia Institute of Technology in Atlanta, and Yudai Takarada, professor of sport sciences at Waseda University in Japan, showing that lifting even lighter weights—as light as 20 percent of one-rep max—while restricting blood flow to the limbs that do the lifting, can in some cases increase both size and strength of muscles as much as high-intensity weight training. The technique of blood-flow restriction training, abbreviated as BFR, involves temporarily applying pressure, as with an elastic band around the top of an arm or leg, to reduce the amount of blood flowing into the exercising limb, and to reduce or briefly stop the blood flowing out of the limb. BFR training seems to work especially well for people with little to no experience of lifting weights, but it seems to be less effective, or ineffective, for well-trained athletes, especially strength athletes such as powerlifters.

Scientists do not completely understand the divergent mechanisms of changing muscle size and strength when people lift lighter or heavier weights. But Jacques Duchateau and his colleagues, in their paper, mentioned one likely difference. They wrote that neural adaptations, engagements of the nervous system and the muscles that increase levels of voluntary muscle activation—levels that "play an important role in improving maximal and functional strength" in the widest range of movements for everyone, from young women breaking powerlifting world records in the barbell squat to old men struggling to stand up from easy chairs—"are not evident after low-load exercise training with or without BFR." In addition, the Belgian team found no clear evidence that blood-flow restriction training could increase muscle power.

This was the big takeaway of the Belgians' research review: New protocols for lifting lighter weights, with or without blood-flow restriction, could do a fine job of building muscles to be bigger and stronger, if perhaps somewhat lacking in the vital attribute of power—but the strength of those muscles might not necessarily be as *integrated* with the workings of a person's nervous system. That was suggested by several measures of nerve and muscle activity and interaction, including electromyography, or EMG, which can record recruitment of those functional neuromuscular entities called motor units that produce muscular movement.

At their laboratory in Brussels, searching for optimal strategies for strength training, Duchateau and his colleagues in recent years have also experimented with a weight training program that can build muscle size, much as a bodybuilder's routine does, but without compromising the neural adaptation of strength; while preserving or building power, too; and at the same time reducing joint stress, as programs of lifting lighter weights and blood-flow restriction training have been intended to do. The program that Duchateau has been investigating is called the three-seven method, with the name written as two integers sitting side by side and separated by a slash: the "3/7 method."

The 3/7 method, as the Belgians described it, "is characterized by five sets of an incremental number of repetitions per set (from three repetitions in the first set to seven repetitions in the last set) with a constant load," about equal to 70 percent of a person's one-repetition maximum. With a rest period of 15 seconds between each set, it can take as little as a minute and a half to complete a single bout of a 3/7 exercise.

One person who read the Belgian group's research on the 3/7 method and was favorably impressed was Roger Enoka, who, as mentioned earlier, literally wrote the book on neural adaptations of muscle, *Neuromechanics of Human Movement*. Enoka was sufficiently impressed that, on behalf of the American College of Sports Medicine, he gave the paper on optimal strength training strategies a prize for being the most significant work in the field of medical exercise science in the year that it was published.

Enoka wrote a commentary on the paper, encouraging Duchateau and other scientists to continue searching for "the optimal combination of training variables" such as set-and-rep schemes and rest periods "for different groups of muscles and training devices," while comparing the effects of these programs not only on muscular attributes of size, strength, and power, but also on functional abilities such as walking, balance, and rising from sitting to standing.

Importantly, too, Roger Enoka gave credit for the 3/7 method to the individual who devised it; and when he did this, he amplified the attribution made by Jacques Duchateau and his colleagues, who had freely, openly, and without a trace of apology or anxiety named the exact source of the strength training protocol they borrowed from athletics and applied in their scientific medical research.

The 3/7 method, as everyone acknowledged, was devised by a coach in France—a coach in the sport of powerlifting—by the name of Emmanuel Legeard.

Here was a ripple effect. The strength training protocol of the French powerlifting coach inspired the Belgian physiologists' academic research that earned the prize from the American medical group—and in late 2022, extending this pattern of insight, investigation, and recognition, the American College of Sports Medicine convened a committee to conduct comprehensive umbrella reviews of current knowledge about optimal strength training strategies, and about using progressive resistance exercise in treatment of chronic diseases including cardiac and peripheral vascular diseases, hypertension, type 2 diabetes, cancer, and obesity.

Maria Fiatarone Singh is on the committee, along with Stuart M. Phillips, professor of kinesiology at McMaster University in Hamilton, in the Canadian province of Ontario. Phillips and his McMaster colleagues have done some of the most influential research on strength and hypertrophy training using lighter weights. As I finished writing this book in 2023, the group was hoping to publish its findings in 2024 and 2025.

In the meantime, while the committee of physicians and physiologists did their work, some of the most knowledgeable and dedicated recreational lifters—people like Charles Stocking—continued the ongoing process of their own personal education in medical and exercise science, reading the research and bringing their athletic experience to bear on it, while perpetually revising their exercise regimens in light of their evolving goals.

"I'm training to be ready to be ninety years old," Stocking tells me.

He looks ahead and sees how mechanisms of movement get worn out: "What are the inevitable injuries that come from old age? Hip replacements, knee replacements, rotator cuff injury." Data backs him up. In recent years in the United States, total joint replacement surgeries have become more common than heart failure. And although many more such surgeries are done in the United States than in any other country, the incidence of total hip replacement surgery, in proportion to population size, is even higher in Australia, France, Germany, and Norway. As populations age, demand for these surgeries will rise—by one estimate, demand for total hip replacement surgeries for Medicare patients in the United States will increase by 659 percent by 2060.

Though some age-related joint degeneration and damage are inevitable, it's not possible to quantify how much. Joint damage may accelerate, though, accompanied by lower back pain, when muscles and connective tissue wither during long days of desk work. As those gradually tightening flexor muscles pull inactive limbs forward, they can curl and hunch our neutral sitting postures almost into fetal positions, a process Stocking calls "*fetal*-i-zation," before he admits, with a chuckle, "I'm making up a term here."

Stocking thinks, "If we can do things from a muscular perspective to prevent those problems, then we should just be doing that every day," and that thought simplifies his training goal.

His goal is: "Stopping that process."

Stocking's workout to maximize the chance of being strong and comfortable and mobile at age ninety involves the same exercises as his workout to minimize the pain and damage of desk work, as well as the main exercise that helped him recover from a back injury when he was a young man—all as described in the first part of this book. The same movements—back extensions, hip thrusts, and rows—are part of Stocking's strategy for building enduring strength and functional alignment of the hips, back, and shoulders.

But his strategy is constantly evolving, based on how he feels on any given day. "Everybody wants the answer to be just one exercise," he says, because "so much of the history of fitness is a history of gimmicks and fetishization." Yet Stocking holds firm to the *kairos* principle, as suggested by Philostratus in the *Gymnasticus*—the principle of choosing exercises "'all improvised for the right time.'" On this he is vehement: There is no universal answer to the question of what exercises will get us to the end of the road in the best possible shape. "Sometimes it's this," he says, and "sometimes it's that."

Sometimes, though, a gimmick may catch on because it speaks to a genuine need in people that has too long gone unfulfilled, and that may be what's happened in recent years, in a lot of gyms and on social media, with the hip thrust exercise. You could probably spend whole days on Instagram watching nothing but videos of beautiful people working to make their backsides more beautiful by doing barbell hip thrusts.

"Ironically," Stocking says, "I bet there's going to be a lot fewer back injuries and hip replacements in the future," even for people whose reasons for glute training are mainly about appearance—because hip exercises that make butts bigger also help to make them stronger, and vice versa. "As a by-product of obsessing over large glutes," Stocking says, a lot of people "will probably not need hip replacements when they're ninety."

Data supports that idea, too. Maria Fiatarone Singh makes a related point when she cites a 2014 randomized controlled trial involving people with hip osteoarthritis that compared long-term outcomes for two groups, in Oslo, Norway. Both groups received classroom education about arthritis; and in addition, one of those groups went through a twelve-week supervised exercise program, involving resistance, flexibility, and functional training. Over the following six years, the exercise program reduced the need for hip replacement surgery by 44 percent. In other trials, older women who did back extensions reduced their risk of osteoporotic fracture in the lumbar spine.

So the quality of your future may depend, in no small part, on what you do for your behind; and Charles Stocking thinks that work could have collateral effects, also, of resolving mind-body dualism.

"Aristotle said that the source of movement is the soul," he says, setting up his own punch line, again with a chuckle: "But biomechanically it's the glutes—so the soul must be located in the glutes."

In the distant future, on that ideal someday when athletes, trainers, doctors, and scientists finally formulate exact exercise prescriptions and regimens for all the individualized performance- and health-related purposes that would make everyone's life better, those theoretically perfect exercise programs still will not give complete, viable answers to the question of what kind of attention muscle needs. Answers to that question are not and will never be only quantitative, technical, prescriptive, and individualized. They are also, as they have always been and always will be, qualitative, philosophical, imaginative, and collaborative.

Training, as I've learned about it from Charles Stocking, Jan Todd, Maria Fiatarone Singh, and many others, is an adventure of improvisation based on knowledge, pursued with help. Finding the right thing to do at the right time, we build and maintain the strength, the tensed and balanced forces of our nature, to meet the challenge of this day and make a difference in the world. Viewed from that perspective, workouts are not extra things to try to fit into your schedule. Workouts are how we get to be here, to enjoy our lives together.

Jan Todd describes the shift of outlook. "You get to that place where lifting is like brushing your teeth."

Maria Fiatarone Singh says, "It actually is as serious, it is as important as anything else you do."

When people make that kind of plain, certain, serious commitment to regimens of training, something sublime can happen. Keeping those commitments creates a new experience of time.

To show what that means, the best I can do is name the most unusual thing I observed about Charles Stocking's practice of training: his articulate awareness of working out for many purposes simultaneously, with those purposes oriented to multiple planes of experience and intention.

Charles Stocking lifts weights partly so his back won't hurt while he does his desk work today; and partly so he'll be less likely to need hip replacement surgery when he's older; and partly because he likes being as strong in his early forties as he was in his early twenties. He lifts weights partly because he likes—and because his wife likes—the way it changes how he looks. He lifts weights because he likes being able to sprint fast—and glute training is key to this, because sprinting speed depends more on the size and strength of glutes than on any other muscles.

All those purposes strive inside Charles Stocking, when he does the desk-proofing workout for future nonagenarians. In that striving, he actively, consciously lives in the past, the present, and the future, all at once.

And no, dear reader, your eyes are not deceived. When I write that Charles Stocking lives in the past, the present, and the future, I am ripping those words right out of *A Christmas Carol*, right out of the mouth of Ebenezer Scrooge. And yes, I am actually saying that, when it comes to exercise, especially lifting weights, we are, every one of us, perpetually due for an awakening like the one that came for Scrooge on his fateful Christmas morning. The awakening about exercise I'm talking about concerns two things above all. Those things are time and generosity.

In physical activity surveys throughout the world, people of many ages, ethnicities, and fitness levels consistently say their biggest obstacle to regular exercise is "lack of time." Everyone knows this feeling. The pull of those *Groundhog Day* time loops, where we get up in the morning and live through another day like yesterday, wondering if life can ever change. Wishing we could do something to make our lives different and better. Trying to find some way to break the repetition.

Repetition, as we have seen, pervades the history of physical culture, too. By the force of example, the lives of some strong people reshape the lives of others—and yet somehow, as practices of strength training drift across the generations, scientific knowledge of the practice fades in and out. To learn this can be as exasperating, and sometimes as heartbreaking, as chasing a mirage.

Sometimes while I was writing this book I found it almost maddening to read about those arguments between doctors and trainers in ancient Rome and to realize that their successors have been fighting about the same things, in almost the same terms—based on false dichotomies of balance and excellence, muscle and mind—right up until today.

As inspiring as it was to learn about the many nineteenth-century girls and women practicing the most challenging resistance exercises during the early years of modern gymnastics movements, it is infuriating to consider how little difference that made to their descendants.

There was something awesome about finding that *The Journal of the American Medical Association* mentioned an octogenarian's muscle-building bicycle training in 1892, almost a hundred years before the *Journal* published "High-Intensity Strength Training in Nonagenarians," and it is enraging to think of all the people in that intervening century growing old and frail and dying in helpless pain because they and their doctors did not know it was not too late for them, they were not past hope, there were things they could have done to help the muscles in their tiny withered legs grow bigger and stronger so they could have another chance at standing on their own two feet.

If we want to stop repeating the same conflicts and conversations about muscle, we have to take the time to make the generous choice to take muscle more seriously. To take muscle seriously not merely as an "instrument of the soul," but as the vital, inextricable, and effective partner of the soul.

The concept of partnership, based on collaboration, describes the link of mind and muscle more accurately than the old metaphor of instrumentality, based on command, grounded in scientific dogma that deemed the brain the *hēgemonikon*, the "ruling part." As the physiologist Richard Lieber has written, "Muscles are designed for a specific function—large excursion, for example. The nervous system provides the signal for the muscle to 'do its thing,' but does not necessarily specify the details of that action. It is as if the nervous system acts as the central control while the muscle interprets the control signal into an external action by virtue of its intrinsic design." Muscle is so well made for doing its thing, Lieber writes, that in movements that make up the act of walking, for example, "the spinal cord can 'learn,'" in collaboration with muscles, "in the absence of influence from higher centers" in the brain.

When I think how these facts of neuromuscular function could reframe our commonsense way of talking about body and soul, mind and muscle, I remember those final words Aristotle wrote about functional movement: "The constitution of an animal must be regarded as resembling that of a well-governed city-state. For when order is once established in a city there is no need of a special ruler with arbitrary powers to be present at every activity, but each individual performs his own task as he is ordered, and one act succeeds another because of custom." I also remember the tradition Aristotle followed when he wrote those words, the tradition going back to Alcmaeon of Croton, the doctor from Milo's hometown, who named health and democracy with the same word. And I wonder: Could imagining the healthy body as a kind of functional body politic help make it easier for more people to consider muscle as a priority on par with mind, when we make decisions about how to care for ourselves and others? Considering muscle on par with mind and conceiving of personal health as a microcosm of democracy might even be steps toward restoring our politics, governments, and societies to health.

Which brings us to generosity. One of the most striking findings in the research on why people do or do not get into the habit of lifting weights is that starting and maintaining the practice often requires social support. This seems to be especially true of people in some of the groups that most acutely need some of the benefits that progressive resistance training can dependably produce, such as older women and people with type 2 diabetes.

For many millions of us, the simple truth is that we lift weights when the people around us support us in doing it, and we do not lift weights if we do not have social support.

Of course, the whole truth is much more complicated—with vast permutations of other factors, including economic, educational, environmental, and racial inequalities also influencing participation in exercise regimens. But a deciding factor as to whether people keep showing up, keep doing our workouts, and learn to reconcile ourselves with the inherent repetition of lifting weights is our sense of whether other people care if we keep training and will help us when we need help.

The stories of Charles Stocking, Jan Todd, and Maria Fiatarone Singh show that training can be as expressive and generative, as rich and meaningful, as orderly and adaptable, and as fundamental to many kinds of human connection as language.

The dialect of that language known as progressive resistance training helps give muscles their best chance to be known in this world, in the most lasting possible way, among the largest number of people. Lifting weights gives muscles the chance to be known so intimately as to be taken for granted—to be habitually forgotten, because of being habitually exercised. Taken for granted in their powers to sustain us in the many complex ways of which even scientists are for now at best only partially aware. And taken for granted in some of their most obvious powers, of which most of us are keenly aware—or will be, in time: the powers that are the prize. Power to help us move as easily as possible and power to act upon the world as effectively as possible—no matter what the fates may bring—for as long as we all shall live.

Epilogue

On the evening of May 22, 2020, Jan Todd celebrated her sixty-eighth birthday at a cookout with a couple of friends. After supper, Todd climbed into the back seat of her friends' sport utility vehicle, and they went out for a drive.

They were on a ranch road, with no light poles alongside the road, and in the darkness the driver could not see the sounder of wild boars until they were in the beams of the headlights. A hard swerve; the SUV flipped. Todd was thrown out of the vehicle as it rolled over, and the vehicle landed on top of her, crushing and breaking more than a dozen bones, and tearing muscles, tendons, and ligaments. She was trapped under the car, and fully conscious, for more than an hour before paramedics arrived.

Following the accident, Todd underwent multiple surgeries and other invasive procedures. Her family, friends, colleagues, students, and admirers rallied around her, supporting her as best as they could; but with the world in lockdown because of the COVID-19 pandemic, a lot of the caretaking happened from a distance. One day Arnold Schwarzenegger called her on FaceTime, and he gently scolded her for not letting him know right away about what happened. "Of course I'm worried about you," he told her. "You are family."

Almost one year after the accident, Todd decided to test her strength by building three large, raised seedbeds for an organic garden, each bed surrounded by concrete blocks. She moved the blocks and dirt by herself, and then planted what she called a "victory garden." In the year after that, she started lifting weights again. She began with weights that were very light, compared to what she had lifted as a younger woman. She deadlifted 115 pounds.

"It feels good," she told me, "and gardening feels good. What I really like is to *work*, you know?"

She also got a puppy. A bullmastiff, a female. She named the puppy Pudgy, after the most famous female lifter of the 1940s, Pudgy Stockton.

One day when we were talking on the phone and about to say goodbye, she remembered one more thing she wanted to say: "And did I tell you what else I'm doing?" she asked. "I call it my 'Jan of Crotona' workout."

She paused for effect and then went on: "I lift my puppy up every day. She was twenty-two pounds when I got her, and she's fifty-six pounds now. And pretty soon she's going to be so big I can't lift her anymore." And the last part, about facing her limit, came out in a voice so bright, it took me a minute to get that she was joking.

Acknowledgments

The roots of this project go back many years, and really took hold when I was lucky enough to meet a trainer, Charles Maturana, and two expert kayakers, Eric Stiller and Jessie Stone, who in various ways changed my experience of strength and muscle.

When my agent Todd Shuster heard about what I'd been learning, he instantly saw the possibility of this book, and then he always had my back as the project took shape.

The team at Dutton who acquired the book, including Christine Ball, Ben Sevier, and especially Stephen Morrow, gave me the greatest possible enthusiasm and support. They set a tone that others sustained as time went on: John Parsley, Ivan Held, and Jill Schwartzman, to whom I am especially grateful for her dependable, shrewd, and sure-footed guidance. But no editor did more to shape and protect this book than Grace Layer, whom I wholeheartedly thank for her meticulous, prudent readings and searching reflections that improved the manuscript through every draft, and for her astute, effective attention to every detail of the book's overall appearance and presentation.

Whatever life is in these pages owes its origin mainly to the generosity, trust, patience, and good humor of Charles Stocking, Jan Todd, and Maria Fiatarone Singh—and of Catherine Pratt and the late Terry Todd and Nalin Singh. It would be impossible to overstate my gratitude for their advice, encouragements, and cautions, for their answers to my innumerable questions, for the many references they provided and facts that they checked, and for introductions to their colleagues, friends, and families.

Research for this book entailed meeting, calling, and corresponding with hundreds of people. Only a few of them are named in these pages, but they have all shaped the whole of the project. For brevity's sake, I've made the hard choice to refrain here from acknowledging anyone who is named in the main text except for a very few who helped in the most defining ways.

No one brings the classics to life like Daniel Mendelsohn, and I was spectacularly lucky to have his inspired and faithful help as I began to learn about ancient Greek athletics and literature. Stephen G. Miller took it from there, first in long hours of phone conversations and then in person, both in Berkeley and at Nemea. At the Mayo Clinic, Michael Joyner welcomed me with energetic support for this project and has consistently been a magnanimous counselor and connector and a reliable sounding board. At the University of Copenhagen, Jesper Løvind Andersen and his colleagues at the Centre for Muscle Research offered warm hospitality in a dark December, and the head of the university's department of nutrition, exercise, and sports, Arne Astrup, also made helpful introductions. Richard Winett helped me start to understand the sociology of strength training research, among many other topics. At annual meetings of the American College of Sports Medicine, Per Aagaard, Marcas Bamman, D.C. Lee, Rob Morton, Stu Phillips, and Katie Schmitz, among many others, engaged me in lengthy and stimulating discussions. A residency at the MacDowell Colony in Peterborough, New Hampshire, gave me time and space to turn the first stage of my research into early draft chapters.

Later, Charles Stocking invited me to present a portion of my manuscript as a lecture at the International Olympic Academy, at a symposium he organized for Harvard University's Center for Hellenic Studies, and then Damian Stocking invited me to make a similar presentation at Occidental College. At both events, feedback from faculty and students helped to clarify some key ideas in the book. By introducing me to Gregory Nagy and Zoie Lafis, Charles Stocking also opened the door for me to spend a couple of days on campus at the Center for Hellenic Studies in Washington, D.C., where Aileen Das sharpened my questions about ancient Greek medicine, and Lanah

Koelle and her colleagues in the library lit the way toward some answers to those questions.

Among the host of librarians who helped organize my inquiries and gather reference materials, probably no one gave more time and effort than Cindy Slater of the H.J. Lutcher Stark Center for Physical Culture and Sports in Austin. Over the course of several visits to the center, I learned a lot from Kim Beckwith and Tolga Ozyurtcu, and Kyle R. Martin gave me the benefit of his creative diligence. Among Jan and Terry Todd's friends and family in Austin who offered insights about the couple's life and work, I especially thank Connie Todd, Waneen Spirduso, and Mark and Jana Henry. As part of my research on the Todds, for several years I also attended the Arnold Strongman Classic competition at the Arnold Sports Festival in Columbus, Ohio, where Bill Henniger and Caity Matter Henniger of Rogue Fitness, and also William J. Kraemer of the Ohio State University, allowed me to witness some unforgettable events.

Among Maria Fiatarone Singh's colleagues who helped me understand her work while I was in Sydney, Michael Inskip and Shelley Kay made special efforts on my behalf. I found Shelley Kay's work with cancer patients as an exercise oncologist at Chris O'Brien Lifehouse in Sydney to be so compelling that, at her suggestion, I also visited one of the leading researchers in that field, Rob Newton, at the Exercise Medicine Research Institute at Edith Cowan University in Joondalup, West Australia, where Nicolas Hart, Mary Kennedy, and Ciaran Fairman were also especially helpful.

For hunting down nearly every reference I could think to ask for, I am grateful to the New York Public Library and its librarians, especially Paul Friedman and Melanie Locay, and the staff of the research study rooms in the Center for Research in the Humanities, as well as the Billy Rose Theatre Division of the New York Public Library for the Performing Arts. At NewYork-Presbyterian/Weill Cornell Medical Center, the Myra Mahon Patient Resource Center and the Samuel J. Wood Library helped with medical and scientific research, and Arlene Shaner, the historical collections librarian at the New York Academy of Medicine Library, showed me some wonderful, surprising treasures.

For help with other queries, I am also grateful to librarians and archivists at the Thomas J. Watson Library at the Metropolitan Museum of Art, the Center for the History of Medicine at the Countway Library at the Harvard Medical School, the UA Libraries Special Collections at the University of Alabama at Birmingham, the Massachusetts General Hospital archives, the Massachusetts Historical Society, the Boston Athenaeum, the Augustus C. Long Health Sciences Library at Columbia University Irving Medical Center, the Franklin D. Roosevelt Presidential Library, the Dwight D. Eisenhower Presidential Library, and the Western Illinois University Library.

For research assistance on a few specific questions, I thank Ben Bonner and Abby McCreary.

Many of the scholars mentioned in these pages read the passages that quote their work to check for accuracy and offered constructive criticisms. Two who are not mentioned but deserve special thanks for their corrections are David L. Morgan and Thomas Scanlon. Ben Kalin fact-checked most of the manuscript. Any errors that remain are my own.

Many friends and colleagues, too, read all or part of the manuscript, and their comments much improved it. Thanks to Elizabeth Bejarano, Matthew Cushing, Paul Elie, Eve Gerber, Michael Joyner, William Langewiesche, Michael Lowenthal, Andrea Malin, Cullen Murphy, Jim Shepard, Karen Shepard, and Ladd Spiegel. For especially close and repeated readings, I am particularly grateful to Amy Whitaker and Dan Josefson, both of whom supported this book, and me, with absolute and unwavering confidence from start to finish.

Other colleagues, friends, and friends of friends whose help was pivotal in a variety of ways—from making one quick phone call on my behalf, to sitting for dozens of hours of interviews over periods of years, to giving me a place to sleep while I was traveling or a dedicated place in which to work when I got sick of writing at the dining room table—include Thomas Rode Andersen and the staff of the Kurhotel Skodsborg, the late Olivia Barker, Rain Bennett, Lisa Boshard, Ed Checo and the Barstarzz, Lyn Christian, Patricia Clark, Curtis Cole, Ben Court, Hope Cushing, Jason Furman, Alison and Andrew Good, Sue Good, Steve Horney, Oleg Ikhelson, E. J. Jackson, Gregory

Jordan, Ethan and Sally Kline, Lauren Liebow, Mark Lotto, Rob MacDonald, Richard McGheehan, George Mayhew, Claus Meyer, Stelios Mormoris, Christian Noble, Mary Parsons, Erin Pauwels, Conrad Rippy, Veronica Roberts, Grayson Schaffer, Ted Smith, Rory Toser, Mark Twight, Diana Untermeyer, Patrick Vignona, Thayer Walker, and Susan Williams—and especially Fitz Boissevain.

The many dear and stalwart friends who don't quite fit into any of the above categories, whose companionship and good cheer meant the world to me through the whole odyssey of making this book, also include Jon Barrett, John Case, Romain Frugé, Sean Keilen, John Kim, David Masenheimer, and Dan Polsby.

For believing in me and my writing, thanks to my whole family—especially Angela, Mac and Karen, Linda and David, Joanne and Phil, Jennifer and Matt, and Sheila and David.

My deepest gratitude is to my husband, Steve, who did more than any person could reasonably have been expected to do to make it possible for me to write this book, and then always found some way of doing more. He is simply the best.

Source Notes

PROLOGUE

On the story of Milo and the myths that grew from his life, see Michael B. Poliakoff, *Combat Sports in the Ancient World: Competition, Violence, and Culture* (New Haven: Yale University Press, 1987), 117–19; also Jean-Manuel Roubineau, *Milon de Crotone ou l'invention du sport* (Paris: Presses Universitaires de France, 2016).

On Thomas DeLorme's early experience of weight training, see Bob Collins, "His Fight Was Against a Frail, Weak Physique," *Birmingham Post*, July 21, 1939.

On the longer arc of DeLorme's life and career, see Janice S. Todd, Jason P. Shurley, and Terry C. Todd, "Thomas L. DeLorme and the Science of Progressive Resistance Exercise," *Journal of Strength and Conditioning Research* 26, no. 11 (2012): 2913–23.

DeLorme's first publication on heavy-resistance exercises for rehabilitating injured soldiers is Thomas L. DeLorme, "Restoration of Muscle Power by Heavy-Resistance Exercises," *Journal of Bone and Joint Surgery* 27, no. 4 (1945): 645–67.

For Delorme's basic progressive resistance training protocol, see Thomas L. DeLorme and Arthur L. Watkins, *Progressive Resistance Exercise: Technic and Medical Application* (New York: Appleton-Century-Crofts, 1951), 23–30. This is the best source for information about the development and early applications of progressive resistance training as a part of medical treatment.

DeLorme's techniques were validated for people with polio, for adolescents, and for women in the following papers: Thomas L. DeLorme et al., "The Response of the Quadriceps Femoris to Progressive-Resistance Exercises in Poliomyelitis Patients," *Journal of Bone and Joint Surgery* 30, no. 4 (1948): 834–47.

J. Roswell Gallagher and Thomas L. DeLorme, "The Use of the Technique of Progressive-Resistance Exercise in Adolescence," *Journal of Bone and Joint Surgery* 31, no. 4 (1949): 847–58.

Sara Jane Houtz, Annie M. Parrish, and F. A. Hellebrandt, "The Influence of Heavy Resistance Exercise on Strength," *Physiotherapy Review* 26, no. 6 (1946): 299–304.

DeLorme tips his cap to Milo in DeLorme and Watkins, *Progressive Resistance Exercise,* 19.

INTRODUCTION

On the proportion of muscle tissue in the human body, see Ian Janssen et al., "Skeletal Muscle Mass and Distribution in 468 Men and Women Aged 18–88 Yr," *Journal of Applied Physiology* 89, no. 1 (2000): 81–88, especially 83; and for the estimate of 40 percent, see Walter R. Frontera and Julien Ochala, "Skeletal Muscle: A Brief Review of Structure and Function," *Calcified Tissue International* 96, no. 3 (2015): 183–95, especially 183. The highest recorded measurement of whole-body muscle mass, at the time of this writing, was published in Enrique N. Moreno et al., "Skeletal Muscle Mass in Competitive Physique-Based Athletes (Bodybuilding, 212 Bodybuilding, Bikini, and Physique Divisions): A Case Series," *American Journal of Human Biology* 36, no. 1 (2023): e23978.

On the variety of roles that muscle plays in human life and health, see Robert R. Wolfe, "The Underappreciated Role of Muscle in Health and Disease," *American Journal of Clinical Nutrition* 84, no. 3 (2006): 475–82.

On myokines, see Bente Klarlund Pedersen, "Muscles and Their Myokines," *Journal of Experimental Biology* 214 (2011): 337–46.

Insightful overviews of research on progressive resistance training and health include Stuart M. Phillips, Jasmin K. Ma, and Eric S. Rawson, "The Coming of Age of Resistance Exercise as a Primary Form of Exercise for Health," *ACSM's Health & Fitness Journal* 27, no. 6 (2023): 19–25.

Jordan J. Smith et al., "The Health Benefits of Muscular Fitness for Children and Adolescents: A Systematic Review and Meta-Analysis," *Sports Medicine* 44, no. 9 (2014): 1209–23.

James Steele et al., "A Higher Effort-Based Paradigm in Physical Activity and Exercise for Public Health: Making the Case for a Greater Emphasis on Resistance Training," *BMC Public Health* 17, no. 1 (2017): 300.

Jonathan C. Mcleod, Tanner Stokes, and Stuart M. Phillips, "Resistance Exercise Training as a Primary Countermeasure to Age-Related Chronic Disease," *Frontiers in Physiology* 10, article 645 (2019).

On the CDC survey that revealed the growing popularity of weight training, see Matthew A. Ladwig et al., "When American Adults Do Move, How Do They Do So? Trends in Physical Activity Intensity, Type, and Modality: 1988–2017," *Journal of Physical Activity and Health* 18, no. 10 (2021): 1181–98.

On the majorities of national populations that do little to no strength training, see, for example, Jason A. Bennie et al., "Muscle-Strengthening Exercise Among 397,423 U.S. Adults: Prevalence, Correlates, and Associations with Health Conditions," *American Journal of Preventive Medicine* 55, no. 6 (2018): 864–74, which reports that 57.8 percent of Americans do no muscle-strengthening exercise. In Europe, only 17.3 percent of adults meet physical activity guidelines for muscle-strengthening exercise, although prevalence varies widely by country, with strength training being most prevalent in Iceland (51.5 percent) and least prevalent in Romania (0.7 percent), according to Jason A. Bennie et al., "The Epidemiology of Muscle-Strengthening Exercise in Europe: A 28-Country Comparison Including 280,605 Adults," *PLoS One* 15, no. 11 (2020): e0242220. Just 3.9 percent of Japanese adults meet physical activity recommendations for strength training, according to Yoshio Nakamura and Kazuhiro

Harada, "Promotion of Strength Training," in *Physical Activity, Exercise, Sedentary Behavior and Health*, ed. Kazuyuki Kanosue et al. (Tokyo: Springer Japan, 2015), 29–42. But physical activity surveys in Japan "did not give a specific definition for strength training," these authors write, and the same can be said of most physical activity surveys, which tend to define strength training very broadly, encompassing everything from sit-ups to deadlifts.

On "the importance of muscular contraction to us," see Charles Sherrington, *Man on His Nature* (Cambridge, UK: Cambridge University Press, 1940, reprinted 1946), 116.

For the finding that weight-lifting exercise increases the size of the brain's posterior cingulate cortex, see Kathryn M. Broadhouse, Maria Fiatarone Singh, et al., "Hippocampal Plasticity Underpins Long-Term Cognitive Gains from Exercise in MCI," *NeuroImage: Clinical* 25 (2020): 102182.

For the finding that aerobic exercise increases the size of the hippocampus, see Kirk I. Erickson et al., "Exercise Training Increases Size of Hippocampus and Improves Memory," *Proceedings of the National Academy of Sciences USA* 108, no. 7 (2011): 3017–22.

On the specific effects of aerobic and resistance exercise, see Angelique G. Brellenthin, Jason A. Bennie, and Duck-Chul Lee, "Aerobic or Muscle-Strengthening Physical Activity: Which Is Better for Health?" *Current Sports Medicine Reports* 21, no. 8 (2022): 272–79.

Regarding training the calf muscles, Maria Fiatarone Singh adds, "Body weight is not sufficient for long. Even if you use just one leg to lift your body," progression of resistance in the form of external weights fairly quickly becomes necessary.

CHAPTER 1

On the historical rarity of mass participation in sport, see Paul Christesen, *Sport and Democracy in the Ancient and Modern Worlds* (Cambridge, UK: Cambridge University Press, 2012).

The claim that ancient Greek athletes "knew about the muscle-building process" comes from Nigel Spivey, *Understanding Greek Sculpture: Ancient Meanings, Modern Readings* (London: Thames and Hudson, 1996), 39.

For Charles Stocking's powerlifting record, see https://www.openpowerlifting.org/u/charlesstocking.

On the glutes, see Françoise K. Jouffroy and Monique F. Médina, "A Hallmark of Humankind: The Gluteus Maximus Muscle: Its Form, Action, and Function," in *Human Origins and Environmental Backgrounds*, ed. Hidemi Ishida et al. (Boston: Springer, 2006), 135–48.

On gluteal amnesia, see Stuart McGill, *Ultimate Back Fitness and Performance*, 6th ed. (Canada: Backfitpro Incorporated, 2017), 108; and Stephanie Freeman, Anthony Mascia, and Stuart McGill, "Arthrogenic Neuromusculature Inhibition: A Foundational Investigation of Existence in the Hip Joint," *Clinical Biomechanics* 28, no. 2 (2013): 171–77.

On hip thrusts, see Bret Contreras and Glen Cordoza, *Glute Lab: The Art and Science of Strength and Physique Training* (Canada: Victory Belt Publishing, 2019), especially 66–67,

306–39. Contreras describes and demonstrates good form for the hip thrust exercise on this web page, which includes a video: https://bretcontreras.com/10-steps-to-the-perfect-hip-thrust/. For more on the exercise, see Walter Krause Neto, Thais Lima Vieira, and Eliane Florencio Gama, "Barbell Hip Thrust, Muscular Activation and Performance: A Systematic Review," *Journal of Sports Science and Medicine* 18 (2019): 198–206.

On strigils and gloios, see Stephen G. Miller, *Ancient Greek Athletics* (New Haven: Yale University Press, 2004), 15–16.

On the situation and dimensions of the site of ancient athletic competitions at Olympia, see Panos Valavanis, *Games and Sanctuaries in Ancient Greece: Olympia, Delphi, Isthmia, Nemea, Athens,* trans. David Hardy (Los Angeles: Getty Publications, 2004); and for a map of the archaeological site, see https://whc.unesco.org/en/documents/102335/.

On evidence of early athletic prizes at Olympia, including bronze tripods from the ninth century BC, see Zinon Papakonstantinou, "Prizes in Early Archaic Greek Sport," *Nikephoros* 15 (2002): 63–65.

On the relation between *agōn* and assembly, see Gregory Nagy, *Pindar's Homer: The Lyric Possession of an Epic Past* (Baltimore: Johns Hopkins University Press, 1990), especially 145 and 287. The full text of Nagy's book is accessible online at http://nrs.harvard.edu/urn-3:hul.ebook:CHS_Nagy.Pindars_Homer.1990.

On the date of early ritual competitions at Olympia, see Alfred Mallwitz, "Cult and Competition Locations at Olympia," in *The Archaeology of the Olympics,* ed. Wendy J. Raschke (Madison: University of Wisconsin Press, 1988), 79–118, especially 99.

The description of the earliest ritual stadion at Olympia draws from Walter Burkert, *Homo Necans: The Anthropology of Ancient Greek Sacrificial Ritual and Myth,* trans. Peter Bing (Berkeley: University of California Press, 1983), especially 93–103; and Nagy, *Pindar's Homer,* especially 123–25.

On the meaning of sacrifice in ancient Greece and specifically at Olympia, see Mark Golden, *Sport and Society in Ancient Greece* (Cambridge, UK: Cambridge University Press, 1998), 17–23; and for discussion of why ancient Greek competitive athletics took place only in a religious context, see Elizabeth Pemberton, "Agones Hieroi: Greek Athletic Contests in Their Religious Context," *Nikephoros* 13 (2000): 111–23.

Gregory Nagy is the scholar who describes ancient athletic competitions as "rituals of compensation for the catastrophe of death." The line comes from *Pindar's Homer,* 118.

On the consumption of meat in ancient Greece, see Burkert, *Homo Necans,* 101–102; also Miller, *Ancient Greek Athletics,* 213; and Jane M. Renfrew, "Food for Athletes and Gods: A Classical Diet," in *The Archaeology of the Olympics,* 176. When it did become common for athletes to eat meat during their training (probably by the end of the fourth century BC), they developed a reputation for carnivorous gluttony, on which see Golden, *Sport and Society in Ancient Greece,* 157–58; also Nigel Nicolson, "Representations of Sport in Greek Literature," in *A Companion to Sport and Spectacle in Greek and Roman Antiquity,* eds. Paul Christesen and Donald G. Kyle (Chichester, UK: Wiley-Blackwell, 2014), 74.

"The gods are there" is Stocking's translation of a line from Ulrich von Wilamowitz-Moellendorff, *Der Glaube der Hellenen* (*The Religious Beliefs of the Greeks*), first published in 1931–1932.

SOURCE NOTES 373

On the bodies of Greek gods, see Jean-Paul Vernant, "Dim Body, Dazzling Body," *Fragments for a History of the Human Body, Part One,* trans. Anne M. Wilson, ed. Michel Feher et al. (Princeton, NJ: Zone, 1989), 19–47.

Describing the atmosphere at ancient Olympia, Damian Stocking paraphrases the treatise "On Providence" by Epictetus, who wrote: "Some things are unseemly and difficult in life. Are there not difficulties at Olympia? Do you not suffer from the heat? Are you not closed in by the crowd? Do you not bathe badly? Are you not soaked when it rains? Do you not take delight in the noise and shouts and other difficulties? But I suppose these things are compensated for by the worthiness of the spectacle and you bear and endure it." The translation is from Charles H. Stocking and Susan A. Stephens, *Ancient Greek Athletics: Primary Sources in Translation* (Oxford: Oxford University Press, 2021), (36b), 171.

The teenaged Charles Stocking's punk band Vox Pop was unrelated to the earlier Los Angeles punk band of the same name, and he had not heard of the earlier band until I asked him about it.

On the labors of Hercules, see Emma Stafford, *Herakles* (London: Routledge, 2012).

On the reasons Hercules was said to have founded the Olympic Games, see Lysias, "*Olympic Oration 1–2: Purpose and Value of Olympic Games,*" in Stocking and Stephens, *Ancient Greek Athletics: Primary Sources* (16b), 108. This is one of several founding myths of the Olympic contests, on which see Golden, *Sport and Society in Ancient Greece,* 12–14.

On ancient Greek heroes, see Gregory Nagy, *The Ancient Greek Hero in 24 Hours,* abridged edition (Cambridge: Harvard University Press, 2019). The book's full text is accessible online at http://nrs.harvard.edu/urn-3:hul.ebook:CHS_NagyG.The_Ancient_Greek_Hero_in_24_Hours.2013.

For Nagy's argument that "the ordeals of heroes, as myths, are analogous to the ordeals of athletes, as rituals," and his observation that "When athletes win or lose in an athletic event, they 'live' or 'die' like heroes," see *Pindar's Homer,* 141.

For Jean-Pierre Vernant's description of "the dual nature of the games," see "From the 'Presentification' of the Invisible to the Imitation of Appearance," in *Mortals and Immortals: Collected Essays,* ed. Froma I. Zeitlin (Princeton, NJ: Princeton University Press, 1991), 160.

Schwarzenegger recommended that beginning lifters should do three sets of eight to ten repetitions of most exercises in Arnold Schwarzenegger and Douglas Kent Hall, *Arnold: The Education of a Bodybuilder* (New York: Simon & Schuster, 1977), 181.

On powerlifting competitions, see USA Powerlifting, "Technical Rules," Version 2021.1, "General Rules of Powerlifting," 4–6, https://www.usapowerlifting.com/wp-content/uploads/2021/04/USAPL-Rulebook-v2021.1.pdf.

On biomechanics of the squat, bench press, and deadlift, and on doing the exercises with good form, see Mark Rippetoe and Lon Kilgore, *Starting Strength: Basic Barbell Training,* 2nd ed. (Wichita Falls: Aasgaard Company, 2007).

For Schwarzenegger's description of the United States as "truly a country without limits," see "The America I Love Needs to Do Better," *The Atlantic,* May 31, 2020, https://www.theatlantic.com/ideas/archive/2020/05/arnold-schwarzenegger-america-needs-do-better/612442/.

For "Don't go there" and for Schwarzenegger's reason for training, see Ebenezer Samuel, "Arnold Schwarzenegger's Secret to Lasting Influence," *Men's Health,* July/August 2023, https://www.menshealth.com/fitness/a43894719/arnold-schwarzenegger-fubar-interview.

On the word *nēdus*, see Charles H. Stocking, *The Politics of Sacrifice in Early Greek Myth and Poetry* (Cambridge, UK: Cambridge University Press, 2017).

On *psychē* and *sōma*, see Charles H. Stocking, "Minds, Bodies and Identities," in *A Cultural History of Sport in Antiquity,* eds. Paul Christesen and Charles H. Stocking (London: Bloomsbury Academic, 2022), 159–77, especially 161; Charles H. Stocking, *Homer's* Iliad *and the Problem of Force* (Oxford: Oxford University Press, 2023); and David C. Young, "*Mens Sana in Corpore Sano*? Body and Mind in Ancient Greece," *The International Journal of the History of Sport* 22, no. 1 (2005): 22–41, especially 25.

On *sarx* and *mys*, see E. Bastholm, "The History of Muscle Physiology: From the Natural Philosophers to Albrecht von Haller; A Study of the History of Medicine," *Acta Historica Scientiarum Naturalium et Medicinalium* 7 (Copenhagen: Ejnar Munksgaard, 1950), 19–20.

The lines from Homer that mention muscles come from *The Iliad: A New Translation by Peter Green* (Berkeley: University of California Press, 2015), 302. (Book XVI, lines 314–316 and lines 323–325.)

On the depiction of dying in Homer as a "loosening" of limbs, see Damian Stocking, "*Res Agens*: Towards an Ontology of the Homeric Self," *College Literature* 34, no. 2 (2007): 56–84.

On the primacy of the joints in Homeric depictions of bodies, see Guillemette Bolens, "Homeric Joints and the Marrow in Plato's *Timaeus*: Two Logics of the Body," *Multilingua* 18, no. 2/3 (1999): 149–57.

On the importance of the knees in Homer, see R. B. Onians, *The Origins of European Thought: About the Body, the Mind, the Soul, the World, Time, and Fate* (Cambridge, UK: Cambridge University Press, 1951, reprinted 1991), especially 174–99. Important.

In the *Iliad*, it is typical for warriors to die after falling "with a thud," according to Alex Purves, "Falling into Time in Homer's *Iliad*," *Classical Antiquity* 25, no. 1 (2006): 179–209. This essay also discusses Nestor's experience of loosening of limbs in older age.

The "unstrung" motif, depicting the loss of tension, appears frequently in Richard Lattimore's translation of *The Iliad of Homer* (Chicago: University of Chicago Press, 1951).

On Homer's words for strength, see Stocking, *Homer's* Iliad *and the Problem of Force.* One of the most detailed treatments of Homer's words for strength prior to Stocking's is Bruno Snell, *The Discovery of the Mind: The Greek Origins of European Thought,* trans. T. G. Rosenmeyer (Cambridge: Harvard University Press, 1953).

On the Olympian temple's relief sculptures, called metopes, depicting Hercules performing his labors, see Judith M. Barringer, "The Temple of Zeus at Olympia, Heroes, and Athletes," *Hesperia* 74 (2005): 211–12; and her book *Olympia: A Cultural History* (Princeton, NJ: Princeton University Press, 2023), especially 128–37, and 197, where Barringer argues that sculptures such as these of Hercules "provided a model for athletes to emulate at Olympia."

For Stocking's formulation of the Homeric person as an "intersubjective dividual," see *Homer's* Iliad *and the Problem of Force*, 220.

CHAPTER 2

The story of Cleobis and Biton comes from Herodotus, *Histories*, 1.31. I quote from *The Landmark Herodotus: The Histories*, trans. Andrea L. Purvis, ed. Robert B. Strassler (New York: Pantheon Books, 2007), 20.

For Miller's observation about Cleobis and Biton, see Stephen G. Miller, *Ancient Greek Athletics* (New Haven: Yale University Press, 2004), 98.

On Boris Sheiko, see Greg Nuckols, "Interview with Coach Boris Sheiko," *Stronger by Science*, July 30, 2014, https://www.strongerbyscience.com/interview-with-coach-boris-sheiko/.

On strength as tension, see Pavel Tsatsouline, *Power to the People! Russian Strength Training Secrets for Every American* (St. Paul: Dragon Door Publications, 2000), especially 12–13, 24.

Charles Sherrington, *Man on His Nature* (Harmondsworth: Penguin Books, 1955, Pelican Books edition), especially 211.

C. S. Sherrington, "Note on the History of the Word 'Tonus' as a Physiological Term," in *Contributions to Medical and Biological Research Dedicated to Sir William Osler . . . In Honour of His Seventieth Birthday, June 12, 1919*, vol. 1 (New York: Paul B. Hoeber, 1919).

Galen's observation about tonic contraction, in which muscles are "active while appearing not to move in the least," comes from the translation of Charles Mayo Goss, "On Movement of Muscles by Galen of Pergamon," *American Journal of Anatomy* 123, no. 1 (1968): 9.

On high-tension contractions for building muscle strength, see Richard L. Lieber, *Skeletal Muscle Structure, Function, and Plasticity*, 3rd ed. (Philadelphia: Wolters Kluwer, 2010), 164.

On early Soviet research regarding periodized training, or cycling, see Tsatsouline, *Power to the People!*, especially 50–57.

On what's wrong with the myth of Milo, see Yuri Verkhoshansky and Mel Siff, *Supertraining*, 6th ed., expanded version (Rome: Verkhoshansky SSTM, 2009), 89–90.

On training as adaptation to stress, see Verkhoshansky and Siff, *Supertraining*, 82–84, 313–15.

Hans Selye, *The Stress of Life*, rev. ed. (New York: McGraw-Hill, 1976, paperback edition 1978), especially 36–40 (on the phases of the general adaptation syndrome), and 74 (for the definition of stress).

On supercompensation, see Verkhoshansky and Siff, *Supertraining*, 83–85.

On Stocking's training before setting the squat record, see Charles Stocking, "How to Be at Your Best When It Counts the Most: Peak Your Strength by 'Greasing the Groove,'" *Hard-Style: Hard Core Tools for Hard Living Types* 2, no. 1 (2005): 59–60.

On Dikon's strigil, see "Dedicated Strigil with Inscription, c. 500 BCE (*CEG* 387)," in Charles H. Stocking and Susan A. Stephens, *Ancient Greek Athletics: Primary Sources in Translation* (Oxford: Oxford University Press, 2021), (I1c), 309.

The essential review of evidence on lifting weights in ancient Greece is Nigel B. Crowther, "Weightlifting in Antiquity: Achievement and Training," *Greece & Rome* 24, no. 2 (1977): 111–20.

The article is reprinted, with some updates, in Nigel B. Crowther, *Athletika: Studies on the Olympic Games and Greek Athletics* (Hildesheim: Georg Olms Verlag, 2004), 269–80.

For inscriptions on lifting stones, see "Rock of Bybon, mid-sixth century (*IvO* 717)," in Stocking and Stephens, *Ancient Greek Athletics: Primary Sources* (I1a), 307.

E. Norman Gardiner, *Athletics in the Ancient World* (Oxford: Oxford University Press, 1955; republished Mineola: Dover, 2002), 54, translates the inscription on the stone of Eumastas.

The Guinness World Record deadlift at the time of this writing in 2023 comes from https://www.guinnessworldrecords.com/world-records/heaviest-deadlift.

Regarding the throwing-stone of Xenareus, see H. A. Harris, *Sport in Greece and Rome* (Ithaca, NY: Cornell University Press, 1972), 146–47.

In contrast to the scarcity of specific evidence of weight training and weight lifting in ancient Greece, there is more evidence of weight training and weight lifting in early centuries of the Common Era, in the Roman empire. Crowther's essay "Weightlifting in Antiquity" says, for instance, that in the second century AD, Epictetus wrote about systematic training with heavy weights; and in the fourth century AD, Jerome wrote about progressive resistance training, even describing some of the movements involved.

Regarding depictions of weight lifting in the art of antiquity, see Christine Sourvinou-Inwood, "Theseus Lifting the Rock and a Cup Near the Pithos Painter," *Journal of Hellenic Studies* 91 (1971): 96, for a list of "four representations of young workmen or athletes lifting round objects."

The best-known of these objects is probably Louvre G56, a wine cup fragment showing a boy attempting to move a large stone, discussed in E. Norman Gardiner, "Throwing the Diskos," *Journal of Hellenic Studies* 27 (1907): 2.

On Spartan women's education, including exercise and athletics, see Sarah B. Pomeroy, *Spartan Women* (Oxford: Oxford University Press, 2002), 3–32, especially 12.

Ellen Millender, "Athenian Ideology and the Empowered Spartan Woman" in *Sparta: New Perspectives,* eds. Stephen Hopkinson and Anton Powell (London: Duckworth, 1999), 355–91.

Paul Christesen, "Athletics and Social Order in Sparta in the Classical Period," *Classical Antiquity* 31, no. 2 (October 2012): 193–255.

The Spartan lawgiver Lycurgus decreed "that females exercise their bodies no less than males" according to Xenophon, "*Constitution of the Lacedaemonians* 1.1–4: Spartan Rearing of Girls," in Stocking and Stephens, *Ancient Greek Athletics: Primary Sources* (10c), 86.

The earliest surviving source that calls Spartan women "thigh-flashers" is a fragment by the poet Ibycus (fr. 339 *PMGF*).

The lines from Euripides's *Andromache* (595–601) are translated in Stocking and Stephens, *Ancient Greek Athletics: Primary Sources* (12b), 98.

The line "I jump up and kick myself in the butt" is spoken by the character Lampito in *Lysistrata,* a comedy by Aristophanes. The translation is Stephen Miller's, in *Ancient Greek Athletics,* 157.

On the bronze statuettes of girls and young women, see Andrew Stewart, *Art, Desire, and the Body in Ancient Greece* (Cambridge, UK: Cambridge University Press, 1997), 108–18; and Thomas F. Scanlon, *Eros and Greek Athletics* (Oxford: Oxford University Press, 2002), especially 101–105 and 121–38.

On the muscle-building effects of plyometric training, see Jozo Grgic et al., "Effects of Plyometric vs. Resistance Training on Skeletal Muscle Hypertrophy: A Review," *Journal of Sport and Health Science* 10, no. 5 (2021): 530–36.

"The wrestler should not break a finger" is from "Wrestling Regulation, 525–500 BCE (*SEG* 48.541)" in Stocking and Stephens, *Ancient Greek Athletics: Primary Sources* (I1b), 307.

On the myth of Milo's training: Many scholars have identified the source of the story as Quintilian 1.9.5. For Stocking's discussion of the source, see Charles H. Stocking, "The Use and Abuse of Training 'Science' in Philostratus' *Gymnasticus*," *Classical Antiquity* 35, no. 1 (2016): 90–91.

On the evolution of the Olympic program, see Donald G. Kyle, "Greek Athletic Competitions: The Ancient Olympics and More," in *A Companion to Sport and Spectacle in Greek and Roman Antiquity,* eds. Paul Christesen and Donald G. Kyle (Chichester, UK: Wiley-Blackwell, 2014), 22–35; and Miller, *Ancient Greek Athletics,* 31–84.

On wrestling in ancient Greece, see Michael B. Poliakoff, *Combat Sports in the Ancient World: Competition, Violence, and Culture* (New Haven: Yale University Press, 1987), 23–53.

On the special popularity of wrestling in ancient Greece, see John Zilcosky, "Wrestling, or the Art of Disentangling Bodies," in *The Allure of Sports in Western Culture,* eds. John Zilcosky and Marlo A. Burks (Toronto: University of Toronto Press, 2019), 79–118.

On age classes in athletic competition at Olympia, see Nigel B. Crowther, "The Age-Category of Boys at Olympia," *Phoenix* 42, no. 4 (1988): 304–308; and Kyle, "Greek Athletic Competitions," 25.

On the heavy events, see Mark Golden, *Sport and Society in Ancient Greece* (Cambridge, UK: Cambridge University Press, 1998), 37–38.

On the prestigious circuit of athletic contests and the development of many other local athletic contests and on the proliferating rewards and honors for victory, see Donald G. Kyle, *Sport and Spectacle in the Ancient World,* 2nd ed. (Chichester, UK: Wiley-Blackwell, 2015), 70–80.

On the sacred truce, see Golden, *Sport and Society in Ancient Greece,* 16–17.

On athletic participation and social privilege, see Paul Christesen, "The Transformation of Athletics in Sixth-Century Greece," in *Onward to the Olympics: Historical Perspectives on the Olympic Games,* eds. Gerald P. Schaus and Stephen R. Wenn (Waterloo, Canada: Wilfrid Laurier University Press, 2007), 59–68, especially 61.

On the Athenian meal rewards, called *sitēsis,* see "*Sitēsis* for Athenian Victors . . . ," in Stocking and Stephens, *Ancient Greek Athletics: Primary Sources* (14e), 329–30.

On Olympic victor statues, see R.R.R. Smith, "Pindar, Athletes, and the Early Greek Statue Habit," in *Pindar's Poetry, Patrons, and Festivals: From Archaic Greece to the Roman Empire*, eds. Simon Hornblower and Catherine Morgan (Oxford: Oxford University Press, 2007), 83–139.

On the earliest of these victor statues, see Pausanias, *Description of Greece* 6.18.7. The passage appears in "Bk. 6.1.1.—18.7 (with omissions): Statues, Shrines, and Stories About Victors," in Stocking and Stephens, *Ancient Greek Athletics: Primary Sources* (37q), 219.

On the size of victor statues, see Lucian, "*Portraiture Defended* 11: Victor Statues at Olympia," in Stocking and Stephens, *Ancient Greek Athletics: Primary Sources* (38d), 243–44; and "Cleombrotus Inscription, sixth century (*CEG* 394)," in Stocking and Stephens, *Ancient Greek Athletics: Primary Sources* (5.15a), 335–36.

On the story of Milo carrying his own statue, see Poliakoff, *Combat Sports in the Ancient World*, 9 and 118.

For Aristotle's definition of strength as the ability to move another body, see Aristotle, "*Rhetoric* bk. 1, 1361b: Bodily Excellence," in Stocking and Stephens, *Ancient Greek Athletics: Primary Sources* (23a), 129–30.

Plato compared his dialogues to the *agōn*, or athletic contest. In Plato's *Theaetetus*, for example, a student of Socrates calls the teacher out for being contentious, telling him, "You do not allow anyone to pass by until you force them to strip down and wrestle with your arguments." For this passage, see Plato, "*Theaetetus* 169a-169c, Training and Socrates' Agonistic Method," in Stocking and Stephens, *Ancient Greek Athletics: Primary Sources* (22c), 119. See also Charles Stocking, "Minds, Bodies, and Identities," in *A Cultural History of Sport in Antiquity*, eds. Paul Christesen and Charles Stocking (New York: Bloomsbury, 2022), 165.

On the Greek gymnasium, see Miller, *Ancient Greek Athletics*, 176–95; David Potter, *The Victor's Crown: A History of Ancient Sport from Homer to Byzantium* (London: Quercus, 2011), 109–36, especially 110–11; Michael Scott, "The Social Life of Greek Athletic Facilities (other than Stadia)," *A Companion to Sport and Spectacle in Greek and Roman Antiquity*, 297–301; and Werner Petermandl, "Growing Up with Greek Sport: Education and Athletics," *A Companion to Sport and Spectacle in Greek and Roman Antiquity*, 238–40.

For a classic study of athletic nudity in ancient Greek art, see Larissa Bonfante, "Nudity as a Costume in Classical Art," *American Journal of Archaeology* 93, no. 4 (1989): 543–70.

For evidence for various theories about athletic nudity, including its origins, see Nigel B. Crowther, "Nudity and Flogging at the Games," in *Athletika*, 135–40.

Stocking's observation about the possible democratizing effects of ancient Greek athletic nudity (from his essay "Minds, Bodies, Identities," 165) builds on Paul Christesen's research, especially concerning Sparta, for instance in *Sport and Democracy in the Ancient and Modern Worlds* (Cambridge, UK: Cambridge University Press, 2012), 164–83, and "Sport and Democratization in Ancient Greece," in *A Cultural History of Sport in Antiquity*, 226–28. Stephen G. Miller makes a similar argument about athletic nudity in Athens in "Naked Democracy," *Polis and Politics: Studies in Ancient Greek History Presented to Mogens Herman Hansen on His Sixtieth Birthday*, ed. Pernille Flensted-Jensen et al. (Copenhagen: Museum Tusculanum Press, 2000), 277–96. Antecedents to

Christesen's and Miller's arguments on nudity and democracy are discussed in Kyle, *Sport and Spectacle in the Ancient World*, 82–85.

On an exclusionary aspect of athletic nudity—based on "the ancient equivalent of a 'farmer's tan'"—see Christesen, *Sport and Democracy in the Ancient and Modern Worlds*, 177.

On class and athletic participation, see David Pritchard, "Athletics, Education and Participation in Classical Athens," *Sport and Festival in the Ancient Greek World*, eds. David J. Phillips and David Pritchard (Swansea: Classical Press of Wales, 2003), 293–344; Golden, *Sport and Society in Ancient Greece*, 141–69; and H. W. Pleket, "Games, Prizes, Athletes and Ideology: Some Aspects of the History of Sport in the Greco-Roman World," *Stadion. Zeitschrift Für Geschichte des Sports und der Korperkultur* 1, no. 1 (1975): 60–61 and 71–74.

On physical training for the elderly, see Nigel B. Crowther, "Old Age, Exercise, and Athletics in the Ancient World," in *Athletika*, 255–66, especially 257.

On Alcmaeon of Croton, see W.K.C. Guthrie, "Alcmaeon," in *A History of Greek Philosophy: Volume 1, The Earlier Presocratics and the Pythagoreans* (Cambridge, UK: Cambridge University Press, 1962), 341–59, especially 344–45; Robin Lane Fox, *The Invention of Medicine: From Homer to Hippocrates* (New York: Basic Books, 2020), 56–62, especially 60–61; and Gregory Vlastos, "Isonomia," *The American Journal of Philology* 74, no. 4 (1953): 337–66.

On Alcmaeon and the Hippocratic definition of health in terms of balance, see Jacques Jouanna, "Politics and Medicine: The Problem of Change in *Acute Diseases* and Thucydides (Book 6)," in *Greek Medicine from Hippocrates to Galen: Selected Papers*, trans. Neil Allies (Leiden: Brill, 2012), 23–24, especially n. 6.

On Aristotle's understanding of muscle and movement, see Aristotle, *Parts of Animals, Movement of Animals, Progression of Animals*, trans. A. L. Peck and E. S. Forster, Loeb Classical Library 323 (Cambridge: Harvard University Press, 1937), especially 463–65.

Klaus Corcilius and Pavel Gregorić, "Aristotle's Model of Animal Motion," *Phronesis* 58 (2013): 52–97.

Pavel Gregorić and Martin Kuhar, "Aristotle's Physiology of Animal Motion: On Neura and Muscles," *Apeiron* 47, no. 1 (2014): 94–115.

Pavel Gregorić, "The Origin and the Instrument of Animal Motion—*De Motu Animalium* Chapters 9 and 10," in Aristotle's *De motu animalium: Symposium Aristotelicum*, eds. Christof Rapp and Oliver Primavesi (Oxford, 2020; online edition, Oxford Academic), 416–44.

Eyvind Bastholm, *The History of Muscle Physiology: From the Natural Philosophers to Albrecht Von Haller* (Copenhagen: Ejnar Munksgaard, 1950), 41–52.

On the history of the concept of *pneuma*, see Bastholm, *The History of Muscle Physiology*, 27–29.

One of the most extensive treatments of Greek insensibility to muscle is Robin Osborne, *The History Written on the Classical Greek Body* (Cambridge, UK: Cambridge University Press, 2011), 27–54.

On the long absence of ancient literary mentions of muscle, see Tyson Sukava, "The Expanding Body: Anatomical Vocabulary and Its Dissemination in Classical Athens" (PhD dissertation, University of British Columbia, Vancouver, 2014), 111.

On the Greek fascination with jointedness, see Shigehisa Kuriyama, *The Expressiveness of the Body and the Divergence of Greek and Chinese Medicine* (New York: Zone Books, 1999), especially 129, 143–45.

On kouroi, and on the statues of Cleobis and Biton, see Brunilde Sismondo Ridgway, *The Archaic Style in Greek Sculpture* (Princeton, NJ: Princeton University Press, 1977), 45–83, especially 63–64, 70.

G.M.A. Richter, *Kouroi: Archaic Greek Youths: A Study of the Development of the Kouros Type in Greek Sculpture*, 3rd ed. (London: Phaidon, 1970), especially 32, 49–50.

Herodotus mentions the statues of Cleobis and Biton in *Histories* 1.31.5 (translation from *The Landmark Herodotus*, 20).

Aristotle writes that "all the other parts exist for the sake of the organ of touch," in *Parts of Animals*, II.7, 653b (Loeb edition, 159).

For Aristotle's understanding of the purposes of what we call the pectoral and gluteal muscles, see *Parts of Animals* IV.10, 687b and 689b (Loeb edition, 377 and 387).

Aristotle's views of muscle follow Plato's. In the *Timaeus* (74B–C, 296–97; quote and further details below), Plato described flesh as padding and insulation, and as a damper on perception and intelligence; and Plato, like Aristotle, saw no connection between flesh and movement.

Aristotle mentioned "inner parts" that "change from solid to supple and from supple to solid" in *Movement of Animals* 8, 702a7–10 (Loeb edition, 467). Corcilius and Gregorić discuss this passage in "Aristotle's Model of Animal Motion," 70. Their observations on Aristotle's understanding of movement and the soul come from the same essay, 86. Bastholm's remark on this topic is in *The History of Muscle Physiology*, 51.

For Aristotle's analogy of functional movement to functional leadership of a community, see *Movement of Animals*, x, 30 (Loeb edition, 475). On this passage, see Gregorić, "The Origin and Instrument of Animal Motion," 24–30.

On cultural prohibition of dissection in ancient Greece, see Heinrich Von Staden, "The Discovery of the Body: Human Dissection and Its Cultural Contexts in Ancient Greece," *Yale Journal of Biology and Medicine* 65 (1992): 225–31.

On the "momentous development" of anatomy in Alexandria, see Vivian Nutton, "Hellenistic and Roman Medicine," in *The Cambridge History of Science, Vol. I: Ancient Science*, eds. Alexander Jones and Liba Taub (Cambridge, UK: Cambridge University Press, 2018), 322–31, esp. 323–24.

On the view of motion that emerged from Alexandria, see Heinrich Von Staden, "Body, Soul, and Nerves: Epicurus, Herophilus, Erasistratus, and Galen," in *Psyche and Soma: Physicians and Metaphysicians on the Mind-Body Problem from Antiquity to Enlightenment*, eds. John P. Wright and Paul Potter (Oxford: Clarendon Press, 2000), 79–116, especially 95.

David Leith, "The Pneumatic Theories of Erasistratus and Asclepiades," in *The Concept of Pneuma After Aristotle*, ed. Sean Coughlin et al., *Berlin Studies of the Ancient World 61* (Berlin: Edition Topoi, 2020), 131–54.

Orly Lewis and David Leith, "Ideas of Pneuma in Early Hellenistic Medical Writers," in *The Concept of Pneuma After Aristotle*, 93–129.

For some of Galen's general statements about movement and muscle and for his explanation of tonic muscular action with reference to the bird in flight, see Goss, "On Movement of Muscles by Galen of Pergamon," especially 1–2 and 10.

For Kuriyama's comment about muscle with reference to Greek and Chinese cultures, see *The Expressiveness of the Body*, especially 111; and for his interpretation of Galen, see especially 144–45.

On how psychic pneuma was understood to be made, see Julia Trompeter, "The Actions of Spirit and Appetite," *Phronesis* 63, no. 2 (2018), especially 194; and Heinrich Von Staden, "Body, Soul, and Nerves: Epicurus, Herophilus, Erasistratus, the Stoics, and Galen," in *Psyche and Soma*, especially 108, 113–16.

On what Galen meant by muscle, see *Galen on the Usefulness of the Parts of the Body*, trans. Margaret Tallmadge May, 2 vols. (New York: Classics of Medicine Library, 1996), especially vol. 1, 61–62; and vol. 2, 552–54. Galen wrote that Plato, in the *Timaeus*, had best described "the uses of flesh in general." He quoted from that text: "Flesh is a protection from the burning heat of the sun and a defense against the winter's cold and also against falls, yielding softly and easily, like articles made of felt, to the objects it encounters. It has within it a warm moisture which in summer exudes and makes the surface damp, thus imparting a natural coolness to the whole body. And again in winter, by aid of this internal warmth, the flesh affords moderate protection from the cold which envelops and attacks it from without." (*Galen on the Usefulness of the Parts of the Body*, vol. I, 85; I.27). The passage quoted is from *Timaeus* 74 (in Plato, *The Dialogues* . . . trans. B. Jowett, 2 vols. [New York: Bigelow, Brown, 1920], vol. II, 52).

On the place of *Usefulness of the Parts* in Galen's work overall and the work's intended readership, see Susan P. Mattern, *The Prince of Medicine: Galen in the Roman Empire* (Oxford: Oxford University Press, 2013), 167–69.

On Galen's authority and influence, see Owsei Temkin, *Galenism: Rise and Decline of a Medical Philosophy* (Ithaca, NY: Cornell University Press, 1973).

For Stocking's account of the exchange between the athlete and the coach who said, "Your body belongs to the university," see "Ancient Greece, the Olympic Revival, and the Modern Student Athlete," in *Looking Towards the Future with Hope: 60th IOA Anniversary*, eds. D. Gangas and K. Georgiadis (International Olympic Academy, 2021), 96–107, especially 103–104.

On the frequency of ACL tears in women's college soccer, see Avinash Chandran et al., "Epidemiology of Injuries in National Collegiate Athletic Association Women's Soccer: 2014–2015 through 2018–2019," *Journal of Athletic Training* 56, no. 7 (2021): 651–58.

On ACL tears and missed practices and games in women's soccer, see Randall Dick et al., "Descriptive Epidemiology of Collegiate Women's Soccer Injuries: National Collegiate Athletic Association Injury Surveillance System, 1988–1989 through 2002–2003," *Journal of Athletic Training* 42, no. 2 (2007): 278–85, especially 281–82.

On the increased risk of recurrent injury following an ACL tear, see Allison M. Ezzat, "New or Recurrent Knee Injury, Physical Activity, and Osteoarthritis Beliefs in a Cohort of Female Athletes 2 to 3 Years After ACL Reconstruction and Matched Healthy Peers," *Sports Health* 14, no. 6 (2022): 842–48.

On the increased risk of osteoarthritis of the knee following an ACL tear, see Erik Poulsen et al., "Knee Osteoarthritis Risk Is Increased 4-6 fold After Knee Injury—A Systematic Review and Meta-Analysis," *British Journal of Sports Medicine* 53, no. 23 (2019): 1454–63.

Female athletes who suffer knee injuries such as ACL tears more than double their risk of needing knee replacement surgery within fifteen years, according to Ilana N. Ackerman et al., "Likelihood of Knee Replacement Surgery up to 15 Years after Sports Injury: A Population-Level Data Linkage Study," *Journal of Science and Medicine in Sport* 22 (2019): 629–34.

This review of research on training to reduce the risk of ACL injuries summarizes the type of work that Charles Stocking did with the UCLA women's soccer team: Yu Song et al., "Trunk Neuromuscular Function and Anterior Cruciate Ligament Injuries: A Narrative Review of Trunk Strength, Endurance, and Dynamic Control," *Strength and Conditioning Journal* 44, no. 6 (2022): 82–93.

On Stuart McGill, see his *Low Back Disorders: Evidence-Based Prevention and Rehabilitation*, 3rd ed. (Champaign, IL: Human Kinetics, 2016); and *Ultimate Back Fitness and Performance*, 6th ed., especially 49, 122, 142–43; and *Back Mechanic: The Secrets to a Healthy Spine Your Doctor Isn't Telling You: The Step by Step McGill Method to Fix Back Pain* (Gravenhurst, Canada: Backfitpro Incorporated, 2015), especially 22, 31, 95.

On the normal ratios of endurance times for trunk muscles, see Stuart M. McGill et al., "Endurance Times for Low Back Stabilization Exercises: Clinical Targets for Testing and Training from a Normal Database," *Archives of Physical Medicine and Rehabilitation* 80, no. 8 (1999): 941–44.

On extensor strength as a benchmark of lower back health, see Stuart McGill et al., "Previous History of LBP with Work Loss Is Related to Lingering Deficits in Biomechanical, Physiological, Personal, Psychosocial and Motor Control Characteristics," *Ergonomics* 46, no. 7 (2003): 731–46.

CHAPTER 3

On the stadium at Olympia, see Alfred Mallwitz, "Cult and Competition Locations at Olympia," in *The Archaeology of the Olympics*, ed. Wendy J. Raschke (Madison: University of Wisconsin Press, 1988), especially 93–101; and Panos Valavanis, *Games and Sanctuaries in Ancient Greece: Olympia, Delphi, Isthmia, Nemea, Athens*, trans. David Hardy (Los Angeles: Getty Publications, 2004), 100–103.

On Olympic stadium crowd capacity, see Thomas Heine Nielsen, "Hellenic Interaction at Olympia," in *Olympia and the Classical Hellenic City-State Culture*, Historisk-filosofiske Meddelelser 96 (Copenhagen: Royal Danish Academy of Sciences and Letters, 2007), 55–58.

On the priestess of Demeter, see Ulrich Sinn, *Olympia: Cult, Sport, and Ancient Festival* (Princeton, NJ: Markus Wiener, 1996), 74.

On the topic of women's presence or absence at Olympia, evidence is late, and it is ambiguous. For a good brief review of the evidence, see Donald G. Kyle, "Greek Female Sport," in *A Companion to Sport and Spectacle in Greek and Roman Antiquity*, eds. Paul Christesen and Donald G. Kyle (Chichester, UK: Wiley-Blackwell, 2014), 266–67.

Mark Golden, *Sport and Society in Ancient Greece* (Cambridge, UK: Cambridge University Press, 1998), 132–33, makes a slightly different reading of the evidence.

On Pausanias and his place in Roman culture during this period, which scholars call the Second Sophistic, see Jason König, *Athletics and Literature in the Roman Empire* (Cambridge, UK: Cambridge University Press, 2005), especially 158–86.

Zahra Newby, *Greek Athletics in the Roman World: Victory and Virtue* (Oxford: Oxford University Press, 2005), especially 202–28.

On the status of Greek culture in the Roman empire, see Tim Whitmarsh, *The Second Sophistic* (Oxford: Oxford University Press, 2005), especially 1–22.

On competition between trainers and doctors in Roman gymnasium education, see König, *Athletics and Literature in the Roman Empire*, 261, showing how "athletic training acted as a battleground for opposing disciplines," and "for opposing views of how contemporary life should be lived in relation to earlier Greek traditions."

On Hysmon, see Pausanias "Bk. 6.1.1–18.7 (with omissions): Statues, Shrines, and Stories About Victors" (6.3.9–10), in Charles H. Stocking and Susan A. Stephens, *Ancient Greek Athletics: Primary Sources in Translation* (Oxford: Oxford University Press, 2021), (37q), 198.

On the Panathenaic amphora that Stocking mentions (Accession number 14.13.12 in the collection of the Metropolitan Museum of Art in New York City)—the amphora pictured in this chapter—see Mary B. Moore, "'Nikias Made Me': An Early Panathenaic Prize Amphora in the Metropolitan Museum of Art," *Metropolitan Museum Journal* 34 (1999): 37–56.

For the value of Panathenaic prize amphoras and their contents, see the formula in David C. Young, *The Olympic Myth of Greek Amateur Athletics* (Chicago: Ares, 1984), 115–33.

A succinct updated estimate of value (based on 2001 currency) is in David C. Young, *A Brief History of the Olympic Games* (Oxford: Blackwell, 2004), 97–101, and 172 (n. 3).

Young's estimates were based on evidence that men's stadion victors won 100 amphoras, but Julia L. Shear, "The Prizes of Athens: The List of Panathenaic Victors and the Sacred Oil," *Zeitschrift für Papyrologie und Epigraphik* 142 (2003): 87–108, especially 95, 105, presents evidence that men's stadion victors won 80 amphoras, not 100.

For the value estimate of approximately $185,000, I began with Young's stadion victory prize value of $135,000 for 100 amphoras of olive oil, then reduced it by 20 percent (to account for Shear's new evidence that the prize was 80 amphoras), yielding $108,480. In those computations, I followed Donald G. Kyle, *Sport and Spectacle in the Ancient World*, 2nd ed. (Chichester, UK: Wiley-Blackwell, 2015), 151 and 170 (n. 3). Then I used the U.S. Bureau of Labor Statistics CPI Inflation Calculator to convert January 2001 dollars to January 2023 dollars.

On sprinting technique, see Daniel E. Lieberman et al., "Foot Strike Patterns and Collision Forces in Habitually Barefoot Versus Shod Runners," *Nature* 463, no. 7280 (2010): 531.

On the problem of sources in the study of ancient Greek athletics, see Golden, *Sport and Society in Ancient Greece*, 46–56.

On the ancient crib sheet of wrestling moves, see Michael B. Poliakoff, *Combat Sports in the Ancient World: Competition, Violence, and Culture* (New Haven: Yale University Press, 1987), 51–53.

On ancient training manuals, see König, *Athletics and Literature in the Roman Empire*, 314.

Jason König, "Introduction" to *Gymnasticus*, in Philostratus, *Heroicus, Gymnasticus, Discourses 1 and 2*, edited and translated by Jeffrey Rusten and Jason König, Loeb Classical Library (Cambridge: Harvard University Press, 2014), especially 372.

On the *Gymnasticus* and its manuscript tradition and translations, see König, "Introduction" to *Gymnasticus*, especially 382–85.

Translations of the *Gymnasticus* quoted in this chapter are Stocking's, published in Philostratus, "Gymnasticus," in Stocking and Stephens, *Ancient Greek Athletics: Primary Sources* (40a), 271–92, with two exceptions.

The lines about a boxer's calves and belly, from *Gymnasticus* 34, come from König's translation, published in the Loeb edition.

The line about *kairos*, from *Gymnasticus* 54, comes from a translation Stocking made for his essay "The Use and Abuse of Training 'Science' in Philostratus' *Gymnasticus*," *Classical Antiquity* 35, no. 1 (2016): 86–165, 101. (He made a slightly different translation of the same line in *Ancient Greek Athletics: Primary Sources*.)

After Stocking's lecture at Olympia, he developed some of its ideas in his essay "Minds, Bodies, and Identities," in *A Cultural History of Sport in Antiquity*, eds. Paul Christesen and Charles Stocking (New York: Bloomsbury, 2022). His essay contains much more detail than my summary of his lecture, with many references. To supplement my summary of Stocking's lecture and essay in this chapter, I've added some context concerning a few topics.

The headline Stocking read at the start of his lecture in Olympia comes from Ben Martynoga, "How Physical Exercise Makes Your Brain Work Better," *The Guardian*, June 18, 2016, https://www.theguardian.com/education/2016/jun/18/how-physical-exercise-makes-your-brain-work-better.

On Hippocrates and the Hippocratic corpus, see Robin Lane Fox, *The Invention of Medicine: From Homer to Hippocrates* (New York: Basic Books, 2020), especially 72–78, 205–208.

Stocking's translation of Seneca, *Letters* 15.2, quoted in this chapter, is published in part in "Minds, Bodies, and Identities," 168. Stocking read the rest of it in his talk at Olympia in 2019.

On the number of Galen's extant and lost works, see Vivian Nutton, "Hellenistic and Roman Medicine," in *The Cambridge History of Science, Vol. I: Ancient Science*, eds. Alexander Jones and Liba Taub (Cambridge, UK: Cambridge University Press, 2018), 341.

On the roots and legacy of Galen's belief that health was balance, see Jacques Jouanna, "The Legacy of the Hippocratic Treatise The Nature of Man: The Theory of the Four Humours," in *Greek Medicine from Hippocrates to Galen*, ed. Philip van der Eijk (Leiden: Brill, 2012), especially 335, 359.

On doctors' and trainers' competing claims to own the field of health, see Stocking, "Minds, Bodies, and Identities," 169.

For Galen's thoughts about the faults of trainers, especially their promotion of building body mass, and for his thoughts about athletes with big, fleshy bodies, see König, *Athletics and Literature in the Roman Empire*, 274–300, especially 282–83 and 295–96.

Galen's remarks on building mass quoted in this chapter come from *Thrasybulus, or Whether Health Is Part of Athletics*, in Stocking and Stephens, *Ancient Greek Athletics: Primary Sources* (39b), 254–66, especially 260, and from *Protrepticus, or Exhortation to Acquire Expertise*, Stocking and Stephens, *Ancient Greek Athletics: Primary Sources* (39a), 246–54, especially 250.

On Galen's "famously ambiguous" writings about the soul, see Stocking, "Minds, Bodies, and Identities," 169.

On Galen's views of mind or soul and body, see also Vivian Nutton, *Galen: A Thinking Doctor in Imperial Rome* (London: Routledge, 2020), 86–90.

On movement and the soul's immortality, see R. J. Hankinson, "Greek Medical Models of Mind," in *Psychology (Companions to Ancient Thought: 2)*, ed. Stephen Everson (Cambridge, UK: Cambridge University Press, 1991), 195–96.

On how "Exercise with a Small Ball" suggests some contours of Galen's views of physical activity, see Stocking, "Minds, Bodies, Identities," 170; for a translation of the text, see Galen, "Exercise with a Small Ball," in Stocking and Stephens, *Ancient Greek Athletics: Primary Sources* (39c), 267–71.

On Galen's work with gladiators, see Susan P. Mattern, *The Prince of Medicine: Galen in the Roman Empire* (Oxford: Oxford University Press, 2013), 81–97.

John Scarborough, "Galen and the Gladiators," *Episteme* 5, no. 2 (1971): 98–111.

On "barley boys" as a name for gladiators, see Gregory S. Aldrete, "Material Evidence for Roman Spectacle and Sport" in *A Companion to Sport and Spectacle*, 448. The term was used by Pliny the Elder in *The Natural History*, Book XVIII, 14.

Archaeological findings in Ephesus in Turkey, in a graveyard for gladiators, provide material evidence that supports ancient accounts of the gladiators' diet. See Sandra Lösch et al., "Stable Isotope and Trace Element Studies on Gladiators and Contemporary Romans from Ephesus (Turkey, 2nd and 3rd Ct. AD)—Implications for Differences in Diet," *PLoS One* 9, no. 10 (2014): e110489.

On Galen's wrestling and his wrestling injury, see Mattern, *The Prince of Medicine*, 179–82.

Susan P. Mattern, "Galen's Ideal Patient," in *Asklepios: Studies on Ancient Medicine*, ed. Louise Cilliers (Bloemfontein: Classical Association of South Africa, 2008), 116–30.

On the end of ancient Greek athletics, see Sofie Remijsen, *The End of Greek Athletics in Late Antiquity* (Cambridge, UK: Cambridge University Press, 2015), especially 47, 173, 167.

Thomas F. Scanlon, *Eros and Greek Athletics* (Oxford: Oxford University Press, 2002), 59.

David Potter, *The Victor's Crown: A History of Ancient Sport from Homer to Byzantium* (London: Quercus, 2011), 308–20, performs a rigorous, brief dissection of the myth that Christianity caused the end of ancient Greek athletics.

On the fate of Galen's writings and his reputation, see Owsei Temkin, *Galenism: Rise and Decline of a Medical Philosophy* (Ithaca, NY: Cornell University Press, 1973).

For general information on Flavius Philostratus, see Ewen Bowie, "Philostratus: The Life of a Sophist" in *Philostratus*, eds. Ewen Bowie and Jaś Elner (Cambridge, UK: Cambridge University Press, 2009).

Graeme Miles, "Philostratus," in *The Oxford Handbook to the Second Sophistic*, eds. Daniel S. Richter and William A. Johnson (Oxford: Oxford University Press, 2017).

Sources informing this chapter's discussion of the *Gymnasticus* include Charles Heiko Stocking, "Greek Ideal as Hyperreal: Greco-Roman Sculpture and the Athletic Male Body," *Arion: A Journal of Humanities and the Classics* 21, no. 3 (2014): 45–74.

Charles Stocking, "The Allure and Ethics of Ancient Aesthetics: Hellenism in the Modern Olympic Movement," in *The Allure of Sports in Western Culture*, eds. John Zilcosky and Marlo A. Burks (Toronto: University of Toronto Press, 2019), 130–53.

Stocking, "The Use and Abuse of Training 'Science' in Philostratus' *Gymnasticus*."

Jason König, "Training Athletes and Interpreting the Past in Philostratus' *Gymnasticus*," in *Philostratus*, eds. Ewan Bowie and Jaś Elsner (Cambridge, UK: Cambridge University Press, 2009), 251–83.

For Aristotle's views of the predominant purposes of beauty in various stages of life, see his "*Rhetoric* bk. 1, 1361b: Bodily Excellence," in Stocking and Stephens, *Ancient Greek Athletics: Primary Sources* (23a), 129–30.

The most famous Weary Hercules statue is also called the Farnese Hercules, after Alessandro Farnese (Pope Paul III), the first custodian of the statue after it was excavated in the sixteenth century. On the original situation of this statue in the baths of Caracalla, and on the cultural significance of baths in Rome, see Zahra Newby, *Greek Athletics in the Roman World: Victory and Virtue* (Oxford: Oxford University Press, 2005), 67–76.

Miranda Marvin, "Freestanding Sculptures from the Baths of Caracalla," *American Journal of Archaeology* 87, no. 3 (1983): 347–84, especially 347–48 and 355.

Janet DeLaine, *The Baths of Caracalla: A Study in the Design, Construction, and Economics of Large-Scale Building Projects in Imperial Rome* (Journal of Roman Archaeology Supplementary Series 25) (Portsmouth, RI: Journal of Roman Archaeology, 1997), especially 80–81.

On the late-fourth-century BC origin of this style of statue of Hercules, see Emma Stafford, *Herakles* (London: Routledge, 2012), 129.

For more on the earlier Greek models of the influential later Roman copies, such as the Farnese Hercules, see Cornelius Vermeule, "The Weary Herakles of Lysippos," *American Journal of Archaeology* 79, no. 4 (1975): 323–32.

For a brief but richly detailed account of the popularity of the Farnese Hercules, see Francis Haskell and Nicholas Penny, *Taste and the Antique: The Lure of Classical Sculpture 1500–1900* (New Haven: Yale University Press, 1981), 229–32, especially 229 (for Napoleon's desire to have the sculpture for the Louvre).

On Eugen Sandow, see David L. Chapman, *Sandow the Magnificent: Eugen Sandow and the Beginnings of Bodybuilding* (Urbana: University of Illinois Press, 1994), especially xi–xii.

For Joe Weider's remark about the Weary Hercules statue, and for an overview of the statue's history and significance, see Jan Todd, "The History of Cardinal Farnese's 'Weary Hercules,'" *Iron Game History* 9, no. 1 (2005): 29–34, especially 29.

On *kairos*, see Catherine Eskin, "Hippocrates, Kairos, and Writing the Sciences," in *Rhetoric and Kairos: Essays in History, Theory, and Praxis*, eds. Phillip Sipiora and James S. Baumlin (Albany: State University of New York Press, 2002), 97–113.

Debra Hawhee, *Bodily Arts: Rhetoric and Athletics in Ancient Greece* (Austin: University of Texas Press, 2004).

CHAPTER 4

On the early years of Jan and Terry Todd's life together, and on Jan's early training, see Jan Todd and Terry Todd, *Lift Your Way to Youthful Fitness: The Comprehensive Guide to Weight Training* (Boston: Little, Brown, 1985); and Terry Todd, *Inside Powerlifting* (Chicago: Contemporary Books, 1978). On Jan Todd's childhood, see also Jan Suffolk Todd, "Father to a Strongwoman," *Aethlon* 19, no. 2 (2002): 15–23. On the couple's life in Nova Scotia, see Sarah Pileggi, "The Pleasure of Being the World's Strongest Woman," *Sports Illustrated* 47, no. 20 (November 14, 1977): 60–71.

For sociological research on women and weight training, see Rachel I. Roth and Bobbi A. Knapp, "Gender Negotiations of Female Collegiate Athletes in the Strength and Conditioning Environment," *Women in Sport and Physical Activity Journal* 25, no. 1 (2017): 50–59, especially 55; Shari L. Dworkin, "A Woman's Place Is in the . . . Cardiovascular Room?? Gender Relations, the Body, and the Gym," in *Athletic Intruders: Ethnographic Research on Women, Culture, and Exercise*, eds. Anne Bolin and Jane E. Granskog (Albany: State University of New York Press, 2003): 131–58. (Though published in 2003, this work refers to fieldwork conducted in 1995–1996.) Shari L. Dworkin, "'Holding Back': Negotiating a Glass Ceiling on Women's Muscular Strength," *Sociological Perspectives* 44, no. 3 (2001): 333–50; and Chris Shilling and Tanya Bunsell, "The Female Bodybuilder as a Gender Outlaw," *Qualitative Research in Sport and Exercise* 1, no. 2 (2009): 141–59.

On Katie Sandwina, see Jan Todd, "Center Ring: Katie Sandwina and the Construction of Celebrity," *Iron Game History* 10, no. 1 (2007): 4–13; and John Fair, "Kati Sandwina: 'HERCULES CAN BE A LADY,'" *Iron Game History* 9, no. 2 (2005): 4–7.

On Jan Todd's connection with Sandwina and on the uncertain origins of Sandwina's name, see Jan Todd, "An Unexpected Image . . . Sandwina at 17," on the H.J. Lutcher Stark Center blog, August 5, 2020, https://starkcenter.org/2020/08/its-a-good-life-2/.

The ROGUE Legends documentary series episode 4, "Sandwina," directed by Todd Sansom, features commentary by Jan and Terry Todd, who were also among the film's producers. It may be viewed at https://www.roguefitness.com/theindex/documentaries/sandwina.

The first physiological study of women's strength training mentioned in this chapter is Jack H. Wilmore, "Alterations in Strength, Body Composition and Anthropometric Measurements Consequent to a 10-Week Weight Training Program," *Medicine & Science in Sports & Exercise* 6, no. 2 (1974): 133–38. The study compared untrained women and untrained men—two groups of people who had not previously lifted weights.

The second publication mentioned here, about the smaller differences between trained men and women, is Jack H. Wilmore, "The Application of Science to Sport: Physiological Profiles of Male and Female Athletes," *Canadian Journal of Applied Sport Sciences* 4, no. 2 (1979): 103–15, especially 103.

The magazine article about Wilmore's experience teaching weight training for college women is "The 'Weaker Sex' Comes On Strong," *Life* 72, no. 19 (May 19, 1972): 67–70.

On the etymology, meaning, and usage of the word *muscle*, I consulted the *Oxford English Dictionary* online, *The Chambers Dictionary*, Samuel Johnson's *Dictionary of the English Language* (1755), the database Literature Online, and James P. Mallory and Douglas Q. Adams, *The Oxford Introduction to Proto-Indo-European and the Proto-Indo-European World* (Oxford: Oxford University Press, 2006), especially 193–97, 66. The quoted definition of muscle comes from the OED.

On short- and long-term effects of sedentary behavior in youth, see Gregory D. Myer et al., "Exercise Deficit Disorder in Youth: A Paradigm Shift Toward Disease Prevention and Comprehensive Care," *Current Sports Medicine Reports* 12, no. 4 (2013): 248–55, especially 251.

On locomotor skills versus object control proficiency, see Lisa M. Barnett et al., "Childhood Motor Skill Proficiency as a Predictor of Adolescent Physical Activity," *Journal of Adolescent Health* 44, no. 3 (2009): 252–59, especially 257.

On the decline in physical activity between childhood and adolescence, especially between the ages of thirteen and eighteen, see James F. Sallis, "Age-Related Decline in Physical Activity: A Synthesis of Human and Animal Studies," *Medicine & Science in Sports & Exercise* 32, no. 9 (2000): 1598–600.

On physical activity measured by accelerometers, see Esther M. F. van Sluijs et al., "Physical Activity Behaviours in Adolescence: Current Evidence and Opportunities for Intervention," *The Lancet* 398, no. 10298 (2021): 429–42.

For more on Jan Todd's unusual handgrip strength, see Terry Todd, "Mac and Jan," *Iron Game History* 3, no. 6 (1995): 17–19.

On self-efficacy, see Albert Bandura, "Self-Efficacy: Toward a Unifying Theory of Behavioral Change," *Psychological Review* 84, no. 2 (1977): 191–215. On self-efficacy as "both a cause and effect of performance," see Sandra E. Moritz et al., "The Relation of Self-Efficacy Measures to Sport Performance: A Meta-Analytic Review," *Research Quarterly for Exercise and Sport* 71, no. 3 (2000): 280–94, especially 289.

The first study of heavy weight training and self-efficacy in young women was Jean Barrett Holloway, Anne Beuter, and Joan L. Duda, "Self-Efficacy and Training for Strength in Adolescent Girls 1," *Journal of Applied Social Psychology* 18, no. 8, pt. 2 (1988): 699–719. Many other studies have explored connections between self-efficacy and weight training, such as Todd A. Gilson et al., "Self-Efficacy and Athletic Squat Performance: Positive or Negative Influences at the Within- and Between-Levels of Analysis," *Journal of Applied Social Psychology* 42, no. 6 (2012): 1467–85.

On efficacy in sport and its possible influence on political efficacy, see Paul Christesen, *Sport and Democracy in the Ancient and Modern Worlds* (Cambridge, UK: Cambridge University Press, 2012), 85–91.

Other studies measuring self-efficacy and physical activity mentioned in this chapter are Rod K. Dishman et al., "Self-Efficacy Partially Mediates the Effect of a School-Based Physical-Activity Intervention Among Adolescent Girls," *Preventive Medicine* 38, no. 5 (2004): 628–36; Rod K. Dishman et al., "Responses to Preferred Intensities of Exertion in Men Differing in Activity Levels," *Medicine & Science in Sports & Exercise* 26, no. 6 (1994): 783–90; and Nalin A. Singh et al., "Effects of High-Intensity Progressive Resistance Training and Targeted Multidisciplinary Treatment of Frailty on

Mortality and Nursing Home Admissions After Hip Fracture: A Randomized Controlled Trial," *Journal of the American Medical Directors Association* 13, no. 1 (2012): 24–30.

On Jan Todd's first Guinness record and her goal of lifting a total of 1,000 pounds, see Danny Thom, "394 Pounds: Woman Sets Weightlifting Record," *The Macon News* (Georgia), May 6, 1975, 1A.

On the neuromuscular phenomenon of strength and on the relation between muscle size and strength, see Roger M. Enoka, *Neuromechanics of Human Movement*, 5th ed. (Champaign, IL: Human Kinetics, 2015), especially 397–405.

On the motor unit, see also this classic paper: C. J. Heckman and Roger M. Enoka, "Motor Unit," *Comprehensive Physiology* 2, no. 4 (2012): 2629–82.

The large study of how resistance training affected muscular size and strength in the arms of men and women is Monica J. Hubal et al., "Variability in Muscle Size and Strength Gain After Unilateral Resistance Training," *Medicine & Science in Sports & Exercise* 37, no. 6 (2005): 964–72. On subsequent research supporting the findings of Hubal and colleagues, see Brandon M. Roberts, Greg Nuckols, and James W. Krieger, "Sex Differences in Resistance Training: A Systematic Review and Meta-Analysis," *Journal of Strength and Conditioning Research* 34, no. 5 (2020): 1448–60, especially 1456.

On attentional focus, see Brad Schoenfeld, *Science and Development of Muscle Hypertrophy*, 2nd ed. (Champaign, IL: Human Kinetics, 2016), 118–19; and Gabriele Wulf, "Attentional Focus and Motor Learning: A Review of 15 Years," *International Review of Sport and Exercise Psychology* 6, no. 1 (2013): 77–104.

On the findings from twin studies regarding heredity and baseline muscle mass, and on the declining influence of genes on muscle-building potential in older age, see Schoenfeld, *Science and Development of Muscle Hypertrophy*, 105.

On some challenges that many female athletes experience in connection with training to build muscle, see Vikki Krane et al., "Living the Paradox: Female Athletes Negotiate Femininity and Muscularity," *Sex Roles* 50, nos. 5–6 (2004): 315–29; and Jason Shurley et al., "Historical and Social Considerations of Strength Training for Female Athletes," *Strength and Conditioning Journal* 42, no. 4 (2020): 22–35.

On the genesis and publication of *Pumping Iron*, see "Paperback Talk," *The New York Times*, April 20, 1975, 135; and Ray Walters, "Paperback Talk," *The New York Times*, February 14, 1982, 114–15.

On stereotypes of bodybuilders in the 1970s, see Charles Gaines, *Pumping Iron: The Art and Science of Bodybuilding* (New York: Simon & Schuster, 1974), 8. An early review of *Pumping Iron* that reflects some of those stereotypes is Albert Nussbaum, "Nussbaum on Books," *Lubbock Avalanche-Journal* (Lubbock, Texas), January 12, 1975, 63.

For the comment about Schwarzenegger by Jacqueline Kennedy Onassis, see Joyce Maynard, "Carter's Sister and Mr. Universe Make the Chic Scene in New York," *The New York Times*, January 13, 1977, 38.

On the first powerlifting meet for women, see "Powerlift Meet Set for Women," *Nashua Telegraph*, April 11, 1977, 25.

On Jan Todd's record-setting powerlifts in Newfoundland, see "Jan Todd Sets Record for Lift at Her Weight," *The Globe and Mail*, December 13, 1977, 36. (The article, though published in December, reports results of a powerlifting meet in June of that year.)

Terry Todd's thoughts about the social significance of Jan Todd's lifting come from *Inside Powerlifting*, 101.

CHAPTER 5

This chapter draws on press clippings from Todd's early years as a powerlifter, including Sarah Pileggi's profile of Jan Todd: "The Pleasure of Being the World's Strongest Woman," *Sports Illustrated* 47, no. 20 (November 14, 1977): 60–71; "Pumping Iron with the Strongest Woman in the World," *The Bulletin* (Bridgewater, Nova Scotia), November 30, 1977; Glenn Leiper, "Jan Todd Sets World Weightlifting Record," *Chronicle-Herald* (Halifax, Nova Scotia), December 13, 1977, 9; and Nora McCabe, "Nova Scotia's Pretty Powerhouse," *Globe and Mail* (Toronto), December 3, 1977, 57–58.

Jan Todd appeared on *The Tonight Show* with Johnny Carson on February 2, 1978. The episode can be seen online at https://www.youtube.com/watch?v=d5EEoc3ZXR4. Her appearance begins around the 52:00 mark of the video.

On anabolic steroids, this chapter's main references are the most recent position statements on the topic issued by the American College of Sports Medicine and the National Strength and Conditioning Association: Shalender Bhasin et al., "Anabolic-Androgenic Steroid Use in Sports, Health, and Society," *Medicine & Science in Sports & Exercise* 53, no. 8 (2021): 1778–94; and Jay R. Hoffman et al., "Position Stand on Androgen and Human Growth Hormone Use," *Journal of Strength and Conditioning Research* 23, no. 5 suppl. (2009): S1–S59.

For the definition of a hormone, see Hans Selye, *The Stress of Life*, rev. ed. (New York: McGraw-Hill, 1976, paperback edition 1978), 21.

On the history of testosterone, see Erica R. Freeman, David A. Bloom, and Edward J. McGuire, "A Brief History of Testosterone," *Journal of Urology* 165, no. 2 (2001): 371–73; and John M. Hoberman and Charles E. Yesalis, "The History of Synthetic Testosterone," *Scientific American* 272, no. 2 (1995): 76–81.

John Hoberman, *Testosterone Dreams: Rejuvenation, Aphrodisia, Doping* (Berkeley: University of California Press, 2005).

Charles E. Yesalis, ed., *Anabolic Steroids in Sport and Exercise*, 2nd ed. (Champaign, IL: Human Kinetics, 2000).

Many of this chapter's details about the use of anabolic drugs in sport come from these three articles:

Terry Todd, "Anabolic Steroids: The Gremlins of Sport," *Journal of Sport History* 14, no. 1 (1987): 87–107, at 93–94.

Jan Todd and Terry Todd, "Significant Events in the History of Drug Testing and the Olympic Movement: 1960–1999," in *Doping in Elite Sport: The Politics of Drugs in the Olympic Movement*, eds. Wayne Wilson and Edward Derse (Champaign, IL: Human Kinetics, 2001), 65–128.

Terry Todd, "The Steroid Predicament," *Sports Illustrated* 59, no. 5 (1983): 62–78.

For Tom Waddell's observation about drug use among Olympic athletes training for the 1968 games, see Jack Scott, "It's Not How You Play the Game, But What Pill You Take," *The New York Times Magazine*, October 17, 1971, 40.

On early uses of anabolics, see also "Cancer Pain Relief Claimed," *New York Herald Tribune* [European Edition], October 21, 1939, p. [1].

Paul de Kruif, *The Male Hormone* (New York: Harcourt, Brace, 1945), especially 226.

The anecdote about Bill Pearl comes from John D. Fair, *Mr. America: The Tragic History of a Bodybuilding Icon* (Austin: University of Texas Press, 2015), 152–53, 396 (n. 73).

Among bodybuilders who won the Mr. America contest in the pre-steroid era, the average age at time of death was about seventy-six; and average age at time of death for winners of the contest in the steroid era was fifty-three—a difference of about twenty-three years, according to Fair, *Mr. America*, 354–55, and 422, n. 22. Fair's numbers are based largely on statistical compilations by Joe Roark at IronHistory.com.

On the primacy of muscle size in bodybuilding, see Fair, *Mr. America*, especially 229–30, 302, 324, 335. According to Fair, "the person most responsible for the emphasis on mass and muscularity, with a legacy going back to the 1950s, was Joe Weider." Fair cites *Muscle and Fitness* (December 1997, 96) for Weider's statement that "mass should take precedence over all else" (*Mr. America*, 302, 415, n. 103).

For a nuanced overview of evidence regarding side effects of anabolic/androgenic drug use, see Bhasin, "Anabolic-Androgenic Steroid Use in Sports, Health, and Society," 1786–88.

The story of Jan Todd's work as a performing strongwoman, including her affiliation with Guinness and her appearance at the Multnomah County Fair, draws on press clippings including Kelso F. Sutton, "Letter from the Publisher," *Sports Illustrated*, October 9, 1978, 6.

Amanda Bennett, "Superwoman," *Weekend Magazine*, January 13, 1979.

Steve Erickson, "'Strongest Woman' Lifts Weights, Life's Limitations," *Oregonian* (Portland), August 3, 1978, C1.

"Champion Woman Lifter Practices with Hay Bales," *Toronto Star*, February 8, 1978.

Jan Todd's powerlifting total, squat, and deadlift records during this period are cited in Terry Todd, "A Legend in the Making," *Sports Illustrated*, November 5, 1979, 46.

On physicians' ridicule of heavy musculature and great strength during the late nineteenth and early twentieth centuries, see James C. Whorton, "'Athletes' Heart': The Medical Debate over Athleticism, 1870–1920," in *Sport and Exercise Science: Essays in the History of Sports Medicine*, eds. Jack W. Berryman and Roberta J. Park (Urbana: University of Illinois Press, 1992), 109–36, especially 114.

"Discrimination in Physical Culture Methods," *Journal of the American Medical Association*, 39, no. 13 (1902): 775.

Michael Foster, *A Text Book of Physiology* (London: Macmillan, 1877), especially 7, 490. (Italics of the quoted line are in the original.)

John B. Hamilton, "Physical Culture a Necessity: Introductory Lecture delivered at Rush Medical College, Chicago, Sept. 29, 1891," *Journal of the American Medical Association* 17, no. 14 (1891): 513–19. (At the time of this lecture, Hamilton had recently ended his term as surgeon general; and a little more than a year later, he became editor of *JAMA*.)

In the *Journal of the American Medical Association*, the earliest specific mention of gymnastic exercise as treatment for disease that I found is "Bromide of Potassium in the Treatment of Diabetes," *Journal of the American Medical Association* 1, no. 14 (1883): 426–27. The report concerns a French doctor, identified as "M. Felizet," who reportedly cured patients of diabetes "in a few weeks or even in a few days" with a cocktail of medications, primarily "bromide of potassium." But the *Journal*'s reporter found it "difficult to judge accurately the value of the bromide as M. Felizet associated with this treatment as much of gymnastic exercise as was possible." Felizet "found that this drug employed habitually to the extent of 4 grammes per day, produced an intellectual depression and a decided prostration of the general forces, which conditions M. Felizet affirms that he effectually overcomes by his gymnastic exercises." In other words, the drug treatment for diabetes caused unpleasant side effects including depression, which were counteracted by gymnastic exercise—but the *Journal* gave no details about Felizet's exercise prescription.

One of the *Journal*'s earliest articles mentioning resistance exercise concerns low-intensity resistance exercise: "The Treatment of Chronic Heart-Disease by Means of Baths and Gymnastic Exercise," *Journal of the American Medical Association* 6, no. 3 (1886), describes a German gymnastic technique in which "the arms, trunk and legs are successively extended, flexed and rotated against slight resistance. This resistance is preferably obtained through another person, who exerts gentle pressure with the hand upon the extremity in a direction opposite to that in which it is to be moved. The exercise must never be carried to the point of noticeably accelerating the respiration."

On exercise as treatment for heart disease and obesity, see "Oertel's Treatment of Disorders of the Circulation," *Journal of the American Medical Association* 5, no. 17 (1885): 462–63.

"The Reduction of Superabundant Fat," *Journal of the American Medical Association* 5, no. 18 (1885): 487–88.

On exercise for older people, see J. Madison Taylor, "The Influence of Bodily Exercises upon Length of Life," *Journal of the American Medical Association* 18, no. 23 (1892): 705–10.

On the leadership and ethos of nineteenth-century physical education departments, see James C. Whorton, *Crusaders for Fitness: The History of American Health Reformers* (Princeton, NJ: Princeton University Press, 1982), 270–303, especially 285 (on their Greek idealism).

Roberta J. Park, "Physiologists, Physicians, and Physical Educators: Nineteenth Century Biology and Exercise, *Hygienic* and *Educative*," in *Sport and Exercise Science: Essays in the History of Sports Medicine*, eds. Berryman and Park, 137–81, especially 162.

On the religious mission of nineteenth-century American universities, see George M. Marsden, *The Soul of the American University: From Protestant Establishment to Established Nonbelief* (Oxford: Oxford University Press, 1994), 3–9.

On Dudley Allen Sargent and anthropometry, see Roberta J. Park, "Muscles, Symmetry and Action: 'Do You Measure Up?' Defining Masculinity in Britain and America from the 1860s to the Early 1900s," *International Journal of the History of Sport* 22, no. 3 (May 2005): 365–95.

Park wrote extensively about Sargent in other essays, including "Physiologists, Physicians, and Physical Educators" and "Physiology and Anatomy Are Destiny!?: Brains, Bodies and Exercise in Nineteenth-Century American Thought," *Journal of Sport History* 18, no. 1 (1991), 31–63. But I find Park's interpretation of Sargent to be tendentious and not always rooted in a close reading of his writings.

For an evenhanded and broadly informed analysis of Sargent's work, see Carolyn Thomas de la Peña, *The Body Electric: How Strange Machines Built the Modern American* (New York: New York University Press, 2003).

On Sargent's anthropometry and race, see Carolyn de la Peña, "Dudley Allen Sargent: Health Machines and the Energized Male Body," *Iron Game History* 8, no. 2 (2003): 3–19, especially 13.

Dudley Allen Sargent: An Autobiography, ed. Ledyard Sargent (Philadelphia: Lea & Febiger, 1927) describes his examination (174), his machines (182–88), and his thoughts on "unbalanced development" (116, 129).

Sargent also discussed this last idea—the concept that Charles Stocking calls "the sport-specific paradox"—in "The Physical Proportions of the Typical Man," in *Athletic Sports* (New York: Charles Scribner's Sons, 1897), especially 11, 43.

Sargent's comments about "[t]he object of muscular exercise" come from "Editor's Open Window," *Outing* 5, no. 2 (November 1884), 138, in *Outing and the Wheelman Vol. 5* (Boston: The Wheelman Company, 1885).

On Sargent's influence, see Bruce L. Bennett, "Contributions of Dr. Sargent to Physical Education," *Research Quarterly* 19, no. 2 (1948): 77–92.

On Sargent and Eugen Sandow, see "Eugene Sandow: Modern Marvel of Physical Development: Ambition to Become Strong Aroused by Statues of Demi-Gods: Skin Fair as a Woman's Conceals the Muscles of a Hercules," *Boston Daily Globe*, July 2, 1893, 16.

Eugen Sandow, *Sandow on Physical Training: A Study in the Perfect Type of the Human Form* (London: J. S. Tait and Sons, 1894), 120–28. Sargent's anthropometric chart showing Sandow's measurements is on pages 240–41.

On Sargent's announcement of Sandow's gift of the statue to Harvard, see "Statue of Sandow for Harvard," *Harvard Crimson*, February 5, 1902.

On the statue's installation in Harvard's gym, see "University Notes," *Harvard Graduates' Magazine* 10, no. 15 (June 1902): 658.

Mark Twain's joke about gymnastics is from "Mark Twain to Pay All," *San Francisco Examiner*, August 17, 1895.

On Sandow's extensive marketing endeavors, including Sandow's Curative Institute, see Dominic G. Morais, "Branding Iron: Eugen Sandow's 'Modern' Marketing Strategies 1887–1925," *Journal of Sport History* 40, no. 2 (2013): 193–214.

The *JAMA* editorial proposing that doctors should prescribe exercise and medical students should study exercise is "Physical Culture and Medicine," *Journal of the American Medical Association* 37, no. 24 (1901): 1612–13.

On the term *physical culture* and its identification with Bernarr Macfadden, see Terry Todd, "Physical Culture," blog, H.J. Lutcher Stark Center for Physical Culture and Sports, September 23, 2009, https://starkcenter.org/2009/09/physical-culture/.

The best biography of Bernarr Macfadden is Mark Adams, *Mr. America: How Muscular Millionaire Bernarr Macfadden Transformed the Nation through Sex, Salad, and the Ultimate Starvation Diet* (New York: HarperCollins, 2009).

The H.J. Lutcher Stark Center for Physical Culture and Sport maintains an online archive of Macfadden's *Physical Culture* magazine: https://archives.starkcenter.org/omeka/items/browse?collection=6.

On Macfadden, see also Donald J. Mrozek, "The Scientific Quest for Physical Culture and the Persistent Appeal of Quackery," in *Sport and Exercise Science*, eds. Berryman and Park, 283–96, especially 288.

Robert Lewis Taylor, "Physical Culture I: Sink or Swim," *New Yorker* 26 (October 14, 1950), 39–51.

Robert Lewis Taylor, "Physical Culture II: Weakness Is a Crime," *New Yorker* 26 (October 21, 1950), 39–52.

Robert Lewis Taylor, "Physical Culture III: Physician, Heal Thyself!" *New Yorker* 26 (October 28, 1950), 37–51.

Edward E. Evans, "Macfaddism and a Starvation Death," *Journal of the American Medical Association* 84, no. 2 (1925): 136.

"Body Love," *Time* 14, no. 15 (October 7, 1929), 64–66.

For Macfadden's equation of physical culture with muscular strength, see Whorton, *Crusaders for Fitness*, 298.

On Macfadden's connection to Charles Atlas, see Jacqueline Reich, "The World's Most Perfectly Developed Man: Charles Atlas, Physical Culture, and the Inscription of American Masculinity," *Men and Masculinities* 12, no. 4 (2010): 444–61.

On the extreme ideological amplitude of *Physical Culture* magazine, see Andrea Dale Lapin, "A Body of Text: *Physical Culture* and the Marketing of Mobility," PhD dissertation, University of Pittsburgh, 2013.

On the medical profession's struggles for legitimacy during the time of *Physical Culture*'s popularity, see Paul Starr, *The Social Transformation of American Medicine* (New York: Basic Books, 1982), 115–16 (regarding the changes at Johns Hopkins Medical School) and 260 (describing how medicine and government together "took up the war against 'quackery'").

For brief accounts of how medicine changed during this period, in broader narratives of subsequent changes to the profession, see Walter M. Bortz II, *Next Medicine: The Science and Civics of Health* (Oxford: Oxford University Press, 2011), 58–63.

Michael L. Millenson, *Demanding Medical Excellence: Doctors and Accountability in the Information Age* (Chicago: University of Chicago Press, 1999), 30–51.

Jack W. Berryman, "Exercise Is Medicine: A Historical Perspective," *Current Sports Medicine Reports* 9, no. 4 (2010): 195–201.

For the background and context of Macfadden's confrontations with the American Medical Association, see Lapin, "A Body of Text," 81–91.

Adams, *Mr. America*, 128–34, especially 133 (on the AMA's shift in strategy from denunciation to silence).

Morris Fishbein, *The Medical Follies: An Analysis of the Foibles of Some Healing Cults, Including Osteopathy, Homeopathy, Chiropractic, and the Electronic Reactions of Abrams, with Essays* on the *Antivivisectionists, Health Legislation, Physical Culture, Birth Control, and Rejuvenation* (New York: Boni & Liveright, 1925), includes a chapter on "'Physical Culture' and Bernarr Macfadden" (172–81) and a chapter on "The Big Muscle Boys" (182–203).

Signs of persistent animosity between Macfadden and the AMA are found in "Current Comment: 'Health Advice Should Be as Free as Air and Water,'" *Journal of the American Medical Association* 107, no. 24 (1936): 1972–73.

"'We Are Advertised by Our Loving Friends'" (editorial), *Journal of the American Medical Association* 112, no. 5 (February 5, 1939): 435.

"Current Comment: Mr. Macfadden Discusses Army Medical Service," *Journal of the American Medical Association* 115, no. 22 (1940): 1890.

Progressive resistance exercise was featured in the following *Journal* articles published in the 1950s:

"Unilateral Exercise," *Journal of the American Medical Association* 143, no. 6 (1950): 559.

Sedgwick Mead, "Intermittent Treatment of Poliomyelitis with Progressive Resistance Exercise," *Journal of the American Medical Association* 144, no. 6 (1950): 458–60.

"*Progressive Resistance Exercise: Technic and Medical Application*" (book review), *Journal of the American Medical Association* 146, no. 6 (1951): 606.

Arthur L. Watkins, "Practical Applications of Progressive Resistance Exercises," *Journal of the American Medical Association* 148, no. 6 (1952): 443–46.

T. B. Quigley, "Knee Injuries Incurred in Sport," *Journal of the American Medical Association* 171, no. 12 (1959): 1666–70.

In 1952, the *Journal* also published a letter from H. I. Weiser, a medical doctor in Tel Aviv. He took issue with Arthur Watkins for writing that progressive resistance exercise "cannot be employed in the treatment of infants and young children." Weiser described his own technique for making progressive resistance exercise feasible as part of some treatments for children. The letter was published as "Progressive Resistance Exercises," *Journal of the American Medical Association* 149, no. 4 (1952): 390.

From various angles, the following sources show the cultural dominance of cardiovascular exercise in the mid-twentieth century:

Roger Bannister, *The First Four Minutes* (New York: Putnam, 1955), 112 (for mention of "The waddling gait and breathlessness of the muscle-bound weight-lifter").

Eugene Braunwald, "Cardiovascular Medicine at the Turn of the Millennium: Triumphs, Concerns, and Opportunities," *New England Journal of Medicine* 337, no. 19 (1997): 1360–69. (Braunwald's reference to the first description of myocardial infarction as a pathological entity is James B. Herrick, "Clinical Features of Sudden Obstruction of the Coronary Arteries," *Journal of the American Medical Association* 59, no. 23 [1912]: 2015–22.)

On lifestyle in the Lalonde report, see Marc Lalonde, "A New Perspective on the Health of Canadians" (Ottawa, 1974), 16–17, accessible online at www.phac-aspc.gc.ca/ph-sp/phdd/pdf/perspective.pdf.

Kenneth H. Cooper, *Aerobics* (New York: M. Evans, 1968), especially 29.

American College of Sports Medicine, "Position Statement on the Use and Abuse of Anabolic-Androgenic Steroids in Sports," *Medicine and Science in Sports* 9, no. 4 (Winter 1977): xi–xii.

Regarding early epidemiological research on exercise and the development of exercise guidelines, I was fortunate to hear William Haskell, "Guidelines for Physical Activity and Health: The First 50 Years," D. B. Dill Historical Lecture, American College of Sports Medicine Annual Meeting, 2019.

The study of London transport workers was published in two parts: J. N. Morris et al., "Coronary Heart-Disease and Physical Activity of Work," *The Lancet* 265, no. 6795 (1953): 1053–57; and J. N. Morris et al., "Coronary Heart-Disease and Physical Activity of Work," *The Lancet* 265, no. 6796 (1953): 1111–20.

The study showing that people who exercised were less likely to die of heart disease even if they smoked is E. Cuyler Hammond, "Smoking in Relation to Mortality and Morbidity. Findings in First Thirty-Four Months of Follow-Up in a Prospective Study Started in 1959," *Journal of the National Cancer Institute* 32, no. 5 (1964): 1161–88.

On reform of physical education departments, see Ellsworth R. Buskirk and Charles M. Tipton, "Exercise Physiology," in *The History of Exercise and Sport Science*, eds. John D. Massengale and Richard A. Swanson (Champaign, IL: Human Kinetics, 1997), 367–438, especially 408–14.

The first exercise guideline document published by the American College of Sports Medicine is "American College of Sports Medicine Position Statement on the Recommended Quantity and Quality of Exercise for Developing and Maintaining Fitness in Healthy Adults," *Medicine and Science in Sports* 10, no. 3 (1978): vii–x.

On attitudes toward weight training in athletics, the development of the National Strength and Conditioning Association, and its publications, see Jason P. Shurley, Jan Todd, and Terry Todd, *Strength Coaching in America* (Austin: University of Texas Press, 2019), especially 129–63.

Jason Shurley and Jan Todd, "'If Anyone Gets Slower, You're Fired': Boyd Epley and the Formation of the Strength Coaching Profession," *Iron Game History* 11, no. 3 (2011): 4–18.

Terry Todd, "Historical Perspective: The Myth of the Muscle-Bound Lifter," *Strength and Conditioning Journal* 7, no. 3 (1985): 37–41.

Jean Barrett Holloway et al., "Strength Training for Female Athletes: A Position Paper: Part I," *National Strength and Conditioning Association Journal* 11, no. 4 (1989): 43–51.

Jean Barrett Holloway et al., "Strength Training for Female Athletes: A Position Paper: Part II," *National Strength and Conditioning Association Journal* 11, no. 5 (1989): 29–36.

Jason Shurley and Jan Todd, "Thirty Years On: A Narrative Review of Research on Strength Training for Female Athletes Since the National Strength and Conditioning Association's Position Paper," *Journal of Kinesiology and Wellness* 7, no. 1 (2018): 46–72.

On Ottley Coulter, see Jan Todd and Michael Murphy, "Portrait of a Strongman: The Circus Career of Ottley Russell Coulter 1912–1916," *Iron Game History* 7, no. 1 (2000): 4–22.

John D. Fair, "Ottley Coulter: A Token Remembrance of an Iron Game Pioneer," *Iron Game History* 17, no. 1 (2023): 14–16.

On the world-record-setting "man-woman" deadlift by Jan Todd and Larry Pacifico, see "Liftoff at the Center," *Dayton Daily News*, June 5, 1979.

Jan Todd found Katie Sandwina's picture in Edmond Desbonnet, *Les rois de la force: histoire de tous les hommes forts depuis les temps anciens jusqu'à nos jours: avec 733* *photographies et dessins* (Paris: Libraire Berger-Levrault, 1911), 377. The book is accessible online at https://starkcenter.org/exhibits/type/digital-resources/les-rois-de-la-force/.

The English translation is *The Kings of Strength: A History of All Strong Men from Ancient Times to Our Own: by Edmond Desbonnet*, trans. David Chapman (Jefferson, NC: McFarland Press, 2022).

On the paradoxical bodily experiences of many female athletes, see Krane et al., "Living the Paradox," 317 and 326.

Shurley et al., "Historical and Social Considerations," 30.

Barbara Cox and Shona Thompson, "Multiple Bodies: Sportswomen, Soccer and Sexuality," *International Review for the Sociology of Sport* 35, no. 1 (2000): 5–20.

Mari Kristin Sisjord and Elsa Kristiansen, "Elite Women Wrestlers' Muscles: Physical Strength and a Social Burden," *International Review for the Sociology of Sport* 44, nos. 2–3 (2009): 231–46.

Amber D. Mosewich et al., "Exploring Women Track and Field Athletes' Meanings of Muscularity," *Journal of Applied Sport Psychology* 21, no. 1 (2009): 99–115.

Molly George, "Making Sense of Muscle: The Body Experiences of Collegiate Women Athletes," *Sociological Inquiry* 75, no. 3 (2005): 317–45.

On the squat, see Terry Todd, *Inside Powerlifting*, 9–26.

On Jan Todd's record-setting lifts in Columbus, Georgia, on January 31, 1981, see Bill Orr, "Two Highest Totals Ever Made in West Ga. Open," *The Powerlifter* 1, no. 3 (February 1981), 18–19.

The weights for those lifts given here come from the notarized world record application certificate that was filled out on the day of the meet.

On Todd's experience and advocacy of drug-free lifting, see Thomas M. Hunt and Jan Todd, "Powerlifting's Watershed," *Iron Game History* 9, no. 3 (2007): 7–19.

Terry Todd, "The Steroid Predicament," especially 75ff.

Jan Todd, "'Chaos Can Have Gentle Beginnings': The Early History of the Quest for Drug Testing in American Powerlifting: 1964–1984," *Iron Game History* 8, no. 3 (2004): 3–22, especially 17–18.

Frederick C. Hatfield, *Anabolic Steroids: What Kind and How Many* (Madison, WI: Fitness Systems, 1982).

On Brother Bennett, see Hunt and Todd, "Powerlifting's Watershed," especially 8.

Brother Bennett's remark about what being a Brother is all about comes from the "About Us" section of the USA Powerlifting Mississippi website (https://usapowerliftingms.org/about-us), which attributes it to the *Gulf Pine Catholic* newspaper.

"Drugs offend the concept of fairness" and the remarks following were reported by Jan Todd in "Chaos Can Have Gentle Beginnings," 14.

For Todd's record in the first American Drug Free Powerlifting Federation Championship, see "Todd Sets New Mark," *Opeleika Auburn News*, August 2, 1982.

The weight of her record lift given here comes from the world record application filled out by referees on the day of the meet.

On the Todds' move to Austin and their collection's arrival at the University of Texas, see "The Todd-McLean Collection," *Iron Game History* 1, no. 1 (1990): 10.

Terry Todd's articles for *Sports Illustrated*, in addition to "The Steroid Predicament," include "To the Giant Among Us," a profile of André the Giant (December 21, 1981);

"He Bends but He Doesn't Break," a profile of powerlifter Lamar Gant (October 22, 1984); and "Behold Bulgaria's Vest-Pocket Hercules," a profile of Bulgarian Olympic weightlifter Naim Suleimanov (June 11, 1984).

Jan Todd's articles on performance-enhancing drugs for a general readership include "Steroids Should Be Controlled Substances" (guest editorial), *USA Today* (August 17, 1983), A-18.

Jan Todd, "Former Powerlifter Calls for Steroid Curbs," *Los Angeles Times* and syndicated nationally (September 12, 1983).

Jan Todd, "Pillow Talks," *Strength Training for Beauty* 2 (July 1985): 65–69.

On strength training in older age, see Archie Young, "Exercise Physiology in Geriatric Practice," *Acta Medica Scandinavica* 220, no. S711 (1986): 227–32, especially 227, 231.

Research on strength training in older age reportedly conducted and published in Bulgaria as early as 1980 received little notice in the West. I occasionally see this research cited as P. A. Dobrev, "Complex Experimental Investigations of the Influence of Weight Training on Persons in Middle, Advanced, and Old Age," *Scientific Methodical Bulletin* 3 (1980): 27–28, but I was not able to track it down.

For the first ACSM guidelines to recommend resistance/strength training, see American College of Sports Medicine, "The Recommended Quantity and Quality of Exercise for Developing and Maintaining Cardiorespiratory and Muscular Fitness in Healthy Adults," *Medicine & Science in Sports & Exercise* 22 (1990): 265–74, especially 269–70.

On the purpose of *Iron Game History*, see Terry Todd and Jan Todd, "A Statement of Purpose." *Iron Game History* 1, no. 1 (February 1990): 1.

Terry Todd, "Steroids: An Historical Perspective." *Iron Game History* 1, no. 2 (1990): 1–3.

CHAPTER 6

The main source for this chapter is Jan Todd, *Physical Culture and the Body Beautiful: Purposive Exercise in the Lives of American Women, 1800–1870* (Macon, GA: Mercer University Press, 1998). The book's introduction (1–9) discusses its genesis.

Jan Todd's prizewinning paper, "Bernarr Macfadden: Reformer of Feminine Form," North American Society for Sport History, Vancouver Canada, May 25, 1986, was published in *Journal of Sport History* 14, no. 1 (1987): 61–75. The paper is also included in *Sport and Exercise Science: Essays in the History of Sports Medicine*, eds. Jack W. Berryman and Roberta J. Park (Urbana: University of Illinois Press, 1992), 213–32.

On the "republic of letters" and the cultural context of German gymnastics in nineteenth-century Boston, see Mark Peterson, *The City-State of Boston: The Rise and Fall of an Atlantic Power, 1630–1865* (Princeton, NJ: Princeton University Press, 2020), especially 515–23, 530–39.

Paul Christesen, *Sport and Democracy in the Ancient and Modern Worlds* (Cambridge, UK: Cambridge University Press, 2012), 220–23, briefly summarizes the history of German gymnastics, emphasizing its political and social effects and influence.

Allan Guttmann, *Games and Empires: Modern Sports and Cultural Imperialism* (New York: Columbia University Press, 1994), 141–56, provides more detail, especially on the international influence and adoption of German gymnastics, on which see also the note below.

Todd, *Physical Culture and the Body Beautiful*, also includes a brief sketch of the development of German gymnastics (33–36) and of the regimen's adoption in New England, especially Boston (78–81).

For a detailed account of the early days of German gymnastics in Boston, see Erich Geldbach, "The Beginning of German Gymnastics in America," *Journal of Sport History* 3, no. 3 (1976): 236–72.

On Todd's discovery of the first physical education program in North America, see chapter 3 of *Physical Culture and the Body Beautiful*, "William Fowle's Monitorial School for Girls: The First American Gymnastics Experiment," 71–88.

The principal's report is William B. Fowle, "Gymnastic* Exercise for Females," *American Journal of Education* 1 (1826): 698–99.

On the development and opening of Boston's public gymnasium in 1826, see Geldbach, "The Beginning of German Gymnastics in America."

Edward Warren, *The Life of John Collins Warren, M.D.: Compiled Chiefly from His Autobiography and Journals*, vol. 1 (Boston: Ticknor and Fields, 1859), 222–26.

In 1855, the year before Warren died, he wrote the eponymous introduction to Paton Stewart Jr., *Warren's Recommendations of Gymnastics* (Boston: Published by the Proprietor of Boylston Gymnasium, 1856).

Warren and many other prominent white men in Boston were members of a Boston gym owned by Paton Stewart, a Black man. The historian Louis Moore writes that Stewart's was "the only gymnasium for middle-class clients" in Boston between 1849 and 1853. On Stewart and some of his peers, see Moore's essay, "Fit for Citizenship: Black Sparring Masters, Gymnasium Owners, and the White Body, 1825–1886," *Journal of African American History* 96, no. 4 (2011): 448–73.

On physical education and the "Laws of Health," see Jack W. Berryman, "Exercise Is Medicine: A Historical Perspective," *Current Sports Medicine Reports* 9, no. 4 (2010): 195–201, especially 196.

On nineteenth-century views of menses as a disability and the topic of exercise during menstruation, see Patricia A. Vertinsky, *The Eternally Wounded Woman: Women, Doctors, and Exercise in the Late Nineteenth Century* (Urbana: University of Illinois Press, 1994), 39–68, especially 39, 53.

On exercise for menopausal and older women in the nineteenth century, see Vertinsky, *Eternally Wounded Woman*, 88–108, especially 99.

On the sufficiency of lifting light weights, see Diocletian Lewis, "The New Gymnastics," *The Atlantic Monthly* 10, no. 58 (August 1862), 129–48, especially 133–35.

On the old philosophical ideals that inspired new exercise regimens in the nineteenth century, see chapter 1 of Todd's *Physical Culture and the Body Beautiful*, "Majestic Womanhood vs. 'Baby-Faced Dolls': The Debate over Women's Exercise," 11–30.

The full text of J. A. Beaujeu, *A Treatise on Gymnastic Exercises, Or Calisthenics, For the Use of Young Ladies. Introduced at the Royal Hibernian Military School, Also at the Seminary for the Education of Young Ladies, Under the Direction of Miss Hincks, in 1824*

(Dublin: R. Milliken, 1828), is accessible online at https://nrs.lib.harvard.edu/urn-3:gse.libr:24906818.

Todd wrote about the *Treatise*, its author, and his wife in "'Souls of Fire in Iron Hearts': The 1820's Gymnastics Revolution," chapter 2 of *Physical Culture and the Body Beautiful*, 33–69, especially 44–55.

The Beaujeus' story as presented here builds on Todd's work, including her documentation of some of Madame Beaujeu's American endeavors (*Physical Culture and the Body Beautiful*, 52–53).

On the social context of the Beaujeus' work in Dublin, see Conor Heffernan, *The History of Physical Culture in Ireland* (London: Palgrave Macmillan, 2020), 91–98.

On proper form for the pull-up, see Beaujeu, *A Treatise*, 90; and Todd's interpretation of that passage in *Physical Culture and the Body Beautiful*, 49 and 65 (n. 64). I also quote from Todd's "Class Lecture Notes for Tas in PED 106C-Weight Training," 20.

On William Rowan Hamilton's workouts, see Robert Perceval Graves, "Our Portrait Gallery.—NO. XXVI. Sir William R. Hamilton, Professor of Astronomy in the University of Dublin, Astronomer Royal for Ireland, President of the Royal Irish Academy, &c. &c.," *Dublin University Magazine* 19 (1842), 94–110. (Anne van Weerden, author of *A Victorian Marriage: Sir William Rowan Hamilton* [Stedum: J. Fransje van Weerden, 2017], brought this article and a few others about the Beaujeus to Jan Todd's attention, and Todd passed them along to me.)

The Beaujeus advertised expansion of their business in "Mr. Beaujeu, Professor of Gymnastics to the Hibernian Military School," *Saunders's News-Letter* (Dublin), September 25, 1828, 3.

On J. A. Beaujeu's death, see "With feelings of regret we have to announce the death of Mons. J.A. Beaujeu . . . ," *Saunders's News-Letter* (Dublin), October 28, 1838.

"Gymnastics," *Liverpool Mercury*, January 23, 1829, 3.

For the letter from the German prince, see Hermann Fürst von Pückler-Muskau, *Tour in England, Ireland, and France in the Years 1826, 1827, 1828, and 1829. With Remarks on the Manners and Customs of the Inhabitants, and Anecdotes of Distinguished Public Characters. In a Series of Letters. By a German Prince.* (Philadelphia: Carey, Lea & Blanchard, 1833), 443–44.

Madame Beaujeu advertised her Dublin "Academy" in "Madame Beaujeu begs leave to announce . . . ," *Saunders's News-Letter* (Dublin), November 3, 1829, 3.

On Madame Beaujeu's New York City calisthenics classes at William Fuller's gymnasium, see "Calisthenics," *New-York Mirror*, 14, no. 30 (January 21, 1837), 240.

On William Fuller's gymnasium, see Thomas W. Gilbert, "'Medical Fellows' and the New York Game: Columbia Physicians and the Origins of Baseball," *Columbia Medicine* 36, no. 2 (2016): 17–18.

On Madame Beaujeu's time in Boston, see "Calisthenic Exercises," *Boston Medical and Surgical Journal* (August 25, 1841), in *Boston Medical and Surgical Journal, volume XXV*, ed. J.V.C. Smith (Boston: D. Clapp Jr., 1842), 53–54.

On Madame Beaujeu's gym in Manhattan in 1846, see Todd, *Physical Culture and the Body Beautiful*, 53, citing Mary S. Gove, *Lectures to Women on Anatomy and Physiology with an Appendix on Water Cure* (New York: Harper & Brothers, 1846), 218–19.

On her teaching in 1855 at the gymnasium of Joseph B. Jones, see *The Brooklyn City and Kings County Record* (Brooklyn: W. H. Smith, 1855), 214.

On Joseph B. Jones, see Gilbert, "'Medical Fellows' and the New York Game," and *How Baseball Happened: Outrageous Lies Exposed! The True Story Revealed* (Boston: David R. Godine, 2020). In Gilbert's book and on his blog, howbaseballhappened.com ("Baseball, Fathers, and Feminism," July 27, 2020), the author writes that Madame Beaujeu Hawley's instruction was controversial, partly because she wore trousers; and he says that the abolitionist minister Henry Ward Beecher wrote a newspaper article to address the controversy, "attesting to the moral probity of gymnastics for women and girls."

The influence of German gymnastics in many countries is described in Guttmann, *Games and Empires*, 146–47 (Belgium, Holland, Switzerland, Italy), 148–51 (Eastern Europe, regarding which Guttmann also notes, "The irony, of course, was that Slavs resorted to a product of German culture in their efforts to liberate themselves from the political shackles of German and Austrian rule"), and 151–52 (Latin and South America).

On the conflicts between regulars and irregulars, and on the formation of the American Medical Association, see Paul Starr, *The Social Transformation of American Medicine* (New York: Basic Books, 1982), 79–144, especially 90–91 and 97–102.

Vertinsky, *The Eternally Wounded Woman*, 114.

For Elizabeth Blackwell's remarks on muscle and exercise, Elizabeth Blackwell, *The Laws of Life: With Special Reference to the Physical Education of Girls* (New York: G.P. Putnam, 1852), 107–108.

On the notion that bodies had limited vitality, see *Physical Culture and the Body Beautiful*, 177.

For more on this theory, see Vertinsky, *The Eternally Wounded Woman*, 5, 46–49, and 69–88.

On Vesalius and his relation to Galen, see Andrew Wear, "Medicine in Early Modern Europe, 1500–1700," in *The Western Medical Tradition: 800 BC to AD 1800*, ed. Lawrence I. Conrad et al. (Cambridge, UK: Cambridge University Press, 1995), 273–79.

On Vesalius's findings about muscle and nerves, see Charles D. O'Malley, *Andreas Vesalius of Brussels, 1514–1564* (Berkeley: University of California Press, 1964), especially 162, 170–71.

Owsei Temkin, "Vesalius on an Immanent Biological Motor Force," *Bulletin of the History of Medicine*, 39, no. 3 (1965): 277–80.

On *spiritus animalis*, see J. F. Fulton, *Muscular Contraction and the Reflex Control of Movement* (Baltimore: Williams & Wilkins, 1926), 3–55, especially 12, 16, 17, 31, 43.

For the comparison of spirits to explosion of gunpowder, see Raymond Hierons and Alfred Meyer, "Willis's Place in the History of Muscle Physiology," *Proceedings of the Royal Society of Medicine* 57, no. 8 (1964): 687–92, especially 688.

On the circumstances of Helmholtz's discovery, see "The Anniversary Meeting of the Royal Society," *Nature* 51, no. 1310 (1894): 132–36, especially 133–34.

On George Barker Windship, see "Autobiographical Sketches of a Strength-Seeker," *The Atlantic Monthly* 9, no. 51 (January 1862), 102–15.

Jan Todd, "Strength Is Health: George Barker Windship and the First American Weight-Training Boom," *Iron Game History* 3, no. 1 (1993): 3–14.

Todd, *Physical Culture and the Body Beautiful*, 186–98.

Joan Paul, "The Health Reformers: George Barker Windship and Boston's Strength Seekers," *Journal of Sport History* 10, no. 3 (1983): 41–57.

On Windship's first public lecture, see "Physical Culture," *Boston Medical and Surgical Journal* (June 16, 1959): 405–406.

For Windship's training principle of "doing the right thing, in the right way, at the right time," see G. B. Windship, "Physical Culture," *The Massachusetts Teacher* 13, no. 4 (1860): 126–32, especially 127. The essay attempts to shape his experience of training into a system—a highly complex system, involving many, many variables, but organized around the central ideas of lifting heavy weights, resting, and eating well.

"You are to keep doing, and are never to overdo" comes from "Editor's Easy Chair," *Harper's New Monthly Magazine* 19, no. 112 (September 1859), 562.

"Dr. Winship Raises Twenty-six Hundred Pounds: How He Does It and What First Gave Him the Idea of Increasing His Strength," *New York Herald*, October 21, 1863, 7.

Thomas Wentworth Higginson, "Gymnastics," *The Atlantic Monthly* 7 (March 1861), 283–302, situates Windship in the context of other contemporary practices of gymnastic exercise.

Jan Todd, "Strength is Health," 8, cites Windship's advertisement of his office and gymnasium's "separate apartment for ladies" to *The Boston Directory* (Boston: Sampson, Davenport & Company, 1870), 839–40.

On Louisa May Alcott's gymnastics classes with Diocletian Lewis, see Todd, *Physical Culture and the Body Beautiful*, 221. Todd cites a letter by Elizabeth Weir, who was governess to Ralph Waldo Emerson's children, saying that another student in the same class was Una Hawthorne, Nathaniel Hawthorne's eldest child.

Belle Moses, *Louisa May Alcott: Dreamer and Worker, A Story of Achievement* (New York and London: D. Appleton and Company, 1909), 124–26, describes Alcott's experience in gymnastics class without naming Lewis.

Around the time when Alcott took classes from Lewis, he wrote highly critical pieces about Windship in his magazine *Lewis' New Gymnastics for Ladies, Gentlemen & Children & Boston Journal of Physical Culture*. See, for example, "Strength and Health" (June 1, 1861), 125; and "Heavy and Light Gymnastics" (November 1860), 10.

Alcott's short story depicting a gymnastic class like the one Lewis led is "The King of Clubs and Queen of Hearts" in *Hospital Sketches, and Camp and Fireside Stories* (Boston: Roberts Brothers, 1885), 99–142, especially 100–101 and 103. Some other published versions of this story do not include all the passages I quote from the story's first publication.

On David P. Butler, see Todd, *Physical Culture and the Body Beautiful*, 189–98, especially 189, 194.

Todd, "Strength is Health," 8–12.

Paul, "The Health Reformers," 49–52, 57.

For Butler's claims about the Health Lift "treating female disease and weaknesses" and the regimen's effects on pregnant women, see D. P. Butler,

Butler's System of Physical Training: The Lifting Cure: An Original, Scientific Application of the Laws of Motion or Mechanical Action to Physical Culture and the Cure of Disease. With a Discussion of True and False Methods of Physical Training (Boston: D. P. Butler, 1868), especially 71–73, 80.

For the comparison of working out with the Health Lift machine to "attuning an instrument," see J. W. Leavitt, *Exercise a Medicine: or, Muscular Action as Related to Organic Life*, 2nd ed. (New York: J. W. Leavitt, 1870), 1.

On Windship's death, see "The Late Dr. Windship," *The New York Times*, September 18, 1876, 5.

Paul, "The Health Reformers," 57, notes that Butler went out of business the year after Windship died.

Some critics of Windship were still penning subtle jabs at his regimen decades after he died—for instance, Edwin Checkley, *A Natural Method of Physical Training: Making Muscle and Reducing Flesh Without Dieting or Apparatus*, 10th ed. (Brooklyn: William C. Bryant & Co., 1890), 56.

Other critics were slightly more direct, such as William Gilbert Anderson, *Anderson's Physical Education: Health and Strength, Grace and Symmetry* (New York: A. D. Dana, 1897), 11.

As more time passed, criticism became more categorical—for example, James D'Wolf Lovett, *Old Boston Boys and the Games They Played* (Boston: Privately printed at the Riverside Press, 1906), 116–17: Windship "carried the heavy weight system . . . to extremes, and died a comparatively young man. As we all know now, this system of exercising is dead wrong, and will probably never again come into vogue, except with professional strong men, or with those working for some particular record in that line."

The remark about Windship lifting a calf until it became an ox comes from Sargent, *Dudley Allen Sargent: An Autobiography*, ed. Ledyard W. Sargent (Philadelphia: Lea & Febiger, 1927), 97–98.

Todd's lecture is available on DVD as Jan Todd, *Women and Weights: An Illustrated History,* videotape lecture produced in cooperation with the American College of Sports Medicine (Monterey: Healthy Learning Videos), 1999.

The document that refuted some earlier concerns about cardiac effects of weight training is Michael L. Pollock et al., "Resistance Exercise in Individuals With and Without Cardiovascular Disease: Benefits, Rationale, Safety, and Prescription An Advisory from the Committee on Exercise, Rehabilitation, and Prevention, Council on Clinical Cardiology, American Heart Association," *Circulation* 101, no. 7 (2000): 828–33, especially 830.

On Kenneth Cooper's weight training, see this web article by Clarence Bass, a longtime patient of Cooper's: https://www.cbass.com/CooperBook.htm.

On "Project Firepower," see Claire Osborn, "Empowering Women to Be Firefighters; Five from UT Strength-Training Class Pass Challenging Physical Test," *Austin American-Statesman*, March 5, 2001, B1.

On the genesis of Schwarzenegger's strongman competition, now called the Arnold Strongman Classic, see Terry Todd, "The Arnold Strength Summit," *Iron*

Game History 7, nos. 2 and 3 (July 2002): 1–21. (In this essay, Todd understates his wife's involvement in those early plans for the Arnold.)

On Jack and Elaine LaLanne, see Benjamin Richard Pollock, *Becoming Jack LaLanne* (PhD dissertation, University of Texas, 2018).

On the Milo Bar-Bell Company, see Kimberly Ayn Beckwith, *Building Strength: Alan Calvert, the Milo Bar-bell Company, and the Modernization of American Weight Training* (PhD dissertation, University of Texas, 2006).

On the H.J. Lutcher Stark Center for Physical Culture and Sport, see Jan Todd and Terry Todd, "The State of the Stark Center," *Iron Game History* 11, no. 1 (2009): 1–3, 34–35.

John D. Fair, "H. J. Lutcher Stark Center for Physical Culture and Sports," *Journal of Sport History* 40, no. 3 (2013): 483–87.

On *Strength* magazine, see Kimberly Beckwith and Jan Todd, "*Strength*: America's First Muscle Magazine: 1914–1935," *Iron Game History* 9, no. 1 (2005): 11–28.

On *Strength and Health*, see John D. Fair, *Muscletown USA: Bob Hoffman and the Manly Culture of York Barbell* (University Park, PA: Pennsylvania State University Press, 1999).

On Pudgy Stockton, see Jan Todd, "The Legacy of Pudgy Stockton," *Iron Game History* 2 (1992): 5–7.

On Tommy Kono, see John D. Fair, *Tommy Kono: The Life of America's Greatest Weightlifter* (Jefferson, NC: McFarland, 2023), 17.

For a variety of articles about Kono, see "Special Issue: Remembering America's Greatest Weightlifter—Tommy Kono," *Iron Game History* 14, nos. 2 and 3 (2007).

On Katie Sandwina, see Jan Todd, "Center Ring: Katie Sandwina and the Construction of Celebrity," *Iron Game History* 10, no. 1 (2007): 4–13.

Jan Todd, "Bring on the Amazons: An Evolutionary History," in *Picturing the Modern Amazon*, eds. Joanna Frueh, Laurie Fierstein, and Judith Stein (New York: Rizzoli, 2000).

For the view of Sandwina as "a living argument in favor of equal franchise," Todd cites "Circus Performers Dance While Executing Dexterous Ring Feats," *New York Evening Journal*, March 29, 1912.

On Terry Todd's life and writings, with extensive commentary by Jan Todd, see the anthology "The Terry Todd Collecteana: A Special Issue," *Iron Game History* 15, no. 1 (2020).

"He Liked Big Things" is the title of an exhibition at the Stark Center about Terry Todd's life: https://starkcenter.org/research/collections/he-liked-big-things/.

DeLorme discusses the shift from "heavy resistance exercise" to "progressive resistance exercise" in Thomas DeLorme and Arthur L. Watkins, *Progressive Resistance Exercise: Technic and Medical Application* (New York: Appleton-Century-Crofts, 1951), 21.

Eleanor DeLorme's suggestion of the name "progressive resistance" is reported in Janice S. Todd, Jason P. Shurley, and Terry C. Todd, "Thomas L. DeLorme and the Science of Progressive Resistance Exercise," *Journal of Strength and Conditioning Research* 26, no. 11 (2012): 2913–23, especially 2919.

The first *Strength and Health* articles in which I found the phrase *progressive resistance* are John M. Hernic, "An American Apollo," *Strength and Health*, August 1935,

72; and George F. Jowett, "Facts in Progressive Training," *Strength and Health*, August 1935, 74.

Other articles from this period that use the language of progressive training include the following:

George F. Jowett, "For Good Results, Try These Exercises," *Strength and Health*, September 1935, 90 ("progressive exercises") and 94 ("progressive methods").

Joe Raymond, "All That I Am, I Owe to Barbell Training," *Strength and Health*, April 1935, 22 ("progressive system with bar bells and dumb bells").

"S&H League News," *Strength and Health*, June 1935, 76 ("Jewish Progressive Club in Atlanta").

Bob Hoffman, "A Brief Story of the Part Dumbbells Play in Physical Training," *Strength and Health*, November 1935, 13 ("progressive system of training").

One of the advertisements for a "double progressive system" based on the principle of "progressive resistance" occupies the back cover of the magazine's February 1937 issue.

***Strength and Health* foreshadows DeLorme's recommendation of ten repetitions at 80 percent of one-repetition maximum** in Bob Hoffman, "The Mat: A Meeting Place for Body Building Enthusiasts," *Strength and Health*, August 1935, 71.

The magazine foreshadows DeLorme's method of performing maximal strength tests in Bob Hoffman, "The Mat: A Meeting Place for Body Building Enthusiasts," *Strength and Health*, September 1935, 90, 94.

CHAPTER 7

On the "Geriatric Olympics," see "First Geriatric Olympics at Center for Aged," *Jewish Advocate*, September 12, 1974, 4.

Debbie Levenson, "Senior Olympians Compete in Hebrew Rehab's Sports Special," *Jewish Advocate*, September 13, 1990, 4.

"HRCA Residents Go for the Gold!" *Jewish Advocate*, July 20, 1989, 15.

"The Thrill of Victory," photo stand-alone 5, *Jewish Advocate*, September 13, 1984, 4.

"HRCA Geriatric Olympics," *Jewish Advocate*, August 18, 1988, 17.

On research at Tufts University's Human Nutrition Research Center on Aging in the 1980s, see William Evans, Irwin H. Rosenberg, and Jacqueline Thompson, *Biomarkers: The 10 Keys to Prolonging Vitality* (New York: Simon & Schuster, 1992).

Regarding neural paradigms for understanding of muscular strength in older age, see Waneen W. Spirduso, *Physical Dimensions of Aging* (Champaign, IL: Human Kinetics, 1995), 135–41.

For the first study of high-intensity weight training that took place in the Tufts lab, see Walter R. Frontera et al., "Strength Conditioning in Older Men: Skeletal Muscle Hypertrophy and Improved Function," *Journal of Applied Physiology* 64, no. 3 (1988): 1038–44.

Frontera found that the older men's whole muscles grew by about 10 percent, and their muscle fiber size grew by approximately 30 percent overall. To the question of how muscle fibers can grow by three times more than the whole muscle of which they are part, Frontera says, "It is not a one-to-one relationship between the growth of whole muscles and the growth of their contractile elements. Also, the techniques to measure whole muscle size and fiber size are different, and whole muscles have other tissue components that may have changed with training."

On physical activity determining peak muscle mass; on the difference between the most and least active people's muscle mass, both in young adulthood and in older age; and on the difference between the muscle mass of the most and least active octogenarians, see Klaas R. Westerterp et al., "Physical Activity and Fat-Free Mass During Growth and in Later Life," *American Journal of Clinical Nutrition* 114, no. 5 (2021): 1583–89.

The estimate that most people lose at least 3 to 5 percent of muscle mass per decade comes from Steven B. Heymsfield and Nicole Fearnbach, "Can Increasing Physical Activity Prevent Aging-Related Loss of Skeletal Muscle?" *American Journal of Clinical Nutrition* 114, no. 5 (2021): 1579–80.

The estimate of muscle loss at a rate of 6 percent per decade, starting in the third decade of life, comes from J. L. Fleg and E. G. Lakatta, "Role of Muscle Loss in the Age-Associated Reduction in VO2 Max," *Journal of Applied Physiology* 65, no. 3 (1988): 1147–51.

Estimates of muscle loss at higher rates in later life, and average rates of loss for women and men, are found in Marjolein Visser, "Epidemiology of Muscle Mass Loss with Age," in *Sarcopenia*, 2nd ed., eds. Alfonso J. Cruz-Jentoft and John E. Morley (Hoboken, NJ: Wiley, 2021), 11–17.

The estimate of muscle loss at higher rates in older men—1.4 percent per year—comes from Walter R. Frontera et al., "Aging of Skeletal Muscle: A 12-Yr Longitudinal Study," *Journal of Applied Physiology* 88, no. 4 (2000): 1321–26.

For evidence that general physical activity does not protect against age-related loss of muscle mass and strength, see Bruno Manfredini Baroni et al., "Functional and Morphological Adaptations to Aging in Knee Extensor Muscles of Physically Active Men," *Journal of Applied Biomechanics* 29, no. 5 (2013): 535–42.

On age-related changes in muscle fibers, muscle mass, and neuromuscular connections, see Per Aagaard et al., "Role of the Nervous System in Sarcopenia and Muscle Atrophy with Aging: Strength Training as a Countermeasure," *Scandinavian Journal of Medicine & Science in Sports* 20, no. 1 (2010): 49–64, especially 55.

For other overviews of the topic, see Roger M. Enoka, *Neuromechanics of Human Movement*, 5th ed. (Champaign, IL: Human Kinetics, 2015), 436–44; and Richard L. Lieber, *Skeletal Muscle Structure, Function, & Plasticity*, 3rd ed. (Philadelphia: Lippincott Williams & Wilkins, 2010), 208–10.

The CT scan photographs of muscles come from René Koopman and Luc J. C. van Loon, "Aging, Exercise, and Muscle Protein Metabolism," *Journal of Applied Physiology* 106, no. 6 (2009): 2040–48.

On body composition and disease and disability in older age, see Evans, Rosenberg, and Thompson, *Biomarkers*, especially 42–60.

The young Maria Fiatarone imagined her own future obituary in "Obituaries: Macabre Essay Assignment for College Class," *The Argus* (Fremont, California), December 26, 1975, 3.

For the survey showing that only 16 percent of medical schools taught students about exercise, see Edmund J. Burke and Phillip B. Hultgren, "Will Physicians of the Future Be Able to Prescribe Exercise?" *Journal of Medical Education and Curricular Development* 50, no. 6 (1975): 624–26.

On the structure of muscles and muscle cells, see Lieber, *Skeletal Muscle Structure, Function, & Plasticity,* 3rd ed., 14–15.

For the pilot study of progressive resistance exercise at the Hebrew Rehabilitation Center for Aged, see Maria A. Fiatarone et al., "High-Intensity Strength Training in Nonagenarians: Effects on Skeletal Muscle," *Journal of the American Medical Association* 263, no. 22 (1990): 3029–34.

On the functional importance of leg extension training, see E. Joan Bassey et al., "Leg Extensor Power and Functional Performance in Very Old Men and Women," *Clinical Science* (London) 82, no. 3 (1992): 321–27.

The concept of sarcopenia was first suggested in a speech published as Irwin H. Rosenberg, "Summary Comments," *American Journal of Clinical Nutrition* 50, no. 5 (1989): 1231–33.

In two papers, Rosenberg later reflected on coining the term *sarcopenia*. Both papers have the same title. The first is Irwin H. Rosenberg, "Sarcopenia: Origins and Clinical Relevance," *Journal of Nutrition* 127, no. 5 (1997): 990S–991S.

The second is Irwin H. Rosenberg, "Sarcopenia: Origins and Clinical Relevance," *Clinics in Geriatric Medicine* 27, no. 3 (2011): 337–39.

Maria Fiatarone Singh spoke about how and why Rosenberg came to name the concept of sarcopenia in *Raising the Bar Sydney* podcast, "Do You Even Lift?" October 21, 2018.

On menopause and the loss of muscle mass and bone mass, see Sarianna Sipilä et al., "Muscle and Bone Mass in Middle-Aged Women: Role of Menopausal Status and Physical Activity," *Journal of Cachexia, Sarcopenia, and Muscle* 11, no. 3 (2020): 698–709.

For a comparison of how the concepts of osteoporosis and of sarcopenia developed, see A. Y. Bijlsma et al., "Chronology of Age-Related Disease Definitions: Osteoporosis and Sarcopenia," *Ageing Research Reviews* 11, no. 2 (2012): 320–24.

Sarcopenia was not officially recognized as a disease until it received an ICD-10 code in 2016. See Stefan D. Anker, John E. Morley, and Stephan von Haehling, "Welcome to the ICD-10 Code for Sarcopenia," *Journal of Cachexia, Sarcopenia, and Muscle* 7 (2016): 512–14.

On some of the first Hebrew Rehab residents who lifted weights, see Steve Rosenberg, "Aged Can Regain Muscle Strength, Says HRCA Study," *Jewish Advocate*, June 14, 1990, 4.

Jean Dietz, "Even at 90, You've Got to 'Use It or Lose It,'" *The Boston Globe*, July 22, 1990, B16.

On the safety of resistance training for older adults, see Maren S. Fragala et al., "Resistance Training for Older Adults: Position Statement from the National Strength and Conditioning Association," *Journal of Strength and Conditioning Research* 33, no. 8 (2019): 2019–52, especially 2021.

In the NSCA Position Statement, Fragala cites Eduardo Lusa Cadore et al., "Effects of Different Exercise Interventions on Risk of Falls, Gait Ability, and Balance in Physically

Frail Older Adults: A Systematic Review," *Rejuvenation Research* 16, no. 2 (2013): 105–14, especially 108, 110.

Fragala also cites Nelson Sousa et al., "Progressive Resistance Strength Training and the Related Injuries in Older Adults: The Susceptibility of the Shoulder," *Aging Clinical and Experimental Research* 26, no. 3 (2014): 235–40.

Sousa's article enumerates the risk factors a bit more clearly than the NSCA's Position Statement. The paper's summary of risk factors affecting the shoulder is as follows: "Improper attention to exercise technique, exercise selection, unfavorable shoulder positioning required on the more common exercises, along with the repetitive nature of lifting heavy weight until failure, increases the likelihood of injury."

On the safety of resistance training for youth, see Rhodri S. Lloyd et al., "Position Statement on Youth Resistance Training: The 2014 International Consensus," *British Journal of Sports Medicine* 48, no. 7 (2014): 498–505.

Lloyd cites Gregory D. Myer et al., "Youth Versus Adult 'Weightlifting' Injuries Presenting to United States Emergency Rooms: Accidental Versus Nonaccidental Injury Mechanisms," *Journal of Strength and Conditioning Research* 23, no. 7 (2009): 2054–60.

On the transience of exercise adaptation in teenagers, see Neil Armstrong and Joanne Welsman, *Young People and Physical Activity* (New York: Oxford University Press, 1997), 100.

On frequency of exercise required for maintaining physical performance capacity, see Barry A. Spiering et al., "Maintaining Physical Performance: The Minimal Dose of Exercise Needed to Preserve Endurance and Strength Over Time," *Journal of Strength and Conditioning Research* 35, no. 5 (2021): 1449–58, especially 1455–56.

Maria Fiatarone Singh quotes a definition of *agapē* that many have attributed to C. S. Lewis, sometimes specifically to his book *The Four Loves*, but I have been unable to find that line in his writings. According to Zach Kincaid, "Lewis on Love" (blog entry, February 13, 2019), cslewis.com, Lewis did once write, in a letter, "Agape is all giving, not getting."

For the protocol of the second strength training study involving residents of Hebrew Rehab, see Maria A. Fiatarone et al., "The Boston FICSIT Study: The Effects of Resistance Training and Nutritional Supplementation on Physical Frailty in the Oldest Old," *Journal of the American Geriatrics Society* 41, no. 3 (1993): 333–37.

On eccentric contraction, exercise-induced muscle damage, and delayed-onset muscle soreness, see Lieber, *Skeletal Muscle Structure, Function, & Plasticity*, 3rd ed., 242–70.

On non-steroidal anti-inflammatory drugs and muscle repair, see Lieber, *Skeletal Muscle Structure, Function, & Plasticity*, 3rd ed., 262–63; and Gabriel Moraes de Oliveira et al., "Is Physical Performance Affected by Non-Steroidal Anti-Inflammatory Drugs Use? A Systematic Review and Meta-Analysis," *The Physician and Sportsmedicine* (2023): 1–10.

For the findings of the second Hebrew Rehab strength training study, see Maria A. Fiatarone et al., "Exercise Training and Nutritional Supplementation for Physical Frailty in Very Elderly People," *New England Journal of Medicine* 330, no. 25 (1994): 1769–75.

On the nuclei of muscle cells, see Lieber, *Skeletal Muscle Structure, Function, & Plasticity,* 3rd ed., 12–13; and on the role of satellite cells in muscle regeneration, see 229–34.

Maria Fiatarone Singh spoke about the muscle biopsies mentioned in this chapter in her lecture, "Fit for Your Life: Exercise Comes of Age," delivered on September 30, 2014, at the annual meeting of the American College of Sports Medicine. The lecture, an excellent general overview of exercise in older age and as a treatment for several chronic diseases, is accessible on YouTube: https://www.youtube.com/watch?v=uxH52foW1ZQ.

The biopsy findings were published in Maria A. Fiatarone Singh et al., "Insulin-Like Growth Factor I in Skeletal Muscle After Weight-Lifting Exercise in Frail Elders," *American Journal of Physiology—Endocrinology and Metabolism* 277, no. 1 (1999): E135–E143.

Similar findings have been reported by others, including A. L. Mackey et al., "Enhanced Satellite Cell Proliferation with Resistance Training in Elderly Men and Women," *Scandinavian Journal of Medicine & Science in Sports* 17, no. 1 (2007): 34–42; and Anders Karlsen et al., "Preserved Capacity for Satellite Cell Proliferation, Regeneration, and Hypertrophy in the Skeletal Muscle of Healthy Elderly Men," *FASEB Journal* 34, no. 5 (2020): 6418–36.

On satellite cells, see Edward Schultz and Kathleen M. McCormick, "Skeletal Muscle Satellite Cells," in *Reviews of Physiology, Biochemistry, and Pharmacology,* ed. M. P. Blaustein et al. (Berlin: Springer-Verlag, 1994), 213–57, especially 222–23 (regarding changes in distribution with age).

Walter Frontera told me, "Evidence shows that we lose satellite cells as we grow older but also that those that survive do not function well, particularly those associated with type II fibers. So it is another example of the loss of quantity and quality typical of older muscles."

For challenge to the notion that satellite cells are lost with age, see Kristian Gundersen and Jo C. Bruusgaard, "Nuclear Domains During Muscle Atrophy: Nuclei Lost or Paradigm Lost?" *Journal of Physiology* 586, no. 11 (2008): 2675–81.

To convey some of the coincidental intersections of contrasting views of muscle, with Maria Fiatarone and Walter Frontera on one side, and Hans and Franz on the other, I drew from sources including the following:

Harrison G. Pope and David L. Katz, "Affective and Psychotic Symptoms Associated with Anabolic Steroid Use," *American Journal of Psychiatry* 145, no. 4 (1988): 487–90.

Jan Todd and Terry Todd, "Significant Events in the History of Drug Testing and the Olympic Movement: 1960–1999," in *Doping in Elite Sport: The Politics of Drugs in the Olympic Movement,* eds. Wayne Wilson and Edward Derse (Champaign, IL: Human Kinetics, 2001), 65–128, especially 91, 93.

William E. Buckley et al., "Estimated Prevalence of Anabolic Steroid Use Among Male High School Seniors," *Journal of the American Medical Association* 260, no. 23 (1988): 3441–45.

Jon Hotten, *Muscle: A Writer's Trip Through a Sport with No Boundaries* (London: Yellow Jersey Press, 2004), 252–55 (names Dorian Yates as the first Mr. Olympia to weigh more than 250 pounds).

Harrison G. Pope Jr. et al., "Muscle Dysmorphia: An Underrecognized Form of Body Dysmorphic Disorder," *Psychosomatics* 38, no. 6 (1997): 548–57.

ABC News, *Primetime Live*, September 3, 1992.

On psychological studies of muscularity, see J. Kevin Thompson and Guy Cafri, eds., *The Muscular Ideal: Psychological, Social, and Medical Perspectives* (Washington, DC: American Psychological Association, 2007).

On the trend toward greater body dissatisfaction, see Judy Kruger et al., "Body Size Satisfaction and Physical Activity Levels Among Men and Women," *Obesity* 16 (2008): 1976–79.

For a review of research on body dissatisfaction trends over time, see Lauren Fiske et al., "Prevalence of Body Dissatisfaction Among United States Adults: Review and Recommendations for Future Research," *Eating Behaviors* 15, no. 3 (2014): 357–65.

On how weight training affects men's satisfaction with their bodies, see L. A. Tucker, "Effect of Weight Training on Body Attitudes: Who Benefits Most?" *Journal of Sports Medicine and Physical Fitness* 27, no. 1 (1987): 70–78.

L. A. Tucker, "Muscular Strength: A Predictor of Personality in Males," *Journal of Sports Medicine and Physical Fitness* 23, no. 2 (1983): 213–20.

For the comparison of body image among male runners and bodybuilders, see Larry Pasman and J. Kevin Thompson, "Body Image and Eating Disturbance in Obligatory Runners, Obligatory Weightlifters, and Sedentary Individuals," *International Journal of Eating Disorders* 7, no. 6 (1988): 759–69.

On the different ways that weight training changes body image for women and for men, see Nicholas J. SantaBarbara, James W. Whitworth, and Joseph T. Ciccolo, "A Systematic Review of the Effects of Resistance Training on Body Image," *Journal of Strength and Conditioning Research* 31, no. 10 (2017): 2880–88.

For the definition of muscle dysmorphia as a type of body dysmorphic disorder, see *Diagnostic and Statistical Manual of Mental Disorders, Fifth Edition*, American Psychiatric Association (Washington, DC: American Psychiatric Publishing, 2013), 235–36 and 242–47.

For questions about evidence for muscle dysmorphia as a valid construct, see Celso Alves dos Santos Filho et al., "Systematic Review of the Diagnostic Category Muscle Dysmorphia," *Australian and New Zealand Journal of Psychiatry* 50, no. 4 (2016): 322–33.

David Tod, Christian Edwards, and Ieuan Cranswick, "Muscle Dysmorphia: Current Insights," *Psychology Research and Behavior Management* 9 (2016): 179–88.

The first book that presented muscle dysmorphia to a general readership is Harrison Pope Jr., Katharine A. Phillips, and Roberto Olivardia, *The Adonis Complex: The Secret Crisis of Male Body Obsession* (New York: Free Press, 2000).

For estimated prevalence of anabolic steroid use in the United States, see Shalender Bhasin et al., "Anabolic-Androgenic Steroid Use in Sports, Health, and Society," *Medicine & Science in Sports & Exercise* 53, no. 8 (2021): 1778–94, especially 1780.

On prevalence of body dysmorphic disorder, see David Castle et al., "Body Dysmorphic Disorder: A Treatment Synthesis and Consensus on Behalf of the International College of Obsessive-Compulsive Spectrum Disorders and the

Obsessive-Compulsive and Related Disorders Network of the European College of Neuropsychopharmacology," *International Clinical Psychopharmacology* 36, no. 2 (2021): 61–75.

David Veale et al., "Body Dysmorphic Disorder in Different Settings: A Systematic Review and Estimated Prevalence," *Body Image* 18 (2016): 168–86.

For the population of Americans aged sixty or over, see U.S. Census Bureau, American Community Survey (ACS), 2021, Table S0101, AGE AND SEX, accessible at https://data.census.gov/table?q=PEPAGE&t=Age+and+Sex&tid=ACSST1Y2021.S0101.

For the estimate that low levels of muscle mass put 45 percent of older Americans at risk of physical disability, see Ian Janssen et al., "Skeletal Muscle Cutpoints Associated with Elevated Physical Disability Risk in Older Men and Women," *American Journal of Epidemiology* 159, no. 4 (2004): 413–21.

For the estimated global population of older people, see United Nations, Department of Economic and Social Affairs, Population Division, *World Population Prospects 2022: Summary of Results* (New York: United Nations, 2022).

For estimated global prevalence of sarcopenia in otherwise healthy adults, see Gita Shafiee et al., "Prevalence of Sarcopenia in the World: A Systematic Review and Meta-Analysis of General Population Studies," *Journal of Diabetes & Metabolic Disorders* 16, no. 1 (2017): 1–10.

For estimated global prevalence of sarcopenia among people who are hospitalized and among residents of long-term care facilities, see S. K. Papadopoulou et al., "Differences in the Prevalence of Sarcopenia in Community-Dwelling, Nursing Home and Hospitalized Individuals. A Systematic Review and Meta-Analysis," *Journal of Nutrition, Health and Aging* 24, no. 1 (2020): 83–90.

About those numbers, Maria Fiatarone Singh comments: "I think it would be much higher," with "sarcopenic obesity making it harder to detect." She points out that nursing homes routinely measure only body weight—not body composition—so all these figures "are gross underestimates, I would say."

On later research that confirmed and enlarged on Walter Frontera's and Maria Fiatarone's early findings, see Aagaard et al., "Role of the Nervous System in Sarcopenia and Muscle Atrophy with Aging: Strength Training as a Countermeasure," especially 57, 59–60.

The older women in Sweden who doubled the strength of their biceps in six weeks were part of the pilot study for Jan Lexell et al., "Heavy-Resistance Training in Older Scandinavian Men and Women: Short- and Long-Term Effects on Arm and Leg Muscles," *Scandinavian Journal of Medicine & Science in Sports* 5, no. 6 (1995): 329–41, which reported findings similar to Frontera's and Fiatarone's, down to the cellular level. Lexell went on to oversee similar studies in stroke patients. These were among the first studies to show that people with central neurological disorders could respond to strength training. As Lexell recalls, "until then it was almost forbidden" to prescribe high-intensity training after stroke, "as people thought it would lead to a worsening of their spasticity." See Ulla-Britt Flansbjer et al., "Progressive Resistance Strength Training After Stroke: Effects on Muscle Strength, Muscle Tone, Gait Performance and Perceived Participation," *Journal of Rehabilitation Medicine* 40, no. 1 (2008), 42–48; and Ulla-Britt

Flansbjer et al., "Long-Term Benefits of Progressive Resistance Training in Chronic Stroke: A 4-Year Follow-Up," *Journal of Rehabilitation Medicine* 44, no. 3 (2012): 218–21.

The article about lifting weights at Hebrew Rehab that was published ten years after Fiatarone's pilot study is Chana Shavelson, "Exercise Program Has Seniors Pumping Iron: Weight-Training at Hebrew Rehabilitation Center for Aged Increases Muscle While Building Self-Esteem," *Jewish Advocate*, March 18, 1999, 3.

CHAPTER 8

Maria Fiatarone Singh related the anecdote that begins this chapter—about the doctor's vague prescription to the patient—to dramatize the status of exercise as medicine in her lecture, "Fit for Your Life: Exercise Comes of Age," cited in the previous chapter and accessible on YouTube: https://www.youtube.com/watch?v=uxH52foW1ZQ. I have heard several other speakers relate the same anecdote in other settings.

The U.S. Food and Drug Administration's definition of a drug appears in chapter II, section 201, paragraph (g)(1) of the Federal Food, Drug, and Cosmetic Act, enacted December 29, 2022, accessible at https://www.govinfo.gov/content/pkg/COMPS-973/pdf/COMPS-973.pdf.

For detailed summaries of evidence that exercise can be a type of medicine, see Frank W. Booth, Christian K. Roberts, and Matthew J. Laye, "Lack of Exercise Is a Major Cause of Chronic Diseases," *Comprehensive Physiology* 2, no. 2 (2012): 1143–211.

Bente Klarlund Pedersen and Bengt Saltin, "Exercise as Medicine—Evidence for Prescribing Exercise as Therapy in 26 Different Chronic Diseases," *Scandinavian Journal of Medicine & Science in Sports* 25 (2015): 1–72.

2018 Physical Activity Guidelines Advisory Committee, *2018 Physical Activity Guidelines Advisory Committee Scientific Report* (Washington, DC: U.S. Department of Health and Human Services, 2018), accessible at https://health.gov/sites/default/files/2019-09/PAG_Advisory_Committee_Report.pdf.

The classic definition of frailty is Linda P. Fried et al., "Frailty in Older Adults: Evidence for a Phenotype," *Journals of Gerontology Series A: Biological Sciences and Medical Sciences* 56, no. 3 (2001): M146–M156.

See also Andrew Clegg et al., "Frailty in Elderly People," *The Lancet* 381, no. 9868 (2013): 752–62.

On frailty as a function of aging, disease, and disuse, see Walter M. Bortz II, "On Disease . . . Aging . . . and Disuse," *Executive Health* 20 (1983): 1–5. (Bortz continued to develop this idea in several later publications.)

On the study of the prevalence of frailty in sixty-two countries, see Rónán O'Caoimh et al., "Prevalence of Frailty in 62 Countries Across the World: A Systematic Review and Meta-Analysis of Population-Level Studies," *Age and Ageing* 50, no. 1 (2021): 96–104.

On aging as shown by the performance of masters athletes, see A. Barry Baker and Yong Q. Tang, "Aging Performance for Masters Records in Athletics, Swimming, Rowing, Cycling, Triathlon, and Weightlifting," *Experimental Aging Research* 36, no. 4 (2010): 453–77, especially 467.

On disuse as shown by the Harvard alumni study, see Ralph S. Paffenbarger Jr., "Contributions of Epidemiology to Exercise Science and Cardiovascular Health," *Medicine & Science in Sports & Exercise* 20, no. 5 (1988): 426–38.

For more detail on this finding, see also Ralph S. Paffenbarger Jr. et al., "Physical Activity, All-Cause Mortality, and Longevity of College Alumni," *New England Journal of Medicine* 314, no. 10 (1986): 605–13.

Resistance training is recommended as treatment for frailty in Elsa Dent et al., "The Asia-Pacific Clinical Practice Guidelines for the Management of Frailty," *Journal of the American Medical Directors Association* 18, no. 7 (2017): 564–75.

On the early popularity of Prozac, see Sara Rimer, "With Millions Taking Prozac, a Legal Drug Culture Arises," *New York Times*, December 13, 1993, A1.

On antidepressant medication side effects among older patients, as understood in the 1990s, see, for instance, Cal K. Cohn et al., "Double-Blind, Multicenter Comparison of Sertraline and Amitriptyline in Elderly Depressed Patients," *Journal of Clinical Psychiatry* 51, no. 12, suppl. B (1990): 28–33.

Chris Brymer and Carol Hutner Winograd, "Fluoxetine in Elderly Patients: Is There Cause for Concern?" *Journal of the American Geriatrics Society* 40, no. 9 (1992): 902–905.

Sylvia Gerson et al., "Pharmacological and Psychological Treatments for Depressed Older Patients: A Meta-Analysis and Overview of Recent Findings," *Harvard Review of Psychiatry* 7, no. 1 (1999): 1–28, especially 15–18.

On antidepressant medications for nursing home residents in that era, see Jerry Avorn and Jerry H. Gurwitz, "Drug Use in the Nursing Home," *Annals of Internal Medicine* 123 (1995): 195–204, especially 200.

For a more recent review of research on these topics, see Rob M. Kok and Charles F. Reynolds III, "Management of Depression in Older Adults: A Review," *Journal of the American Medical Association* 317, no. 20 (2017): 2114–22, especially 2116.

In 1997, when Nalin Singh and Maria Fiatarone published their research on progressive resistance training as treatment for depression, the only previous randomized controlled trial of exercise as treatment for depression in older people they could identify was J. Kevin McNeil, Esther M. LeBlanc, and Marion Joyner, "The Effect of Exercise on Depressive Symptoms in the Moderately Depressed Elderly," *Psychology and Aging* 6, no. 3 (1991): 487–88.

Regarding other research concerning exercise and depression that preceded Singh and Fiatarone's first collaboration, see the first report of that trial: Nalin A. Singh, Karen M. Clements, and Maria A. Fiatarone, "A Randomized Controlled Trial of Progressive Resistance Training in Depressed Elders," *Journals of Gerontology Series A: Biological Sciences and Medical Sciences* 52, no. 1 (1997): M27–M35.

On prevalence of obesity in early adulthood, see GBD 2015 Obesity Collaborators, "Health Effects of Overweight and Obesity in 195 Countries over 25 Years," *New England Journal of Medicine* 377, no. 1 (2017): 13–27, especially 16.

On the mutually reinforcing link between obesity and depression, see Floriana S. Luppino et al., "Overweight, Obesity, and Depression: A Systematic Review and Meta-Analysis of Longitudinal Studies," *Archives of General Psychiatry* 67, no. 3 (2010): 220–29, especially 225.

For the findings of the second phase of Singh and Fiatarone's trial of strength training as an antidepressant, see Nalin A. Singh, Karen M. Clements, and Maria A. Fiatarone, "The Efficacy of Exercise as a Long-Term Antidepressant in Elderly Subjects: A Randomized, Controlled Trial," *Journals of Gerontology Series A: Biological Sciences and Medical Sciences* 56, no. 8 (2001): M497–M504.

On the effectiveness of low-intensity resistance training as treatment for depression, see Nalin A. Singh et al., "A Randomized Controlled Trial of High Versus Low Intensity Weight Training Versus General Practitioner Care for Clinical Depression in Older Adults," *Journals of Gerontology Series A: Biological Sciences and Medical Sciences* 60, no. 6 (2005): 768–76.

For speculation regarding mechanisms by which resistance training may help to alleviate depression, see Nalin A. Singh and Maria A. Fiatarone Singh, "Exercise and Depression in the Older Adult," *Nutrition in Clinical Care* 3, no. 4 (2000): 197–208.

On how trials of exercise as treatment for depression have disproportionately focused on aerobic training, to the substantial exclusion of resistance training, see Felipe B. Schuch et al., "Exercise as a Treatment for Depression: A Meta-Analysis Adjusting for Publication Bias," *Journal of Psychiatric Research* 77 (2016): 42–51.

2018 Physical Activity Guidelines Advisory Committee Scientific Report, F3–49.

On effectiveness of exercise as treatment for depression, see *2018 Physical Activity Guidelines Advisory Committee Scientific Report*, F3-32 to F3-35.

World Health Organization, *World Health Organization Guidelines on Physical Activity and Sedentary Behavior*, 2020, especially 26 and 34.

For statements in clinical practice guidelines about exercise as treatment for depression, see Gin S. Malhi et al., "The 2020 Royal Australian and New Zealand College of Psychiatrists Clinical Practice Guidelines for Mood Disorders," *Australian and New Zealand Journal of Psychiatry* 55, no. 1 (2021): 7–117, especially 33–35.

American Psychiatric Association, "Practice Guideline for the Treatment of Patients with Major Depressive Disorder," 3rd ed., *American Journal of Psychiatry* 167, suppl. 10 (2010): 9–118, especially 29–30.

American Psychological Association, *Clinical Practice Guideline for the Treatment of Depression Across Three Age Cohorts* (Washington, DC: American Psychological Association, 2019), 11.

Fiatarone and Singh's trial of weight training as treatment for insomnia is published as Nalin A. Singh, Karen M. Clements, and Maria A. Fiatarone, "A Randomized Controlled Trial of the Effect of Exercise on Sleep," *Sleep* 20, no. 2 (1997): 95–101.

On subsequent research regarding the effects of resistance exercise on sleep, see Ana Kovacevic et al., "The Effect of Resistance Exercise on Sleep: A Systematic Review of Randomized Controlled Trials," *Sleep Medicine Reviews* 39 (2018): 52–68.

On type 2 diabetes, see https://www.cdc.gov/diabetes/basics/diabetes.html and https://www.mayoclinic.org/diseases-conditions/type-2-diabetes/symptoms-causes/syc-20351193.

On shifts in body composition that can accompany type 2 diabetes, see Yi Wang et al., "Muscle and Adipose Tissue Biopsy in Older Adults with Type 2 Diabetes," *Journal of Diabetes Mellitus* 1, no. 3 (2011): 27–35.

The study of distance runners, lifters, and sedentary men that showed relations between body composition and glucose metabolism is Hannele Yki-Järvinen and Veikko A. Koivisto, "Effects of Body Composition on Insulin Sensitivity," *Diabetes* 32, no. 10 (1983): 965–69.

On global prevalence of type 2 diabetes, see Hong Sun et al., "IDF Diabetes Atlas: Global, Regional, and Country-Level Diabetes Prevalence Estimates for 2021 and Projections for 2045," *Diabetes Research and Clinical Practice* 183 (2022): 109–19.

On risk factors associated with type 2 diabetes among Swedish male soldiers, see Casey Crump et al., "Physical Fitness Among Swedish Military Conscripts and Long-Term Risk of Type 2 Diabetes: A Cohort Study," *Annals of Internal Medicine* 164, no. 9 (2016): 577–84.

On prevalence of diabetes among people aged seventy-five to seventy-nine, see Sun et al., "IDF Diabetes Atlas," 6.

On prevalence of youth-onset type 2 diabetes in the United States, see Jean M. Lawrence et al., "Trends in Prevalence of Type 1 and Type 2 Diabetes in Children and Adolescents in the US, 2001–2017," *Journal of the American Medical Association* 326, no. 8 (2021): 717–27.

On prevalence of prediabetes among youth in the United States, see Junting Liu et al., "Trends in Prediabetes Among Youths in the US from 1999 Through 2018," *JAMA Pediatrics* 176, no. 6 (2022): 608–10.

On outcomes and comorbidities of type 2 diabetes, see *2018 Physical Activity Guidelines Advisory Committee Scientific Report*, F10–F56.

On average cost of medical care for diabetics, see Juliana C. N. Chan et al., "The *Lancet* Commission on Diabetes: Using Data to Transform Diabetes Care and Patient Lives," *The Lancet Commissions* 396, no. 10267 (2020): P2019–P2082, especially 2030–31 (for costs in Italy); and Fatma Al-Maskari, Mohammed El-Sadig, and Nicholas Nagelkerke, "Assessment of the Direct Medical Costs of Diabetes Mellitus and Its Complications in the United Arab Emirates," *BMC Public Health* 10 (2010): 679.

On growing numbers of patients with diabetes and related strains on the capacities of health care systems, see Moien Abdul Basith Khan et al., "Epidemiology of Type 2 Diabetes—Global Burden of Disease and Forecasted Trends," *Journal of Epidemiology and Global Health*, 10, no. 1 (2020): 107–11, especially 109.

The estimated numbers of people with diabetes in China, Pakistan, and India in 2045 come from *IDF Diabetes Atlas*, 10th ed. (2021), Table 3.4. Adding the numbers for those three countries produces a total of 361.5 million.

The population of Western Europe in 2050, according to the United Nations, *World Population Prospects* (2022), is predicted to be about 196,000,000: https://population.un.org/wpp/Graphs/DemographicProfiles/Line/926.

On large randomized controlled trials of lifestyle changes for diabetes prevention, see Karla I. Galaviz et al., "Lifestyle and the Prevention of Type 2 Diabetes: A Status Report," *American Journal of Lifestyle Medicine* 12, no. 1 (2018): 4–20.

For the findings that metformin and lifestyle changes had different effects on blood sugar control in older and younger people, see Diabetes Prevention Program Research Group, "Reduction in the Incidence of Type 2 Diabetes with Lifestyle Intervention or Metformin," *New England Journal of Medicine* 346, no. 6 (2002): 393–403, especially 398.

On accelerated loss of functional capacity with type 2 diabetes, see Marika Leenders et al., "Patients with Type 2 Diabetes Show a Greater Decline in Muscle Mass, Muscle Strength, and Functional Capacity with Aging," *Journal of the American Medical Directors Association* 14, no. 8 (2013): 585–92.

On accelerated age-related loss of strength and muscle, see Seok Won Park et al., "Excessive Loss of Skeletal Muscle Mass in Older Adults with Type 2 Diabetes," *Diabetes Care* 32, no. 11 (2009): 1993–97.

On elevated risk of vascular complications for undiagnosed diabetics, see Katherine Ogurtsova et al., "IDF Diabetes Atlas: Global Estimates of Undiagnosed Diabetes in Adults for 2021," *Diabetes Research and Clinical Practice* 183 (2022): 109118.

For the American Diabetes Association's recommendation of resistance exercise, see Sheri R. Colberg, "Physical Activity/Exercise and Diabetes: A Position Statement of the American Diabetes Association," *Diabetes Care* 39, no. 11 (2016): 2065–79.

The ADA's Position Statement cites this paper cowritten by Fiatarone Singh: Karen A. Willey and Maria A. Fiatarone Singh, "Battling Insulin Resistance in Elderly Obese People with Type 2 Diabetes: Bring on the Heavy Weights," *Diabetes Care* 26, no. 5 (2003): 1580–88.

On the rarity of resistance training among people with diabetes, see Ronald C. Plotnikoff, "Physical Activity in the Management of Diabetes: Population-Based Perspectives and Strategies," *Canadian Journal of Diabetes* 30, no. 1 (2006): 52–62, especially 57–58.

On "strong evidence" that exercise affects risk of the four major indicators of diabetes progression, see *2018 Physical Activity Guidelines Advisory Committee Scientific Report*, F10-54. Unless otherwise noted, most details about how exercise affects indicators of diabetes progression come from this report. For details on exercise and hypertension, see F5-12 to F-21, especially F5-18 to F-21.

The 2018 report's findings on exercise and hypertension are also summarized in Linda S. Pescatello et al., "Physical Activity to Prevent and Treat Hypertension: A Systematic Review," *Medicine & Science in Sports & Exercise* 51, no. 6 (2019): 1314–23.

On the prevalence and outcomes of hypertension, see Katherine T. Mills et al., "Global Disparities of Hypertension Prevalence and Control: A Systematic Analysis of Population-Based Studies from 90 Countries," *Circulation* 134, no. 6 (2016): 441–50.

On type 2 diabetes and mortality of older adults, see Hwee H. Tan et al., "Diagnosis of Type 2 Diabetes at an Older Age: Effect on Mortality in Men and Women," *Diabetes Care* 27, no. 12 (2004): 2797–99.

For research on exercise and diabetes since 2018, see Alberto J. Alves et al., "Exercise to Treat Hypertension: Late Breaking News on Exercise Prescriptions That FITT," *Current Sports Medicine Reports* 21, no. 8 (2022): 280–88, especially 283.

Huseyin Naci et al., "How Does Exercise Treatment Compare with Antihypertensive Medications? A Network Meta-Analysis of 391 Randomised Controlled Trials Assessing Exercise and Medication Effects on Systolic Blood Pressure," *British Journal of Sports Medicine* 53, no. 14 (2019): 859–69, especially 866. (Like others who have reviewed the research on this topic, Naci and colleagues also found that "the effectiveness of exercise increased" as treatment for hypertension among people with higher levels of blood pressure.)

Pedro L. Valenzuela et al., "Lifestyle Interventions for the Treatment and Prevention of Hypertension," *Nature Reviews Cardiology* 18, no. 4 (2021): 251–75.

Effects of exercise as treatment for hypertension "are not additive or synergistic," but exercise by itself may reduce high blood pressure more than medication by itself, according to Linda S. Pescatello et al., "Do the Combined Blood Pressure Effects of Exercise and Antihypertensive Medications Add Up to the Sum of Their Parts? A Systematic Meta-Review," *BMJ Open Sport & Exercise Medicine* 7, no. 1 (2021): e000895.

For the United States surgeon general's statement that conflicts with those findings, see U.S. Department of Health and Human Services, *The Surgeon General's Call to Action to Control Hypertension* (Washington, DC: U.S. Department of Health and Human Services, Office of the Surgeon General, 2020), 19, accessible online at https://www.cdc.gov/bloodpressure/docs/SG-CTA-HTN-Control-Report-508.pdf.

On effects of reducing HbA1c, see UK Prospective Diabetes Study (UKPDS) Group, "Intensive Blood-Glucose Control with Sulphonylureas or Insulin Compared with Conventional Treatment and Risk of Complications in Patients with Type 2 Diabetes (UKPDS 33)," *The Lancet* 352, no. 9131 (1998): 837–53.

For later analysis of these UKPDS findings, see Marcus Lind et al., "Historical HbA1c Values May Explain the Type 2 Diabetes Legacy Effect: UKDPS 88," *Diabetes Care* 44, no. 10 (2021): 2231–37.

On resistance training intensity and control of blood sugar and insulin, see Yubo Liu et al., "Resistance Exercise Intensity Is Correlated with Attenuation of HbA1c and Insulin in Patients with Type 2 Diabetes: A Systematic Review and Meta-Analysis," *International Journal of Environmental Research and Public Health* 16, no. 1 (2019): 140.

The University of Sydney group's publications on weight training for diabetics include Yorgi Mavros et al., "Changes in Insulin Resistance and HbA1c Are Related to Exercise-Mediated Changes in Body Composition in Older Adults with Type 2 Diabetes: Interim Outcomes from the GREAT2DO trial," *Diabetes Care* 36, no. 8 (2013): 2372–79.

Kylie A. Simpson et al., "Graded Resistance Exercise and Type 2 Diabetes in Older Adults (The GREAT2DO Study): Methods and Baseline Cohort Characteristics of a Randomized Controlled Trial," *Trials* 16, no. 512 (2015): 1–14.

Yorgi Mavros et al., "Reductions in C-Reactive Protein in Older Adults with Type 2 Diabetes Are Related to Improvements in Body Composition Following a Randomized Controlled Trial of Resistance Training," *Journal of Cachexia, Sarcopenia and Muscle* 5, no. 2 (2014): 111–20.

On the prevalence of chronic kidney disease among people with type 2 diabetes, see https://www.cdc.gov/diabetes/data/statistics-report/coexisting-conditions-complications.html.

On chronic kidney disease and its treatment, as well as its links to diabetes and high blood pressure, see Teresa K. Chen, Daphne H. Knicely, and Morgan E. Grams, "Chronic Kidney Disease Diagnosis and Management: A Review," *Journal of the American Medical Association* 322, no. 13 (2019): 1294–304.

On the progression of kidney disease, see Patricia Painter, "Physical Functioning in End-Stage Renal Disease Patients: Update 2005," *Hemodialysis International* 9, no. 3 (2005): 218–35.

For evidence that weight training protects against muscle wasting for people with chronic kidney disease on low-protein diets, see Carmen Castaneda et al., "Resistance Training to Counteract the Catabolism of a Low-Protein Diet in Patients with Chronic Renal Insufficiency: A Randomized, Controlled Trial," *Annals of Internal Medicine* 135, no. 11 (2001): 965–76.

For evidence that weight training during dialysis improves muscle quantity, strength, function, and quality of life, among other outcomes, see Birinder S. B. Cheema et al., "Progressive Resistance Training During Hemodialysis: Rationale and Method of a Randomized-Controlled Trial," *Hemodialysis International* 10, no. 3 (2006): 303–10.

Bobby Cheema et al., "Progressive Exercise for Anabolism in Kidney Disease (PEAK): A Randomized, Controlled Trial of Resistance Training During Hemodialysis," *Journal of the American Society of Nephrology* 18, no. 5 (2007): 1594–601.

On medium-intensity strength training to counteract losses of muscle size and strength in transplant recipients who take corticosteroids, see Randy W. Braith et al., "Resistance Exercise Prevents Glucocorticoid-Induced Myopathy in Heart Transplant Recipients," *Medicine & Science in Sports & Exercise* 30, no. 4 (1998): 483–89. (The equation of a three-month course of prednisone with a decade of aging, in terms of muscle loss, is Maria Fiatarone Singh's reading of this study's results.)

On osteoarthritis of the knee, see https://www.mayoclinic.org/diseases-conditions/osteoarthritis/symptoms-causes/syc-20351925 and https://my.clevelandclinic.org/health/diseases/21750-osteoarthritis-knee.

Galen's description of cartilage comes from *Galen on the Usefulness of the Parts of the Body*, trans. Margaret Tallmadge May, vol. 2 (Ithaca, NY: Cornell University Press, 1968), 552.

Radiographic evidence of osteoarthritis is most commonly found in the hand, but symptomatic osteoarthritis is most commonly found in the knee, as shown by Anna Litwic et al., "Epidemiology and Burden of Osteoarthritis," *British Medical Bulletin* 105, no. 1 (2013): 185–99.

For global and national prevalence estimates of knee arthritis, see Aiyong Cui et al., "Global, Regional Prevalence, Incidence and Risk Factors of Knee Osteoarthritis in Population-Based Studies," *EClinicalMedicine* 29–30 (2020): 100587, 1.

On prevalence with regard to age and gender, see Saeid Safiri et al., "Global, Regional and National Burden of Osteoarthritis 1990–2017: A Systematic Analysis of the Global

Burden of Disease Study 2017," *Annals of the Rheumatic Diseases* 79, no. 6 (2020): 819–28, especially 826.

On global prevalence trends, see Huibin Long et al., "Prevalence Trends of Site-Specific Osteoarthritis from 1990 to 2019: Findings from the Global Burden of Disease Study 2019," *Arthritis and Rheumatology* 74, no. 7 (2022): 1172–83, especially 1174.

On risk factors for osteoarthritis of the knee, see V. Silverwood et al., "Current Evidence on Risk Factors for Knee Osteoarthritis in Older Adults: A Systematic Review and Meta-Analysis," *Osteoarthritis and Cartilage* 23, no. 4 (2015): 507–15.

On muscle weakness as a risk factor, see B. E. Øiestad et al., "Knee Extensor Muscle Weakness Is a Risk Factor for Development of Knee Osteoarthritis. A Systematic Review and Meta-Analysis," *Osteoarthritis and Cartilage* 23, no. 2 (2015): 171–77.

On obesity as a risk factor, see D. Coggon et al., "Knee Osteoarthritis and Obesity," *International Journal of Obesity* 25, no. 5 (2001): 622–27.

On ACL tears as a risk factor, see Erik Poulsen et al., "Knee Osteoarthritis Risk Is Increased 4-6 Fold After Knee Injury—A Systematic Review and Meta-Analysis," *British Journal of Sports Medicine* 53, no. 23 (2019): 1454–63.

The cemetery across the street from the Cumberland Campus of the University of Sydney is Rookwood Cemetery.

On erosion of cartilage and biomechanics of knee alignment, see Nasim Foroughi, Richard Smith, and Benedicte Vanwanseele, "The Association of External Knee Adduction Moment with Biomechanical Variables in Osteoarthritis: A Systematic Review," *The Knee* 16, no. 5 (2009): 303–309.

The main source of clinical guidelines for treatment of osteoarthritis is Osteoarthritis Research Society International. See T. E. McAlindon et al., "OARSI Guidelines for the Non-Surgical Management of Knee Osteoarthritis," *Osteoarthritis and Cartilage* 22, no. 3 (2014): 363–88.

See also Royal Australian College of General Practitioners, "Guideline for the Management of Knee and Hip Osteoarthritis," 2nd ed. (2018): 1–71. The RACGP Guideline reviews evidence of effectiveness for the gamut of treatments enumerated in this chapter. The only evidence favoring any of these therapies reported by the RACGP Guideline concerns arthroscopic surgery to remove torn cartilage in the condition of a "locked" knee, though the guidelines recommend against this surgery except in cases when exercise has failed to fix the problem. For the general review of evidence see 1–33; and on the "locked" knee see 3, 6, 59.

In Maria Fiatarone Singh's view, the following study gives the best evidence of the lack of efficacy of most arthroscopic knee surgeries: J. Bruce Moseley et al., "A Controlled Trial of Arthroscopic Surgery for Osteoarthritis of the Knee," *New England Journal of Medicine* 347, no. 2 (2002): 81–88.

For a comprehensive review of the effectiveness of arthroscopic surgeries for knee arthritis, see also Wiroon Laupattarakasem et al., "Arthroscopic Debridement for Knee Osteoarthritis," *Cochrane Database of Systematic Reviews* 1 (2008), CD05118.

For effect sizes of strength training and walking programs on the outcomes of pain and physical function in randomized controlled trials of land-based exercise for osteoarthritis of the knee, see M. Fransen et al., "Exercise for Osteoarthritis of the Knee (Review)," *Cochrane Database of Systematic Reviews* 1 (2015), CD004376, especially 101.

For effect sizes of various exercise programs for outcomes of pain and physical function in randomized controlled trials of land-based and aquatic exercise for osteoarthritis of the knee, see Olalekan A. Uthman et al., "Exercise for Lower Limb Osteoarthritis: Systematic Review Incorporating Trial Sequential Analysis and Network Meta-Analysis," *British Medical Journal* 347, no. 21 (2013): 10–11 (Table 2 and Table 3).

On weight-loss recommendations for people with knee arthritis, see Yuan Z. Lim et al., "Recommendations for Weight Management in Osteoarthritis: A Systematic Review of Clinical Practice Guidelines," *Osteoarthritis and Cartilage Open* 4, no. 4 (2022): 100298.

On how dietary weight loss affects lean body mass, Fiatarone Singh says the best evidence is found in publications such as the following: In the IDEA randomized clinical trial, involving more than 450 people, eighteen months of diet alone led to weight loss of 8.9 kg, on average, of which 4.2 kg—47 percent—was fat-free mass. Subjects in the same study who lost weight by a combination of diet and exercise (regimens of roughly two-thirds aerobic training and one-third strength training) lost an average of 10.6 kg, of which 4.7 kg—44 percent—was fat-free mass. See Stephen P. Messier et al., "Effects of Intensive Diet and Exercise on Knee Loads, Inflammation, and Clinical Outcomes Among Overweight and Obese Adults with Knee Arthritis: The IDEA Randomized Clinical Trial," *Journal of the American Medical Association* 310, no. 12 (2013): 1263–73, especially 1266 (Table 2).

In the PREVIEW randomized intervention study, involving more than 2,300 people, regimens combining various types of diet and aerobic exercise were associated with losses of, on average, about 8 kg of fat mass and about 2.7 kg of fat-free mass—a ratio of 70 percent to 30 percent—over a period of eight weeks. See Anne Raben et al., "The PREVIEW Intervention Study: Results from a 3-Year Randomized 2x2 Factorial Multinational Trial Investigating the Role of Protein, Glycaemic Index and Physical Activity for Prevention of Type 2 Diabetes," *Diabetes, Obesity and Metabolism* 23, no. 2 (2021): 324–37, especially 331–32 (Table 2).

A systematic review of six studies involving older obese adults who lost weight by diet alone, or by diet in combination with resistance training, found that resistance training made up for almost all of the loss of lean body mass—93.5 percent of it—with similar reductions in fat mass and overall body mass. See Amanda V. Sardeli et al., "Resistance Training Prevents Muscle Loss Induced by Caloric Restriction in Obese Elderly Individuals: A Systematic Review and Meta-Analysis," *Nutrients* 10, no. 4 (2018): 423–33.

For details of the osteoarthritis study in which Ramanee participated, see Yareni Guerrero et al., "Train High Eat Low for Osteoarthritis Study (THE LO Study): Protocol for a Randomized Controlled Trial," *Journal of Physiotherapy* 61, no. 4 (2015): 217.

On how activities of daily life subject the knees to various levels of force, see I. Kutzner et al., "Loading of the Knee Joint During Activities of Daily Living Measured in vivo in Five Subjects," *Journal of Biomechanics* 43 (2010): 2164–73.

On the biomechanics of descending stairs backwards, see Masaki Hasegawa et al., "Effects of Methods of Descending Stairs Forwards Versus Backwards on Knee Joint Force in Patients with Osteoarthritis of the Knee: A Clinical Controlled Study," *Sports Medicine, Arthroscopy, Rehabilitation, Therapy and Technology* 2, no. 14 (2010): 1–7.

The trial that showed strength training works equally well at low and high intensity as treatment for knee arthritis, on which Fiatarone Singh collaborated, is Nasim Foroughi et al., "Lower Limb Muscle Strengthening Does Not Change Frontal Plane Movements in Women with Knee Osteoarthritis: A Randomized Controlled Trial," *Clinical Biomechanics* 26, no. 2 (2011): 167–74.

Other investigations have made similar findings, such as this study that compared very low intensity resistance training (10 percent of 1-RM) to moderate intensity (60 percent of 1-RM): Mei-Hwa Jan et al., "Investigation of Clinical Effects of High- and Low-Resistance Training for Patients with Knee Osteoarthritis: A Randomized Controlled Trial," *Physical Therapy* 88, no. 4 (2008): 427–36.

CHAPTER 9

Andrew M. Briggs et al., "Musculoskeletal Health Conditions Represent a Global Threat to Healthy Aging: A Report for the 2015 World Health Organization World Report on Ageing and Health," *The Gerontologist* 56, suppl. 2 (2016): S243–S255, was prepared for World Health Organization, *World Report on Ageing and Health* (Geneva: World Health Organization, 2015).

Projected estimates of the global population of older people come from United Nations, Department of Economic and Social Affairs, Population Division, *World Population Prospects 2022: Summary of Results* (New York: United Nations, 2022), 7.

On *The Jack LaLanne Show*, see Benjamin Richard Pollock, *Becoming Jack LaLanne* (PhD dissertation, University of Texas, 2018).

On muscle power and general mobility in older adults, see Jonathan F. Bean et al., "A Comparison of Leg Power and Leg Strength Within the InCHIANTI Study: Which Influences Mobility More?" *Journals of Gerontology Series A: Biological Sciences and Medical Sciences* 58, no. 8 (2003): 728–33.

For a broader review of the topic, see Kieran F. Reid and Roger A. Fielding, "Skeletal Muscle Power: A Critical Determinant of Physical Functioning in Older Adults," *Exercise and Sport Science Reviews* 40, no. 1 (2012): 4–12.

On loss of muscle power and strength in older age, see E. Joan Bassey et al., "Leg Extensor Power and Functional Performance in Very Old Men and Women," *Clinical Science* (London) 82, no. 3 (1992): 321–27.

Jeffrey E. Metter, "Age-Associated Loss of Power and Strength in the Upper Extremities in Women and Men," *Journals of Gerontology Series A: Biological Sciences and Medical Sciences* 52, no. 5 (1997): B267–B276, found smaller but very significant age-related losses of strength and power in the muscles of the upper body. Measurements of more than 1,000 participants in the Baltimore Longitudinal Study of Aging showed losses of more than 40 percent of power, and more than 30 percent of strength, between the third decade of life and the eighth decade.

More recently, Brandon M. Roberts et al., "Human Neuromuscular Aging: Sex Differences Revealed at the Myocellular Level," *Experimental Gerontology* 106 (2018): 116–24, reported on tests of strength and power of groups of people at various ages, ranging from twenty-five to seventy-two. By one measurement, the strength of the seventy-two-year-olds was found to be 40 percent lower than that of the twenty-five-year-olds. The decline in power between the two age groups, by another measurement, was almost 60 percent.

The Danish study showing how aerobic and resistance exercise may preserve some youthful qualities of muscle into older age is Henrik Klitgaard et al., "Function, Morphology and Protein Expression of Ageing Skeletal Muscle: A Cross-Sectional Study of Elderly Men with Different Training Backgrounds," *Acta Physiologica Scandinavica* 140, no. 1 (1990): 41–54.

While the study reported similar overall declines of strength, power, and muscle cross-sectional area in the older swimmers, runners, and sedentary men, it also noted a couple of anomalies. The runners preserved a bit more knee extension strength than did swimmers and sedentary men; and the swimmers preserved a great deal more knee extension power than did runners or sedentary men. These comparative advantages were slight, however, compared to the much greater strength, power, and muscle size of the older men who lifted weights.

On the importance of intent in power training, see see David G. Behm and Digby G. Sale, "Intended Rather Than Actual Movement Velocity Determines Velocity-Specific Training Response," *Journal of Applied Physiology* 74, no. 1 (1993): 359–68; and for a review of research that followed lines of inquiry raised by that paper, Timothy B. Davies et al., "Effect of Movement Velocity During Resistance Training on Dynamic Muscular Strength: A Systematic Review and Meta-Analysis," *Sports Medicine* 47, no. 8 (2017): 1603–17.

For a couple of the Singhs' early studies of muscle power, see Nathan J. de Vos et al., "Optimal Load for Increasing Muscle Power During Explosive Resistance Training in Older Adults," *Journals of Gerontology Series A: Biological Sciences and Medical Sciences* 60, no. 5 (2005): 638–47.

Nathan J. de Vos et al., "Effect of Power-Training Intensity on the Contribution of Force and Velocity to Peak Power in Older Adults," *Journal of Aging and Physical Activity* 16, no. 4 (2008): 393–407.

Nalin Singh compiled statistics on the outcomes of his progressive resistance training prescriptions in unpublished reports that he shared with me, including "The Centre for STRONG Medicine Results" (2015).

"Development of a STRONG Medicine Unit, Report on 12 Months Activity" (2000).

On inclusion body myositis (IBM), see Marinos C. Dalakas, "Inflammatory Muscle Diseases," *New England Journal of Medicine* 372, no. 18 (2015): 1734–47.

On prescriptions of the generic version of Lipitor, see Matej Mikulic, "Number of Atorvastatin Prescriptions in the U.S. from 2004 to 2021," Statista.com (2022).

No causal relationship between statin exposure and IBM has been established, but data shows association between statin exposure and idiopathic inflammatory myositis, the class of diseases that includes IBM. See Gillian E. Caughey et al., "Association of Statin Exposure with Histologically Confirmed Idiopathic Inflammatory Myositis in an Australian Population," *JAMA Internal Medicine* 178, no. 9 (2018): 1224–29.

For a relevant case study, see Nicole Daver and Sara Tonini, "Phenotypical Statin-Associated Immune-Mediated Necrotizing Myositis with Histological Features of Inclusion Body Myositis," *Rheumatology & Autoimmunity* 3, no. 1 (2023): 50–55.

For statistics on falls and fractures in Australia, see Australian Institute of Health and Welfare, *Falls in Older Australians 2019–20: Hospitalisations and Deaths Among People Aged 65 and Over* (AIHW, Australian Government, 2022). https://www.aihw.gov.au/reports/injury/falls-in-older-australians-2019-20-hospitalisation/contents/about.

On the likelihood that someone who falls will fall again, perhaps repeatedly, within the following year, see Bradley D. Lloyd et al., "Recurrent and Injurious Falls in the Year Following Hip Fracture: A Prospective Study of Incidence and Risk Factors from the Sarcopenia and Hip Fracture Study," *Journals of Gerontology Series A: Biomedical Sciences and Medical Sciences* 64, no. 5 (2009): 599–609, especially 604.

For more on the likelihood of falls in older age and in long-term care facilities, see World Health Organization, *World Report on Ageing and Health*, September 29, 2015, https://www.who.int/publications/i/item/9789241565042, 64.

On falls as the leading preventable cause of death, see Joseph E. Ibrahim et al., "Premature Deaths of Nursing Home Residents: An Epidemiological Analysis," *Medical Journal of Australia* 206, no. 10 (2017): 442–47.

On health care costs associated with falls in New South Wales, see Wendy L. Watson, Yang Li, and Rebecca J. Mitchell, "Projections of Hospitalised Fall-Related Injury in NSW, Australia: Impacts on the Hospital and Aged Care Sectors," *Journal of Safety Research* 42, no. 6 (2011): 487–92, especially 489–90.

On perception of falls risk, see Stephen R. Lord et al., "An Epidemiological Study of Falls in Older Community-Dwelling Women: The Randwick Falls and Fractures Study," *Australian Journal of Public Health* 17, no. 3 (1993): 240–45, especially 243.

On comparative effectiveness of various modes of exercise for preventing falls, see Lesley D. Gillespie et al., "Interventions for Preventing Falls in Older People Living in the Community," *Cochrane Database of Systematic Reviews* 9 (2012), CD007146.

Catherine Sherrington et al., "Exercise for Preventing Falls in Older People Living in the Community," *Cochrane Database of Systematic Reviews* 1 (2019), CD012424.

For the observation that research "does not support the use of low-intensity walking as a primary mode of physical activity to reduce the risk of fall-related injuries and fractures," see *2018 Physical Activity Guidelines Advisory Committee Scientific Report*, F9-9.

The two walking exercise programs for falls reduction that caused people to fall were Barbara Resnick, "Testing the Effect of the WALC Intervention on Exercise Adherence in Older Adults," *Journal of Gerontological Nursing* 28, no. 6 (2002): 40–49, and Mark A. Pereira et al., "A Randomized Walking Trial in Postmenopausal Women:

Effects on Physical Activity and Health 10 Years Later," *Archives of Internal Medicine* 158, no. 15 (1998): 1695–701.

On the five-year due diligence process that preceded the Singhs' Sarcopenia and Hip Fracture Study, see Maria A. Fiatarone Singh et al., "Methodology and Baseline Characteristics for the Sarcopenia and Hip Fracture Study: A 5-Year Prospective Study," *Journals of Gerontology Series A: Biomedical Sciences and Medical Sciences* 64, no. 5 (2009): 568–74.

For more on this comprehensive assessment of risk factors and treatment modalities for hip fracture, see Maria A. Fiatarone Singh, "Exercise, Nutrition and Managing Hip Fracture in Older Persons," *Current Opinion in Clinical Nutrition and Metabolic Care* 17, no. 1 (2014): 12–24.

For the study's results, see Nalin A. Singh et al., "Effects of High-Intensity Progressive Resistance Training and Targeted Multidisciplinary Treatment of Frailty on Mortality and Nursing Home Admissions After Hip Fracture: A Randomized Controlled Trial," *Journal of the American Medical Directors Association* 13, no. 1 (2012): 24–30.

On building bone density and bone mass in childhood, see Shona L. Bass, "The Prepubertal Years: A Uniquely Opportune Stage of Growth When the Skeleton Is Most Responsive to Exercise?" *Sports Medicine* 30, no. 2 (2000): 73–78.

Katherine B. Gunter, Hawley C. Almstedt, and Kathleen F. Janz, "Physical Activity in Childhood May Be the Key to Optimizing Lifespan Skeletal Health," *Exercise and Sport Science Reviews* 40, no. 1 (2012): 13–21.

Many doctors and scientists write that increasing peak bone mass by 10 percent in early life could delay the later risk of osteoporosis by thirteen years and cut the lifetime risk of fracture in half, citing C. J. Hernandez, G. S. Beaupré, and D. R. Carter, "A Theoretical Analysis of the Relative Influences of Peak BMD, Age-Related Bone Loss and Menopause on the Development of Osteoporosis," *Osteoporosis International* 14, no. 10 (2003): 843–47.

"While there is no experimental evidence from large lifelong studies to test such a notion," the widely cited estimates were made on the basis of high-powered computer modeling programs, and "a growing body of evidence suggests that adaptations to mechanical loading in youth translate to greater bone strength over a lifetime," according to Belinda R. Beck et al., "Exercise and Sports Science Australia (ESSA) Position Statement on Exercise Prescription for the Prevention and Management of Osteoporosis," *Journal of Science and Medicine in Sport* 20, no. 5 (2017): 438–45.

The ESSA position statement provides a good overview of research on what types of exercise and sport do and do not build bone density at various stages in life.

On pharmaceutical industry efforts to develop drugs that could help older people become more muscular, see Daniel Rooks and Ronenn Roubenoff, "Development of Pharmacotherapies for the Treatment of Sarcopenia," *Journal of Frailty and Aging* 8, no. 3 (2019): 120–30.

Some early trials of anabolic drugs for sarcopenia—including drugs now banned by the World Anti-Doping Association, including DHEA and human growth hormone—are mentioned in John E. Morley et al., "Sarcopenia," *Journal of Laboratory and Clinical Medicine* 137, no. 4 (2001): 231–43.

For more on muscle quality, see Luigi Ferrucci et al., "Of Greek Heroes, Wiggling Worms, Mighty Mice, and Old Body Builders," *Journals of Gerontology Series A: Biomedical Sciences and Medical Sciences* 67, no. 1 (2012): 13–16.

On the new emerging technique that selectively estimates contractile proteins, which was devised by Walter Frontera and Maria Fiatarone Singh's former boss Bill Evans and his colleagues, see William J. Evans and Peggy M. Cawthon, "D_3Creatine Dilution as a Direct, Non-Invasive and Accurate Measurement of Muscle Mass for Aging Research," *Calcified Tissue International* 114, no. 1 (2024): 3–8.

On the process of defining sarcopenia, see Alfonso J. Cruz-Jentoft, Beatriz Montero-Errasquin, and John E. Morley, "Definitions of Sarcopenia," in *Sarcopenia*, 2nd ed., eds. Alfonso J. Cruz-Jentoft and John E. Morley (Hoboken, NJ: Wiley, 2021), 1–10.

Alfonso J. Cruz-Jentoft et al., "Sarcopenia: European Consensus on Definition and Diagnosis. Report of the European Working Group on Sarcopenia in Older People," *Age and Ageing* 39, no. 4 (2010): 412–23.

Liang-Kung Chen et al., "Sarcopenia in Asia: Consensus Report of the Asian Working Group for Sarcopenia," *Journal of the American Medical Directors Association* 15, no. 2 (2014): 95–101.

Alfonso J. Cruz-Jentoft et al., "Sarcopenia: Revised European Consensus on Definition and Diagnosis," *Age and Ageing* 48, no. 1 (2019): 16–31.

Liang-Kung Chen et al., "Asian Working Group for Sarcopenia: 2019 Consensus Update on Sarcopenia Diagnosis and Treatment," *Journal of the American Medical Directors Association* 21, no. 3 (2020): 300–307.

On techniques for measuring muscle mass and muscle quality, see Steven B. Heymsfield et al., "Skeletal Muscle Mass and Quality: Evolution of Modern Measurement Concepts in the Context of Sarcopenia," *Proceedings of the Nutrition Society* 74, no. 4 (2015): 355–66.

On low muscle strength and risk of functional decline, see Laura A. Schaap, Annemarie Koster, and Marjolein Visser, "Adiposity, Muscle Mass, and Muscle Strength in Relation to Functional Decline in Older Persons," *Epidemiologic Reviews* 35, no. 1 (2013): 51–65.

Schaap's finding on the risk of disability associated with sarcopenic obesity is based on Richard N. Baumgartner, "Sarcopenic Obesity Predicts Instrumental Activities of Daily Living Disability in the Elderly," *Obesity Research* 12, no. 12 (2004): 1995–2004.

On sarcopenic obesity, see also Sari Stenholm et al., "Sarcopenic Obesity—Definition, Etiology and Consequences," *Current Opinion in Clinical Nutrition and Metabolic Care* 11, no. 6 (2008): 693–700.

On the developmental origins of sarcopenia, see Richard Dodds and Avan Aihie Sayer, "A Lifecourse Approach to Sarcopenia," in *Sarcopenia*, 2nd ed., eds. Cruz-Jentoft and Morley, 77–93.

While Dodds and Sayer mainly focus on developmental origins of sarcopenia in early life, they also say that efforts to prevent sarcopenia "could be particularly relevant in midlife, when assessment of future cardiovascular disease risk is already undertaken in primary care."

For the protocol of the New Zealand trial of high-intensity progressive resistance training for overweight and obese children and adolescents, see Amanda C. Benson, Margaret E. Torode, and Maria A. Fiatarone Singh, "A Rationale and Method for High-Intensity Progressive Resistance Training with Children and Adolescents," *Contemporary Clinical Trials* 28, no. 4 (2007): 442–50.

For trial results, see A. C. Benson et al., "The Effect of High-Intensity Progressive Resistance Training on Adiposity in Children: A Randomized Controlled Trial," *International Journal of Obesity* 32, no. 6 (2008): 1016–27.

On the Borg Scale of Perceived Exertion, see Gunnar A. V. Borg, "Psychophysical Bases of Perceived Exertion," *Medicine & Science in Sports & Exercise* 14, no. 5 (1982): 377–81.

Panteleimon Ekkekakis, "Gunnar A. V. Borg (November 28, 1927–February 2, 2020): A Multilayered Legacy," ACSM Blog, February 18, 2020: https://www.acsm.org/blog-detail/acsm-blog/2020/02/18/gunnar-borg-perceived-exertion-multilayered-legacy.

The American Academy of Pediatrics Guidance on weight training is Paul R. Stricker et al., "Resistance Training for Children and Adolescents," *Pediatrics* 145, no. 6 (2020).

See also Bruno Ribeiro et al., "The Benefits of Resistance Training in Obese Adolescents: A Systematic Review and Meta-Analysis," *Sports Medicine–Open* 8, no. 1 (2022): 109.

For people who want to know more about resistance training for children and adolescents, two good starting points are Rhodri S. Lloyd et al., "Position Statement on Youth Resistance Training: The 2014 International Consensus," *British Journal of Sports Medicine* 48, no. 7 (2014): 498–505; and Avery D. Faigenbaum et al., "Youth Resistance Training: Updated Position Statement Paper from the National Strength and Conditioning Association," *Journal of Strength and Conditioning Research* 23, no. 5 (2009): S60–S79.

When describing her as a kind of younger-elder stateswoman of medical exercise science, I cite Maria A. Fiatarone Singh, Mikel Izquierdo, and John E. Morley, "Physical Fitness and Exercise," in *Pathy's Principles and Practice of Geriatric Medicine*, 6th ed., ed. Alan Sinclair et al. (Hoboken, NJ: Wiley, 2022), 77–107.

Mikel Izquierdo et al., "International Exercise Recommendations in Older Adults (ICFSR): Expert Consensus Guidelines," *Journal of Nutrition, Health and Aging* 25, no. 7 (2021): 824–53, especially 828.

For evidence that high-intensity progressive resistance training can produce roughly the same improvement in older adults' aerobic capacity as moderate intensity walking, see the SMART study results published in Yorgi Mavros et al., "Mediation of Cognitive Function Improvements by Strength Gains After Resistance Training in Older Adults with Mild Cognitive Impairment: Outcomes of the Study of Mental and Resistance Training," *Journal of the American Geriatrics Society* 65, no. 3 (2017): 550–59; and compare with Kirk I. Erickson et al., "Exercise Training Increases Size of Hippocampus and Improves Memory," *Proceedings of the National Academy of Sciences USA* 108, no. 7 (2011): 3017–22.

For the SMART study protocol, see Nicola J. Gates et al., "Study of Mental Activity and Regular Training (SMART) in At Risk Individuals: A Randomised Double Blind, Sham Controlled, Longitudinal Trial," *BMC Geriatrics* 11, no. 19 (2011).

For details on how this intervention improved global cognitive function, see Maria A. Fiatarone Singh et al., "The Study of Mental and Resistance Training (SMART) Study—Resistance Training and/or Cognitive Training in Mild Cognitive Impairment: A Randomized, Double-Blind, Double-Sham Controlled Trial," *Journal of the American Medical Directors Association* 15, no. 12 (2014): 873–80.

Progressive resistance training "has a significant effect on aerobic capacity," according to Chiung-ju Liu and Nancy K. Latham, "Progressive Resistance Strength Training for Improving Physical Function in Older Adults," *Cochrane Database of Systematic Reviews* 3 (2009), CD002759, a review of 121 trials involving 6,700 participants.

For the World Health Organization's statement that "strength and balance training should precede aerobic exercise" and summary of evidence for that pronouncement, see *World Report on Ageing and Health*, 71.

For the assertion that walking does not improve balance, the WHO cites Tracey E. Howe et al., "Exercise for Improving Balance in Older People," *Cochrane Database of Systematic Reviews* 11 (2011), CD004963.

For the assertion that walking has no effect on preventing falls, the WHO cites Catherine Sherrington et al., "Effective Exercise for the Prevention of Falls: A Systematic Review and Meta-Analysis," *Journal of the American Geriatric Society* 56, no. 12 (2008): 2234–43; and Alexander Voukelatos et al., "The Impact of a Home-Based Walking Programme on Falls in Older People: The Easy Steps Randomized Controlled Trial," *Age and Ageing* 44, no. 3 (2015): 377–83.

For the definition of epidemiology, see the abstract of Ralph S. Paffenbarger Jr., "Contributions of Epidemiology to Exercise Science and Cardiovascular Health," *Medicine & Science in Sports & Exercise* 20, no. 5 (1988): 426–38.

On the California longshoremen study, see Ralph S. Paffenbarger Jr. et al., "Work Activity of Longshoremen as Related to Death from Coronary Heart Disease and Stroke," *New England Journal of Medicine* 282, no. 20 (1970): 1110–14.

Regarding the development of epidemiological research on strength training, see Jason A. Bennie, Jane Shakespear-Druery, and Katrien De Cocker, "Muscle Strengthening Exercise Epidemiology: A New Frontier in Chronic Disease Prevention," *Sports Medicine–Open* 6, no. 40 (2020).

Jason A. Bennie et al., "The Epidemiology of Aerobic Physical Activity and Muscle-Strengthening Activity Guideline Adherence Among 383,928 U.S. Adults," *International Journal of Behavioral Nutrition and Physical Activity* 16, no. 34 (2019).

On independent and combined effects of resistance training and aerobic training on all-cause mortality, cardiovascular disease mortality, and cancer mortality, see Prathiyankara Shailendra et al., "Resistance Training and Mortality Risk: A Systematic Review and Meta-Analysis," *American Journal of Preventive Medicine* 63, no. 2 (2022): 277–85.

On difficulties and ambiguities of epidemiological assessment of resistance exercise, see Jane Shakespear-Druery et al., "Assessment of Muscle-Strengthening

Exercise in Public Health Surveillance for Adults: A Systematic Review," *Preventive Medicine* 148 (2021): 106566.

The discussion of exercise and medical school curricula draws from Angela V. Connaughton et al., "Graduating Medical Students' Exercise Prescription Competence as Perceived by Deans and Directors of Medical Education in the United States: Implications for Healthy People 2010," *Public Health Reports* 116, no. 3 (2001): 226–34.

Jeff K. Vallance, Mark Wylie, and Randy MacDonald, "Medical Students' Self-Perceived Competence and Prescription of Patient-Centered Physical Activity," *Preventive Medicine* 48, no. 2 (2009): 164–66.

Kara Solmundson, Michael Koehle, and Donald McKenzie, "Are We Adequately Preparing the Next Generation of Physicians to Prescribe Exercise as Prevention and Treatment? Residents Express the Desire for More Training in Exercise Prescription," *Canadian Medical Education Journal* 7, no. 2 (2016): e79–e86.

Edward Phillips et al., "Including Lifestyle Medicine in Undergraduate Medical Curricula," *Medical Education Online* 20, no. 1 (2015): 26150.

On the tendency of many doctors to prescribe drugs instead of exercise, see L. Denoeud et al., "First Line Treatment of Knee Osteoarthritis in Outpatients in France: Adherence to the EULAR 2000 Recommendations and Factors Influencing Adherence," *Annals of the Rheumatic Diseases* 64, no. 1 (2005): 70–74, especially 72.

On the question of why "the medical community" has "neglected exercise as a standard treatment," see Robert Sallis, "Exercise Is Medicine: A Call to Action for Physicians to Assess and Prescribe Exercise," *The Physician and Sportsmedicine* 43, no. 1 (2015): 23–26.

For more on the Exercise is Medicine initiative, see Walter R. Thompson et al., "Exercise Is Medicine," *American Journal of Lifestyle Medicine* 14, no. 5 (2020): 511–23.

On prescribing exercise as an ethical or Hippocratic imperative, see Mikel Izquierdo and Maria Fiatarone Singh, "Promoting Resilience in the Face of Ageing and Disease: The Central Role of Exercise and Physical Activity," *Ageing Research Reviews* 88 (2023): 101940.

Mikel Izquierdo and Maria Fiatarone Singh, "Urgent Need for Integrating Physical Exercise into Geriatric Medicine: A Call to Action," *British Journal of Sports Medicine* 57, no. 15 (2023): 953–54.

Those essays build on an editorial by Mikel Izquierdo et al., "Is it Ethical Not to Prescribe Physical Activity for the Elderly Frail?" *Journal of the American Medical Directors Association* 17, no. 9 (2016): P779–81.

The study Fiatarone Singh mentions that integrated balance and strength training into activities of daily life is Lindy Clemson et al., "Integration of Balance and Strength Training into Daily Life Activity to Reduce Rate of Falls in Older People (the LiFE Study): Randomised Parallel Trial," *British Medical Journal* 345 (2012): e4547.

CONCLUSION

On the effectiveness of lower-load strength training, see Jacques Duchateau et al., "Strength Training: In Search of Optimal Strategies to Maximize Neuromuscular Performance," *Exercise and Sport Sciences Reviews* 49, no. 1 (2021): 2–14.

Duchateau's paper builds on the work of, among others, Brad J. Schoenfeld et al., "Strength and Hypertrophy Adaptations Between Low- vs. High-Load Resistance Training: A Systematic Review and Meta-Analysis," *Journal of Strength and Conditioning Research* 31, no. 12 (2017): 3508–23.

Minoru Shinohara et al., "Efficacy of Tourniquet Ischemia for Strength Training with Low Resistance," *European Journal of Applied Physiology and Occupational Physiology* 77, no. 1–2 (1998): 189–91.

Yudai Takarada et al., "Effects of Resistance Exercise Combined with Moderate Vascular Occlusion on Muscular Function in Humans," *Journal of Applied Physiology* 88, no. 6 (2000): 2097–106.

Murat Karabulut et al., "The Effects of Low-Intensity Resistance Training with Vascular Restriction on Leg Muscle Strength in Older Men," *European Journal of Applied Physiology* 108, no. 1 (2010): 147–55, in which BFR training produced nearly equal strength gains for one group of healthy older men who lifted weights at 20 percent of 1-RM, and another group who lifted weights at 80 percent of 1-RM.

Apiwan Manimmanakorn et al., "Effects of Low-Load Resistance Training Combined with Blood Flow Restriction or Hypoxia on Muscle Function and Performance in Netball Athletes," *Journal of Science and Medicine in Sport* 16, no. 4 (2013): 337–42, in which BFR training produced nearly equal gains of strength and size for groups of netball players who lifted weights at 20 percent of 1-RM and 80 percent of 1-RM.

For commentary on Duchateau's paper, see Roger M. Enoka, "Exercise and Sports Sciences Reviews: 2021 Paper of the Year," *Exercise and Sports Sciences Reviews* 50, no. 4 (2022): 173–74.

For more on Emmanuel Legeard, see his website http://www.emmanuel-legeard.com.

Stuart M. Phillips, a leading expert on muscle protein synthesis, cowrote two of the clearest brief overviews of research on resistance training for health, focused on evidence of the benefits of lower-load training: Stuart M. Phillips, Jasmin K. Ma, and Eric S. Rawson, "The Coming of Age of Resistance Exercise as a Primary Form of Exercise for Health," *ACSM's Health & Fitness Journal* 27, no. 6 (2023): 19–25; and Stuart M. Phillips and Richard A. Winett, "Uncomplicated Resistance Training and Health-Related Outcomes: Evidence for a Public Health Mandate," *Current Sports Medicine Reports* 9, no. 4 (2010): 208–13.

Phillips has overseen and cowritten many influential studies that show how training at lower loads can make muscles bigger and stronger, including Robert W. Morton et al., "A Systematic Review, Meta-Analysis and Meta-Regression of the Effect of Protein Supplementation on Resistance Training-Induced Gains in Muscle Mass and Strength in Healthy Adults," *British Journal of Sports Medicine* 52, no. 6 (2018): 376–84.

Cameron J. Mitchell et al., "Resistance Exercise Load Does Not Determine Training-Mediated Hypertrophic Gains in Young Men," *Journal of Applied Physiology* 113, no. 1 (2012): 71–77.

Nicholas A. Burd et al., "Low-Load High Volume Resistance Exercise Stimulates Muscle Protein Synthesis More Than High-Load Low Volume Resistance Exercise in Young Men," *PLoS One* 5, no. 8 (2010): e12033.

Robert W. Morton et al., "Neither Load nor Systemic Hormones Determine Resistance Training-Mediated Hypertrophy or Strength Gains in Resistance-Trained Young Men," *Journal of Applied Physiology* 121, no. 1 (2016): 129–38.

On the prevalence of total joint replacement surgeries, see Hilal Maradit Kremers et al., "Prevalence of Total Hip and Knee Replacement in the United States," *Journal of Bone and Joint Surgery* American 97, no. 17 (2015): 1386–97, especially 1393 (for the comparison with prevalence of heart failure).

For international comparisons of volume and incidence of total hip replacement, see Mohammad S. Abdelaal et al., "Global Perspectives on Arthroplasty of Hip and Knee Joints," *Orthopedic Clinics of North America* 51, no. 2 (2020): 169–76, especially 170–71.

For the predicted growth in demand for total hip replacement among Medicare patients in the United States, see Ittai Shichman et al., "Projections and Epidemiology of Primary Hip and Knee Arthroplasty in Medicare Patients to 2040–2060," *Journal of Bone and Joint Surgery Open Access* 8, no. 1 (2023): e22.00112. According to this paper, 17 percent of Medicare patients required full knee or hip replacement in the year 2000, and the demand more than doubled by 2019, when 36 percent of those on Medicare required the surgeries.

The Norwegian study in which exercise reduced or delayed the need for hip replacement surgery is Ida Svege et al., "Exercise Therapy May Postpone Total Hip Replacement Surgery in Patients with Hip Osteoarthritis: A Long-Term Follow-Up of a Randomised Trial," *Annals of the Rheumatic Diseases* 74, no. 1 (2015): 164–69.

On back extensions and risk of osteoporotic fracture in older women, see Mehrsheed Sinaki et al., "Stronger Back Muscles Reduce the Incidence of Vertebral Fractures: A Prospective 10 Year Follow-up of Postmenopausal Women," *Bone* 30, no. 6 (2002): 836–41.

For much more detail and practical guidance on resistance exercise and osteoporosis, see Katherine Brooke-Wavell et al., "Strong, Steady and Straight: UK Consensus Statement on Physical Activity and Exercise for Osteoporosis," *British Journal of Sports Medicine* 56, no. 15 (2022): 837–46.

On the partnership of mind and muscle, see Richard L. Lieber, *Skeletal Muscle, Function & Plasticity*, 3rd ed. (Philadelphia: Lippincott Williams & Wilkins, 2010), 36.

Research on the factors associated with participation in strength training is fairly sparse. Two reviews of this research that I found to be especially helpful are Ryan E. Rhodes et al., "Factors Associated with Participation in Resistance Training: A Systematic Review," *British Journal of Sports Medicine* 51, no. 20 (2017): 1466–72; and Yoshio Nakamura and Kazuhiro Harada, "Promotion of Strength Training," in *Physical Activity, Exercise, Sedentary Behavior and Health*, ed. Kazuyuki Kanosue et al. (Tokyo: Springer, 2015), 29–42.

On social support as an influence on strength training among people with diabetes, see Heather Tulloch et al., "Exercise Facilitators and Barriers from Adoption to Maintenance in the Diabetes Aerobic and Resistance Exercise Trial," *Canadian Journal of Diabetes* 37, no. 6 (2013): 367–74.

Mireia Vilafranca Cartagena, Glòria Tort-Nasarre, and Esther Rubinat Arnaldo, "Barriers and Facilitators for Physical Activity in Adults with Type 2 Diabetes Mellitus:

A Scoping Review," *International Journal of Environmental Research and Public Health* 18, no. 10 (2021): 5359, notes that, for diabetic patients, social support for exercise often diminishes after diagnosis and as the disease progresses.

Among people with cancer: Cynthia C. Forbes et al., "Prevalence and Correlates of Strength Exercise Among Breast, Prostate, and Colorectal Cancer Survivors," *Oncology Nursing Forum* 42, no. 2 (2015): 118–27.

Among older women: Aishwarya Vasudevan and Elizabeth Ford, "Motivational Factors and Barriers Towards Initiating and Maintaining Strength Training in Women: A Systematic Review and Meta-Synthesis," *Prevention Science* 23, no. 4 (2022): 674–95 reviewed twenty studies from four countries and found "the main barriers and facilitators were social factors." In sum: "When friends and family supported women by praising them and accompanying them to the gym, this motivated women to continue [strength training]; however, when friends and family discouraged them, they found it challenging to continue" strength training.

Mark D. Litt, Alison Kleppinger, and James O. Judge, "Initiation and Maintenance of Exercise Behavior in Older Women: Predictors from the Social Learning Model," *Journal of Behavioral Medicine* 25, no. 1 (2002): 83–97.

Among older adults in general: Nancy E. Sherwood and Robert W. Jeffery, "The Behavioral Determinants of Exercise: Implications for Physical Activity Interventions," *Annual Reviews of Nutrition* 20 (2000): 21–44.

Richard A. Winett, David M. Williams, and B. M. Davy, "Initiating and Maintaining Resistance Training in Older Adults: A Social Cognitive Theory-Based Approach," *British Journal of Sports Medicine* 43, no. 2 (2009): 114–19.

List of Illustrations

CHAPTER 1
  9 Bronze strigil (scraper), fifth or fourth century BC. (*The Metropolitan Museum of Art*)
  9 Terra-cotta fragment of a kylix (drinking cup), fifth century BC. (*The Metropolitan Museum of Art*)
 29 Metope of Atlas and the Apples of Hesperides depicting Herakles and Athena, from the Temple of Zeus at Olympia, ca. 470 BC–457 BC. (*Archaeological Museum of Olympia. Photo from Wikimedia Commons*)

CHAPTER 2
 47 Bronze statuette of a girl runner, sixth century BC. (*Kar. 24, National Archaeological Museum, Athens*)
 59 Statues of Cleobis and Biton, ca. 580 BC. (*Delphi Archaeological Museum. Photo by Rob Stoeltje, Wikimedia Commons*)
 60 Closeups of the knees of Cleobis and Biton. (*Photo by Michael Joseph Gross*)

CHAPTER 3
 78 Charles Stocking and Stella Stocking on the stadium track at ancient Olympia, July 2023. (*Photo by Catherine E. Pratt*)
 81 Terra-cotta Panathenaic prize amphora, ca. 530 BC. (*The Metropolitan Museum of Art*)
 81 Detail of terra-cotta Panathenaic prize amphora. (*Image copyright © The Metropolitan Museum of Art. Image source: Art Resource, NY*)
 97 Themistokles Base: relief depicting wrestlers, ca. 510 BC. (*National Archaeological Museum, Athens. Photo by Paolo Villa, Wikimedia Commons*)
 99 Statue of the Weary Hercules after performing the last of his labors, sculpted by Glykon of Athens ca. early third century AD; a copy of a Greek original by Lysippos, late fourth century BC. (*Naples National Archaeological Museum, Farnese Collection. Photo by Marie-Lan Nguyen, Wikimedia Commons*)
101 Eugen Sandow as the Weary Hercules, ca. 1893. (*Photo by Napoleon Sarony*)

CHAPTER 4
118 Jan Suffolk's fourth-grade school photo, ca. 1961. (*Collection of Jan Todd*)
125 Jan Todd at Powerbuilder's Gym in Macon, Georgia, ca. 1975 (*Collection of Jan Todd*)

CHAPTER 5
163 Strongwoman Katie Sandwina, as depicted in *Les rois de la force* by Edmond Desbonnet, 1911.
169 Jan Todd squatting at the U.S. Powerlifting Federation Women's Nationals in Los Angeles, in 1980. (*Photo by Kathy Tuite Leistner*)
171 Jan Todd, moments after becoming the first woman to squat more than 500 pounds—she squatted 507—in Germantown, Tennessee, 1981. (*Collection of Jan Todd*)

CHAPTER 6
181 Illustration of women with dumbbells, from *The New Gymnastics for Men, Women, and Children*, 6th edition, by Dio Lewis, 1864.
187 Illustration of Venus de Milo as an ideal of beauty, from *Physiology and Calisthenics* by Catherine Beecher, 1856. (*Collection of the H.J. Lutcher Stark Center and Archive of Physical Culture and Sport*)
187 Illustration of Venus de Milo as an ideal of beauty, from *Outlines of Physiology* by J. L. Comstock, 1860.
189 Illustration of a woman preparing to do a chin-up, from *A Treatise on Gymnastic Exercises, Or Calisthenics, For the Use of Young Ladies* by J. A. Beaujeu, 1828. (*Collection of Monroe C. Gutman Library, Harvard University*)
191 Illustration of a woman doing a dip between parallel bars, from Beaujeu's 1828 *Treatise*. (*Collection of Monroe C. Gutman Library, Harvard University*)
200 Handbill advertising lecture on physical culture by George Barker Windship, 1859. (*Collection of the Massachusetts Historical Society*)
203 George Barker Windship carte de visite portrait, ca. 1860s. (*Collection of Jan Todd*)
205 Butler's Health Lift advertisement from *The Daily Graphic*, New York, December 17, 1873. (*Collection of Jan Todd*)

CHAPTER 7
224 Cross-sectional images by computed tomography (CT) scan of muscles in the legs of a twenty-five-year-old and an eighty-one-year-old. (*Journal of the American Physiological Society*)
242 Muscle cells before and after eccentric exercise. (*Reproduced with permission by Professor Jan Fridén, MD, PhD*)
244 Helene Freundlich working out in the gym at Hebrew Rehabilitation Center for Aged, Roslindale, Massachusetts, ca. 1994. (*Photo by George R. McLean*)
248 Ed Rosenthal curling a five-pound dumbbell, ca. 1994. (*Photo by Bob Kramer*)

CHAPTER 8

303 Ramanee in the gym at Lidcombe Campus, University of Sydney, September 2019. *(Photo by Michael Joseph Gross)*

EPILOGUE

360 The bullmastiff puppy Pudgy Stockton. *(Photo by Jan Todd)*

Index

ACL (anterior cruciate ligament) injuries, 68–69, 73, 287
Adams, Robert, 154
adolescents and children
   activity levels, 120–121, 333
   anabolic steroid use, prevalence of, 250
   in ancient Olympics, 50 (boys)
   in ancient Sparta, 45 (boys), 46–48 (girls and young women)
   birth weight as predictor of adult muscle strength, 330
   body image and muscularity, 251–252 (boys)
   girls' gymnastic training (nineteeth century), 184, 191, 193, 204, 205
   metabolic health (prevalence of prediabetes and type 2 diabetes), 276
   obesity and overweight, study of progressive resistance training for, 330–333
   sarcopenia, early life risk factors and preventive interventions for, 330 (*see also* Dodds, Richard; Sayer, Avan Aihie)
   strength training, benefits of, xiv, 123 (girls), 252 (boys), 330–333
   strength training, safety of, 233–234
The Adonis Complex (Pope), 254
aerobic activity and exercise
   benefits of, xxiv, 336, 337
   as depression treatment, 267–268, 273–274
   and falls risk, 320, 335
   as high blood pressure treatment, 279, 280
   and mortality risk, 158, 336–338
   for older adults, 175, 234, 266–267, 268, 333–335
   and osteoarthritis, 290–291
   public health recommendations, 159
   and type 2 diabetes, 275, 278, 279, 280, 281
   and weight loss, 291
   for youth, 332, 333
aerobic fitness built by weight training, xv, 175, 333–335
Aerobics (Cooper), xix, 157, 158, 208
aging. *See* older adults
agōn, 11, 53
Alcmaeon of Croton, 54–55, 57, 62, 88, 357
Alcott, Louisa May, 203–204
Alexandria (Egypt), 62–63, 64, 197
alkē, 26
Altis (sanctuary at Olympia), 10–12
Alves, Alberto J., 279, 280
American Academy of Pediatrics, 332–333
American College of Sports Medicine, 159, 161, 176, 208, 255, 279, 321, 341, 350, 351
American Diabetes Association, 277–278
American Drug Free Powerlifting Federation, 173–174
American Heart Association, 208
American Medical Association (AMA), 143–145, 147–151, 153–156, 195, 215, 233, 250, 341
American Psychiatric Association (APA), 253, 273
American Psychological Association, 251, 274

amphora. *See* Panathenaic amphora
anabolic steroids
  historical context, 139–141, 158, 159, 176–177
  and muscle dysmorphia, 254–256
  and Olympic games, modern, 139–141, 250, 325
  and 'roid rage, 250
  Stocking on, 18–19
  testing for, 140–141, 172, 173
  Todd on, 113, 125, 141–142, 166, 172–175, 176–177
Ancient Greek Athletics (Miller), 32
Ancient Greek Athletics: Primary Sources in Translation (Stocking and Stephens), 83
anterior cruciate ligament (ACL) injuries, 68–69, 73, 287
Any Given Sunday (film), 18
Aphrodite. *See* Venus
Aristotle
  on anatomy and movement, 56–57, 60–63, 69, 89, 119, 196, 354, 357
  on beauty, 98
  on excellence, 324
  gymnasium of, 56
  influence on Galen, 65–66, 89, 196
  and Olympic games, ancient, 75
  and Plato, 52, 53, 54, 61, 89
  strength defined by, 52, 55
Armstrong, Neil, 235
Arnold: The Education of a Bodybuilder (Schwarzenegger and Hall), 18, 19
arthritis (osteoarthritis), 68, 73, 228, 262, 276, 278, 284–302, 305–306, 338, 341, 353
Asian Working Group for Sarcopenia, 328
Athens
  perspectives on Sparta, 46, 48
  physical culture of, 51, 52–54, 56, 80–81
athletic nudity, 53–54
Athletics and Literature in the Roman Empire (König), 82
athletics and medicine, tensions between. *See* medicine and athletics, tensions between
athlos, 15–16
Atlas, Charles, 152
"Autobiographical Sketches of a Strength-Seeker" (Windship), 201

back and core strength and endurance, 4–8, 33, 41, 69–72, 74, 352, 353
Back Mechanic (McGill), 70
Baker, A. Barry, 265
Bamman, Marcas, 340, 342
Bannister, Roger, 156
Bastholm, Eyvind, 23, 57, 62
Bean, Jonathan, 311
Beaujeu, J. A., 188–193
Beaujeu, Madame, 192–195
beauty. *See* body aesthetics
bench press, 19, 33, 39, 105, 131, 134, 168, 170, 171, 190
Bennie, Jason, 337–338
Benson, Amanda C., 330–332
Berryman, Jack, 184
bibasis, 48
biceps muscles, xx, 4, 35–36, 71, 116, 127–128, 189, 197, 241, 248–249, 257, 258, 314
biē, 27–28
Bishop, Bennett, 172–174
Biton (ancient Greek athlete), 31–32, 59, 60, 93, 141, 431
Björnsson, Hafþór Júlíus, 44
Blackwell, Elizabeth, 195
blood-flow restriction (BFR) training, 349
body aesthetics
  in ancient Greek physical culture, 45–46, 48, 96–102
  female standards, 112–113, 114, 115, 117, 118, 126, 130–131, 137–138, 164–165, 180, 186–188, 210–211, 212
  in modern physical culture, 145–149
bodybuilding, xvi, 18, 19, 20–21, 35, 100–101, 113, 131, 132, 133, 140, 141, 208, 249, 250, 252, 253, 256, 350
body composition
  in adolescents, 330–332
  and body image perception, 252–253
  and insulin sensitivity, 275
  in older adults, 225–226, 229, 277, 282–283, 326
  and sarcopenia, 277, 326, 328–330
  and type 2 diabetes, 275, 277, 282–283
  and weight-loss diets, 278, 293
body dysmorphic disorder. *See* muscle dysmorphia
bone density, xv, xxv, 231, 278, 301, 321, 322–323

Booth, Frank W., 262
Borg, Gunnar, 331
Borg Scale of Perceived Exertion, 331
Bortz, Walter II, 306
bottle cap bending, 122, 142
Braith, Randy W., 283–284
Braunwald, Eugene, 156
Briggs, Andrew M., 305–306
Brumbach, Katharina, 114. *See also* Sandwina, Katie
Butler, David P., 204–206
Butler, George, 133
Butler Health Lift, 204–206
Butler's System of Physical Training (Butler), 204–206
Bybon, stone of, 43–44, 49–50

Cafri, Guy, 251–252
cancer mortality risk and exercise, 336–337
cardiovascular health, 156–157, 158, 208, 225, 279–281, 334–335, 336–337
Carson, Johnny, xxiii, 138
Castaneda, Carmen, 283
Chapman, David L., 100
Cheema, Bobby, 283
Chen, Liang-Kung, 328
choral dancing, 48
Christesen, Paul
  on ancient precedent for modern physical culture, 3
  on athletic nudity, 53–54
  on Spartan girls and women, 46, 47–48
  on social status and ancient Greek athletics, 51, 53–54
  on Stocking, 21
chronic kidney disease, 283–284
Clemson, Lindy, 344
Cleobis (ancient Greek athlete), 31–32, 59, 60, 93, 141, 431
college and university athletes, 66–69, 72–74, 112–113, 164–165, 265
Conan the Barbarian (film), 20, 174
conservation of energy, 198
Cooper, Kenneth H., xix, 157, 208
Corcilius, Klaus, 56, 57, 61–62, 69
core endurance, 69–72, 74
corticosteroids, 283–284
Coulter, Ottley, 162–164, 179, 181, 210
cross-education, 155

Crowther, Nigel, 44–45, 54
Cruz-Jentoft, Alfonso J., 327

damazō, 27
deadlift, 19–20, 33, 44, 111–112, 115, 122, 124–126, 131, 134, 138, 142, 163, 168, 170, 171, 173, 360
Deca-Durabolin, 125
Defoe, Daniel, 199
De Humani Corporis Fabrica (Vesalius), 196–197
delayed-onset muscle soreness (DOMS), 240–243
DeLorme, Eleanor, 213, 240
DeLorme, Thomas Lanier. *See also* Progressive Resistance Exercise; progressive resistance training
  background and early life, xiii–xiv, 76
  on medical prejudice against muscle and weight training, xviii–xix
  one-repetition maximum test by, xiv, 214–215, 229
  strength training protocol of, xiv, 19, 229, 230, 314, 348
  strength training protocol applied as polio treatment, 155–156
  strength training protocol inspired by *Strength and Health* magazine, 213–216
  strength training studies by, xiv, 222
  Todd on, 212–213, 215–216
Demeter, priestess of, 75
demos, 26
depression, 263, 267–274, 332
Desbonnet, Edmond, 163
Descartes, René, xviii
deskproofing workout, 4–8, 352, 355
de Vos, Nathan J., 313
diabetes. *See* type 2 diabetes
Diabetes Prevention Program, 276–277
diet
  and chronic kidney disease, 283
  dietary protein, 237, 283, 293
  dietary weight loss linked to muscle and bone loss, 291–292
  diets of ancient Greek athletes, 13, 86–87
  diets of Roman gladiators, 90
  and fall risk, 322
  and osteoarthritis, 291–293

diet (*cont.*)
  and type 2 diabetes, 274, 276–277, 278
  and weight training, 237, 245, 291–293, 342
Dikon (ancient Greek athlete), 43
Dinnie, Donald, 145
Dodds, Richard, 330
Drîmbă, Ion, 226
Duchateau, Jacques, 348, 349–351
dunamis, 26, 65, 196, 325
Durant, Will, 324
Dworkin, Shari L., 113

eccentric contractions, 241–243
efficacy, 123, 322
Engleman, Ben, 258–259
Enoka, Roger, 126–127, 350–351
epidemiology of strength training, 233–234, 336–338
Epley, Boyd, 160
Eskin, Catherine, 103
Esposito, Anthony, 119
The Eternally Wounded Woman (Vertinsky), 185, 195
Eumastas, stone of, 44, 49–50
Euripides, 46
European Working Group on Sarcopenia in Older People, 325–326, 327–328, 330
Evans, William J., 225–226
"Exercise is Medicine" initiative, 341, 342
"Exercise with a Small Ball" (Galen), 88–89
Exercise Physiology (Fahey), 36
The Expressiveness of the Body and the Divergence of Greek and Chinese Medicine (Kuriyama), 58, 64

Fabricius, Hieronymus, 36
Fahey, Thomas D., 36
Fair, John D., 141
falls and hip fractures, xv, xvi, xxiv, 233, 266–267, 307–311, 318–323, 334–335
"farmer's tan," 53–54
Farnese Hercules. *See* Weary Hercules (statues)
Farrar, Roger, 209
Fasting for Health (Macfadden), 152
female athletes. *See also* Sandwina, Katie; Todd, Jan
  in ancient Greece, 45–48, 58
  body aesthetics, 112–113, 114, 117, 118, 126, 130–131, 137–138, 164–165, 180, 186–188, 210–211, 212
  gender comparison in exercise studies, 114–115, 128, 161–162
  muscularity and strength of, 112–113, 126, 130–131, 137–138, 161–162, 164–165, 180, 186–188, 210–211, 212
  powerlifting competitions for, 124, 132–134, 136
Ferrucci, Luigi, 324–325, 326
Fiatarone Singh, Maria
  background and early life, 226–227, 307–310
  on benefits of lifting weights (overview), xxiii–xxv
  on body composition, 282–283, 291–292, 328–329, 332–333
  career overview, xxiii–xxv, 220–221, 274, 306–307, 333
  clinical practice at Centre for Strong Medicine, 343
  on falls and hip fractures, 319–323
  on frailty, 234, 263, 266–267, 321–323
  on grip strength, 327
  Hebrew Rehabilitation Center for Aged strength training studies, 227–235 ("High-Intensity Strength Training in Nonagenarians" (*JAMA*) pilot study); 236–240, 244–249, 250 (*New England Journal of Medicine* follow-up study—randomized controlled trial); 257–259 (influence of studies)
  marriage to and research collaborations with Nalin Singh, 274, 282–283, 306–307, 312–314, 321–322
  on medical prescriptions of exercise, 261–264, 265–267, 342, 343–345
  on rarity of strength training prescription in mainstream medicine, 339
  view that failure to prescribe exercise may violate Hippocratic oath, 342–343

on organ recipients and anabolic
    exercise, 284
on osteoarthritis, 287–288, 289–292,
    300–301, 353
personal training regimen of, 344–345
physical fitness and exercise guidelines,
    contributions to, 333–334
on power training, 312–313
on principle of specificity, xxiv
progressive resistance training medical
    research highlights
  chronic kidney disease studies, 283
  depression studies, 267–268,
    269–272, 273
  falls and hip fracture studies,
    321–322
  insomnia study, 274
  knee osteoarthritis studies, 287–288,
    291–293, 295–296, 301
  mild cognitive impairment study,
    finding on progressive resistance
    training and aerobic fitness,
    333–334
  overweight and obese children's and
    adolescents' strength training
    study, 329–332
  type 2 diabetes studies, 274, 277–278,
    282–283
on ratings of perceived exertion,
    331, 345
on research evidence limitations,
    335–336
on weight-lifting exercise and increased
    size of brain's posterior cingulate
    cortex, xvii–xviii, xxiii–xxiv
Filho, Celso Alves dos Santos, 253–254
Fishbein, Morris, 154, 155
Fit for Your Life (organization), 258
football (American), 18, 66, 72–74
football (soccer). *See* soccer
Foster, Michael, 143–144
Fowle, William B., 184
Fox, Robin Lane, 54–55
frailty, xxv, 234, 263, 264–267, 321–323
Freundlich, Helene, 238–239,
    244–245, 259
Frontera, Walter, 221–223, 225–226,
    227, 231, 247, 249, 257, 310–311,
    312, 326
Fuller, William, 193
Fulton, John Farquhar, 198

Gaines, Charles, 133
Gaiter, Denise, 161–162
Galen of Pergamon
  on anatomy, muscles, and movement,
    36, 63–66, 67, 87, 88–91, 92, 93,
    196–197
  on athletics and health, xviii, 86–91,
    92, 93
  authority and influence of, xviii, 36, 64,
    66, 67, 86, 92, 93, 95, 104, 105,
    143, 145, 148, 196–198
  background and career, 86–87, 90, 95
  on cartilage, 286
  injury while wrestling, 91
  value of moderation, xviii
general adaptation syndrome (Selye),
    40, 94
George, Molly, 164–165
Gerenos (ancient wrestler), 103, 104
geriatrics. *See* older adults
Gilstrap, Lori, 161–162
gladiators, 90–91
gloios, 10
gluteal amnesia, 7
gluteal muscles, 4, 6–7, 8, 41, 61, 68, 96,
    169, 352, 353–354, 355
Graded Resistance Exercise and Type 2
    Diabetes in Older Adults
    (GREAT2DO), 282–283
Gregorić, Pavel, 56–57, 61–62, 69
Greviskes, Lindsey, 130–131
grip strength, 122, 326, 327–328
Guerrero, Yareni, 292–293
The Guinness Book of World Records, xxii,
    112, 115–116, 125, 133–134,
    142–143, 164
Guttmann, Allan, 194
gymnasion or *gymnasium/gymnasia*, 52–54,
    56, 76, 84, 86–87, 91, 92
"Gymnastic* Exercise for Females"
    (Fowle), 184
gymnastics, 182–185, 188–195, 195–196,
    198–199, 201–202, 203–204, 356
Gymnasticus (Philostratus)
  aims and authorship, 94–95
  anticipation of modern exercise science
    and training principles, 94, 96,
    102–104, 129, 201, 353
  on athletes' bodies, 96–100, 101–102
  on athletic training as a form of
    wisdom, 94–95, 102

442 INDEX

Gymnasticus (Philostratus) *(cont.)*
  on *kairos*, 103–104, 129, 201, 353
  on the Tetrads, 102–104
  Stocking on, 83, 84, 94–95, 96–98, 100–104, 208, 353
  translations of, 82–83
  on victory, 104–105

Hamilton, John B., 144
Hamilton, William Rowan, 192
handgrip strength, 122, 326, 327–328
Hankinson, R. J., 88
Haskell, William L., 158
Hawley, Madame Beaujeu, 192–195
heart attack, 156–157, 208, 281, 336
"heavy-resistance exercises" (DeLorme), 212–213, 240
Hebrew Rehabilitation Center for Aged, 219–220, 227–231, 232–241, 243–249, 250, 251, 257–260, 262, 263–264, 265
Hebrew SeniorLife, 260
Helmholtz, Hermann von, 198
Hera, 100
Herakles. *See* Hercules.
Hercules, labors of, 16, 17
Hercules, sculptures of, 28, 29, 98–102, 208, 209
Herodotus, 31–32, 60
Higginson, Thomas Wentworth, 202
hip flexors, 4–5, 6, 7, 8, 68–69
hip fractures and falls, xv, xvi, xxiv, 233, 266–267, 307–311, 318–323, 334–335
hip hinge, 19–20, 71, 142
Hippocrates, 32, 86, 102, 143
Hippocratic oath, 342–343
hip replacement surgeries, 352, 353
hip thrust exercise, 7–8, 352–353
Histories (Herodotus), 31–32, 60
The History of Muscle Physiology (Bastholm), 23, 57, 62
The History Written on the Classical Greek Body (Osborne), 57, 58
H.J. Lutcher Stark Center and Archive for Physical Culture and Sports, 209–210, 211
Hoffman, Bob, 139–140
Holloway, Jean Barrett, 161–162

Holmes, Knox, 145
Homer. *See Iliad; Odyssey*
Homer's *Iliad and the Problem of Force* (Stocking), 5, 25–29
Hubal, Monica J., 127–128
Human Nutrition Research Center on Aging (Tufts), 221–223, 225–226
hypertrophy. *See* muscle, hypertrophy
Hysmon (ancient Greek pentathlete), 76

Iliad (Homer). *See also* Homer's *Iliad* and the Problem of Force (Stocking)
  dying as depicted by, 24–25
  funeral games in, 16
  knees prominent in, 25, 59, 88
  mind/body language in (*psychē* and *sōma*), 23, 84
  muscle references in, 23–24
  Stocking's early reading of, 14–15
  strength as depicted by, 24–29, 36, 42, 65, 325
InCHIANTI study, 311, 324, 326
inclusion body myositis (IBM), 317
Inside Powerlifting (Terry Todd), 134–135
insomnia, 274
Intensive Diet and Exercise for Arthritis (IDEA) study, 291
International Olympic Academy, 10, 15
iphi, 26
Iron Game History: The Journal of Physical Culture, 176–177, 208
isonomia, 55
Izquierdo, Mikel, 342–343

The Jack LaLanne Show (TV show), 308–309
Johnson, Dwayne (The Rock), 21
joint replacement surgeries, 68, 352, 353
Jones, Joseph B., 193–194
Joseph, Chris, 72–74
Jouffroy, Françoise K., 6–7
Joyner, Michael J., 158–159

kairos, 103–104, 129, 201, 353
Keiser, Dennis, 239

Keiser Leg Press, 239
Kennedy, Jacqueline, 117, 133 (as Jacqueline Kennedy Onassis)
Khrushchev, Nikita, 38
The Kings of Strength (Desbonnet), 163
Knapp, Bobbi A., 112–113
knee osteoarthritis, 284–302
knee replacement surgery, 68
knees, ancient Greek understanding and depiction of, 25, 58–61
König, Jason, 82, 83
Kono, Tommy, 210
Kraemer, William, 140
Kramer, Peter D., 267
Krane, Vikki, 130, 165
kratos, 26, 42, 325–326
Kuriyama, Shigehisa, 58, 64
Kyle, Donald G., 50–51
kylix, 431

LaLanne, Elaine, 209, 308–309
LaLanne, Jack, 209, 308–309
Lalonde, Marc, 157
latissimus dorsi muscles, 96–97
Laye, Matthew J., 262
Legeard, Emmanuel, 351
Les Rois de la Force (Desbonnet), 163
Lewis, Diocletian, 185–186, 203–204
Lewis, Orly, 65
Lexell, Jan, 257
Lieber, Richard L., 37, 356–357
Lieberman, Daniel E., 81
lifting stones, 43–45, 49–50
Lift Your Way to Youthful Fitness (Todd and Todd), 175
Lipitor, 317
Lipsitz, Lewis A., 239
Listening to Prozac (Kramer), 267
Liu, Yubo, 281–282
Low Back Disorders (McGill), 70
luein, 24–25
Lydston, G. Frank, 147–148

Macfadden, Bernarr, 151–155, 156, 161, 179, 215
"main strength," 199
Majestic Womanhood, 186–188

"Making Sense of Muscle" (George), 164–165
Marvin, Miranda, 100
Massachusetts, nineteenth-century educational reform in, 182–184
Mattern, Susan P., 90–91
Matveyev, Leonid, 38
Mavros, Yorgi, 282–283, 334
McCabe, Nora, 137
McGill, Stuart, 69–72, 74, 107
Mead, Sedgwick, 155
The Medical Follies (Fishbein), 154
medicine and athletics, tensions between. *See also* DeLorme, Thomas Lanier; Galen
    ancient rivalry between doctors and trainers, xviii, xxii, 32, 84–93
    continuation of ancient rivalry in modern era, xviii–xx, 32, 147–155
    episodic easing of tensions in modern era, 143–145, 155–162, 175–176, 351
    Macfadden's influence on rivalry, 150–155
    repetitive intransigence of rivalry, 356
    role of medical education in rivalry's persistence, 340–341
Médina, Monique F., 6–7
menos, 26–27
mental health, xvii–xvix, 123, 251–255, 263, 267–268, 269–274, 332, 337
Millender, Ellen, 45, 46
Miller, Stephen G., 32
Milo Bar-Bell Company, 209, 210
Milo of Croton, xiii, xiv, xv, 3, 33, 38–39, 43, 49, 50, 51–52, 54, 95, 97, 207, 357
MILO: A Journal for Serious Strength Athletes, 35
mind-body dualism, xvii–xix, xx, 83–88, 93–94, 353–354, 356–357
monarchia, 55
mortality risk and exercise, 336–338
Mr. Olympia contest, 250
Mrozek, Donald J., 151, 152
muscle
    in ancient Greek and Roman medical and scientific writings, 57–58, 61–62, 63–66, 89–90
    contraction, 241–243 (eccentric); 69–71, 258 (isometric)

444 INDEX

muscle (*cont.*)
  as depicted by ancient Greek sculptors, 58–61
  fiber types, 224, 312–313
  hypertrophy, 126–129, 130, 222, 231, 243, 245, 324–325, 348–351
  loss of strength and mass with age, 223–225
    named as sarcopenia, 231–235
    popular media coverage of sarcopenia, 254–256
    prevalence and evolving definition of sarcopenia, 255–256, 324–328
    preventing sarcopenia by building muscle in early life, 330
    sarcopenia accelerated by undiagnosed diabetes, 277
    sarcopenia exacerbated by obesity, 329
    studies of progressive resistance training to treat sarcopenia, 228–231, 236–239, 244–248
  and metabolic health, 275–278, 281–283
  misconceptions of, xvi–xx. *See also* medicine and athletics, tensions between *and* female athletes (muscularity and strength of)
  myokines, xv
  as organ of voluntary movement, 62–66, 88–90, 196–198
  power, 19, 247, 310–313, 349–350
  quality, 325, 327
  satellite cells, 246–248
  structure and function, xv, xx–xi, 4–5, 35–36, 69–72, 126–127, 223–226, 241–243, 246–248, 325, 327
  training for size and strength, 126–129
Muscle Builder magazine, 161
muscle dysmorphia (body dysmorphic disorder), 250, 253–256
muscle power, 19, 247, 310–313, 349–350
muscle quality, 325–326, 327
The Muscular Ideal (Thompson and Cafri), 251–253
Mussolini, Benito, 152
mys, 23–24, 57–58

Naci, Huseyin, 279
Nagy, Gregory, 12, 16–17, 25, 107
National Aeronautics and Space Administration, 159
National Strength and Conditioning Association, 160–161
National Strength and Conditioning Association Journal, 161–162
National Strength Coaches' Association, 160
nēdus, 22
Nestor (Homeric hero), 27
Neuromechanics of Human Movement (Enoka), 126–127, 350
Newton, Isaac, 197
Noble, Yian, 293
non-steroidal anti-inflammatory drugs, 243
North American Society for Sport History, 179, 209
nudity. *See* athletic nudity
nutrition. *See* diet
Nutton, Vivian, 62–63

O'Caoimh, Rónán, 264–265
Odyssey (Homer), 14–15, 23, 24–25
Oertel, M. J., 144
older adults
  in ancient Greek gymnasia, 54
  body composition and risk of disease and disability, 225–226, 275, 277, 282–283, 328–331
  depression, studies of exercise as treatment for, 268–274
  falls and hip fractures
    early lifestyle interventions that may reduce risk of, 322–323
    epidemiology and economic costs of, 319
    types of exercise that may reduce risk of, 307, 320–322
  frailty, 264–267, 321–323
  grip strength, 326–328
  muscle hypertrophy, 128, 222, 245, 231, 324–325, 245, 348–351
  muscle power, 247, 310–313
  musculoskeletal conditions, risk of and effects on, 305–306
  myokines, xv
  osteoarthritis, 284–302, 352–354
  sarcopenia. *See* muscle, loss of strength and mass with age
  strength training for aerobic fitness, 333–335

strength training for older adults, early studies
  by Fiatarone at Hebrew Rehabilitation Center for Aged, 219–220, 227–241, 243–249, 257–260
  by Frontera, 222–226
  by Young, 175, 225
 strength training safety, 233–234
 strength training to reverse effects of immobilization, 175
 type 2 diabetes
  elevated risk of, 276
  studies of progressive resistance training as treatment for, 282–283
Olympia (history of the site), 10–13, 28, 51–52, 75–78, 91
 described by Pausanias, 75–76, 103–104
Olympic games, ancient
 events, 11–14, 50, 75–78, 91–92, 103–105
 prizes, 17, 50
 origins of, 10–14
 rules of competition, 48–49, 50
 victor statues, 51–52
 victory, primacy of, 16–18
 waning of, 91–92
Olympic games, modern. See also Kono, Tommy
 and anabolic drugs, 139–141, 250, 325
 and sport science research, 38–39
The Olympic Myth of Greek Amateur Athletics (Young), 80–81
Olympic weightlifting, 21, 40, 132, 139–140, 142, 210, 250, 265
O'Neill, Evelyn, 237–241, 243–245, 258–259
one-repetition maximum strength, 214–215, 222, 229, 230, 233, 257, 270, 282, 301, 314, 331, 345, 348, 350
Onians, R. B., 25
"On Movement of Muscles" (Galen), 36, 63–64
On the Usefulness of the Parts of the Body (Galen), 65–66
organ transplant. See Fiatarone Singh, Maria (on organ recipients and anabolic exercise)
The Origins of European Thought (Onians), 25
Osborne, Robin, 57, 58

osteoarthritis, 68, 276, 284–302, 341, 352–354
Ozempic, 278

Paffenbarger, Ralph S., Jr., 336
Panathenaic amphora, 80–81
Papadopoulou, Sousana K., 255
paradox
 of athletic training, 84–85, 93–94
 for female athletes, 130–131, 137, 164–165
 of muscles, xx–xxi, 3–4, 36, 241–243
 sport-specific paradox, 5–6, 66–69, 93–94, 96–97, 98, 146
 of strength in Homer, xxii, 27–29
 of strength training, 40, 347
 of training for the long term, 74
Park, Roberta J., 145
Park, Seok Won, 277
Pausanias, 75–76, 103–104
Pearl, Bill, 140
pectoral muscles, 61, 86, 190
Pedersen, Bente Klarlund, 262
Pepi, Esther, 244, 246
perceived exertion, ratings of, 331, 345
periodized training regimens, 38–43, 72–73, 94, 102–104, 129–130
Pescatello, Linda S., 280–281
Phillips, Edward, 340
Phillips, Stuart M., 351
philology, 7, 22
Philostratus, Flavius, 95–105, 107, 129, 201, 208–209, 259, 353. See also Gymnasticus
physical culture, ancient. See also medicine and athletics, tensions between; mind-body dualism; Stocking, Charles
 aging and, 54
 anatomy and movement, theories of (see Aristotle; Galen)
 athletic competitions, 10–14, 15–18, 49–52, 75–78, 91–92 (see also Olympic games, ancient)
 athletic nudity, 53–54
 athletic training and competition as religious experience and ritual, 10–14, 17–18, 28, 42–43, 53, 77–78, 80, 106–107

physical culture, ancient. (*cont.*)
  diets of ancient Greek athletes, 13, 86–87
  diets of Roman gladiators, 90
  education and, 45–46, 52–58, 61–62, 76, 84, 86–87, 92, 104
  female athletes, 45–48, 58
  language of, 10, 11, 15–18, 22–29, 43, 55–58, 61, 62, 63 (*see also Iliad; Odyssey*)
  political analogies for health and functional movement, 55, 62, 357
  strength, concepts of, 25–29, 36, 42–43, 52
  strength training and feats of strength, 31–32, 43–50, 52, 82–83
  victory, 13, 16–18, 42–43, 51–52, 80–81, 104–106
Physical Culture and the Body Beautiful (Todd), 182–184, 185, 186–190, 194, 195–196, 199, 201, 202–203, 204, 206, 208
Physical Culture magazine, 150–151, 152–153, 154, 161
physical culture, modern. *See also* anabolic steroids; gymnastics; medicine and athletics, tensions between; Todd, Jan
  definition, xxiii, 150–151
  gender differences, exercise research on, 114–115, 128, 161–162
  gender norms and stereotypes, evolution of, 110–116, 117, 118–119, 126, 130–131, 132–133, 134–136, 137–138, 164–165, 180, 182, 186–191, 210–211, 212
  Macfadden's influence on, 151–155, 156, 161, 179, 215
  muscle and weight training, ambivalence about, 143–146, 147–151, 154–158, 159–162, 164–165, 206–207, 212–216, 220, 249–257, 338–343
  political dimensions, xvii, 182–184, 194, 211, 357
  Sandow's influence on, 100–101, 114, 147–149, 151
  Sandwina's influence on, 114–115, 163–164, 180, 210–211
  Sargent's influence on, 146–147
  Schwarzenegger's influence on, xvii, 20–21, 133–134, 137, 174
  Windship's influence on, 198–203, 204, 206–207
"Physical Culture" (Windship essay), 201
Pileggi, Sarah, 137
Pindar, 17
Plato, 52–53, 54, 56, 84
  on flesh (muscle), 61, 66
  influence on Galen, 86, 89, 92
  and mind-body dualism, xviii, 84
pneuma, 56–57, 63–66, 119, 196–197. *See also* psychic pneuma
Pomeroy, Sarah B., 45
Pope, Harrison G., Jr., 254
Potter, David, 51
power. *See* muscle power
Powerlifting. *See also* Stocking, Charles; Todd, Jan; Todd, Terry
  American Drug Free Powerlifting Federation, 173–174
  anabolic drug use and testing for, 113, 125, 166, 172–174
  definition of sport and competition format, 19–20, 168
  for female athletes, 124–125, 132–135, 166–172
  training protocols, 33–34, 41–42, 130, 350–351
Power to the People! Russian Strength Training Secrets for Every American (Tsatsouline), 34–36, 37–38, 39, 40, 41–42
power training for muscle, 310–313, 349–350
Practice Guideline for the Treatment of Patients with Major Depressive Disorder (APA), 273
Pratt, Catherine, 10–11, 14, 76
prednisone, 283–284
Progressive Resistance Exercise (DeLorme), 156
progressive resistance training. *See also* strength and strength training principles
  barriers to adoption as medical treatment, xviii–xix, 339–342, 357
  benefits (overview), xv–xvi, xxiii–xxv, 220
  DeLorme on, xiv, 155–156, 212–216
  development of name, 213–216

as medical treatment, 208, 220, 222, 225–227, 262, 269–274, 274–284, 296–302, 306–307, 312–318, 321–323, 324, 330–333, 333–335, 353 (*see also* Fiatarone Singh, Maria; Singh, Nalin)
Prozac, 267
psychē, xviii, 23, 84–85, 87–88
psychic pneuma, 63–66, 196–197. *See also* pneuma
Pumping Iron (book by Gaines and documentary film by Butler), 133
purposive exercise, 182–183
Purves, Alex, 27

Ramanee, 284–286, 288–289, 292–302, 305–306
recovery. *See* periodized training regimens
Reid, Heather, 92
Reinhoudt, Cindy, 133
The Religious Beliefs of the Greeks (von Wilamowitz-Moellendorff), 13
Remijsen, Sofie, 91
Rhetoric (Aristotle), 52, 98
rhomboid muscles, 5
Ridgway, Brunilde Sismondo, 59–60
Ritchie, Meg, 161–162
Roberts, Christian K., 262
Rockwood, Kenneth, 264–265
'roid rage, 250
Roosevelt, Franklin Delano, 152
Rosenberg, Irwin, 221, 231–232, 249, 250, 256, 325, 326
Rosenthal, David, 260
Rosenthal, Ed, 244–246, 248–249, 258, 259–260
Roth, Rachel I., 113
Rousseau, Jean-Jacques, 186
row (resistance exercise), 7–8, 352
running and runners, xvi, 6, 14, 18, 54, 70, 75–82, 98, 106, 120, 268, 311–312, 355. *See also* sprinting and sprinters

Saint-Gaudens, Jeanne Marie Céline Torre, 307–310
Sallis, Robert, 341, 342

Saltin, Bengt, 262
Sandow, Eugen, 100–101, 114, 147–149, 151, 153
Sandow's Curative Institute, 149
Sandwina, Katie, 114, 129, 163–164, 174, 180, 207, 210–211
SantaBarbara, Nicholas J., 252–253
sarcopenia (age-related loss of muscle strength and mass), 223–225, 228–235, 237, 250, 254–256, 266, 277, 321, 324–330, 333
sarcopenic obesity, 328–329, 330
Sargent, Dudley Allen, 146–147, 206–207
sarx, 23, 61, 232
Saxon, Arthur, 44
Sayer, Avan Aihie, 330
Schaap, Laura A., 328–329
Schoenfeld, Brad, 128, 348
Schwarzenegger, Arnold
  books by, 18, 19, 35
  films starring, 20, 133, 174
  influence of, xvii, 20–21, 107, 133, 137, 141, 161, 174, 249
  Sandow comparison, 149
  Todd and, 209, 359
Schweitzer, Albert, 226
Science and Development of Muscle Hypertrophy (Schoenfeld), 128
sculpture, Ancient Greek
  beauty standards based on, 97–100
  of female athletes, 45, 46–48, 58
  of Hercules, 28, 98–102, 208, 209
  knees, rendering of, 58–61
  as model for athletes, 96, 97–100
  modern influence of, 100–101, 147, 151, 187–188
  musculature, depiction of, 3–4, 58–61, 98
  of Olympic victors, 51–52
Selye, Hans, 40, 93, 94, 139
semaglutide, 278
Seneca, 85–86
Shailendra, Prathiyankara, 337
Sheiko, Boris, 33–34, 102
Sherrington, Charles Scott, xvii, 36–37
Shinohara, Minoru, 348–349
Shurley, Jason, 130–131, 162, 212–213
Siff, Mel, 39–41
Singh, Harold, 269
Singh, Maria Fiatarone. *See* Fiatarone Singh, Maria.

Singh, Nalin
   background and early life, 268–269
   clinical practice at Centre for Strong
      Medicine, 307, 313–319, 321
   death of, 323–324
   marriage to and research collaborations
      with Maria Fiatarone (Singh), 274,
      282–283, 306–307, 312–314,
      321–322
   medical training (geriatrics), 263
   on progressive resistance training as
      treatment for frailty, 263–266,
      321–322
   progressive resistance training medical
      research highlights
         depression studies, 267–272, 273
         falls and hip fracture studies, 321–322
         insomnia study, 274
         mild cognitive impairment study,
            finding on weight training and
            aerobic fitness, 333–334
         power training studies, 312–313
         type 2 diabetes studies, 274, 282–283
      on research evidence limitations,
         335–336
sitting, exercise regimen to counteract, 4–8
Skeletal Muscle Structure, Function, and
   Plasticity (Lieber), 37
Slater, Cindy, 209
Sluijs, Esther M. F van., 121
soccer, 66, 68–69, 112, 164–165
sōma, xviii, 23, 84–85, 87, 88
sophia, 43, 102
Sparta, physical culture of, 45–48, 53, 58
Spartan Women (Pomeroy), 45
spine health and function, 7, 20, 33–34,
   41, 69–72, 74, 353. See also back
   and core strength and endurance
spiritus animalis, 197–198
Sport and Democracy in the Ancient and
   Modern Worlds (Christesen), 53–54
Sport and Spectacle in the Ancient World
   (Kyle), 50–51
sport-specific paradox, 5–6, 66–69, 93–94,
   96–97, 98, 146
sprinting and sprinters, xxi, 6, 11–13,
   68–69, 70, 75–82, 96, 105–106,
   121, 355. See also running and
   runners
Staden, Heinrich von, 63
stadion, 11–14, 16, 75, 80–81, 95, 207
Stalin, Joseph, 38

Starr, Paul, 154
statues. See sculpture, ancient Greek
Steno, Nicolaus, 197
Stephens, Susan, 83
steroids. See anabolic steroids;
   corticosteroids
Stewart, Andrew, 46
sthenos, 26–27
Stocking, Charles. See also physical
   culture, ancient
   on ancient Greek athletics and
      religion, 11, 12–13, 42–43,
      77–78, 105–107
   on ancient Greek concepts of strength,
      xxii, 25–29 (Homer); 52 (Aristotle)
   on ancient Greek language of athletics
      and physical culture, 10, 11, 15–18,
      22, 23, 25–29, 43, 52–53, 82–89,
      92–93
   on ancient Greek sculptures, 27–28,
      97–98, 100, 101–102
   on athletic training as a matter of life
      and death, 105–107
   background and early life, 5, 14–15,
      18–22
      as athlete (powerlifter), xxii, 19–20,
         33–36, 37–42
      injuries, experience of, 34–35, 41,
         67–69, 72, 94
      as strength coach (UCLA), 5, 22,
         66–69, 72–74
   career overview, xxi–xxii
   on Galen, 86–87, 88
   on the Gymnasticus (Philostratus),
      82–83, 94, 96–98, 101–105
   on Herodotus (story of Cleobis and
      Biton), 31–32
   at International Olympic Academy
      symposium on ancient and modern
      athletics in Olympia, 10–11, 15–18,
      27–28, 76, 79–80, 83–93
   on lifting weights in ancient Greece,
      43–44, 52
   on literary sources regarding ancient
      Greek athletics, 82, 83
   on mind and body in Homer, 23
   on mind-body dualism, 83–87, 88,
      92–93, 353–354
   Pausanias, translation of, 76
   publications by, 5, 25–29, 83, 94
   on sport-specific paradox, 5–6, 67–69,
      93–94, 96–97

on sprinting biomechanics, ancient and modern, 80–82
Todd as influence on, 107, 208, 211
training regimens of, 4–8 (to compensate for sitting at desk); 33–36, 37–42 (powerlifting); 67–69 (as strength coach, to minimize athletes' risk of injuries); 69–74 (as strength coach, to build athletes' core endurance); 76–80 (sprinting on ancient tracks); 352–355 (to prepare for older age)
on victory, 17–18, 42–43, 105–107
Stocking, Damian
on ancient Greek concepts of strength, 28
background and early teaching experience, 14–15, 84
on Homer, 24–25
on Olympic games, ancient, 13–14
Stocking, Stella, 76
Stockton, Abbye "Pudgy," 208, 210, 360
Stoessel, Lynne, 161–162
Stoicism, 85–86, 93
stones. See lifting stones
Strength and Health magazine, xiii–xiv, 161, 210, 213–216
strength and strength training. See also adolescents and children; back and core strength and endurance; medicine and athletics, tensions between; muscle; older adults; periodized training regimens; powerlifting; weight lifting
aerobic fitness built by weight training, xv, 175, 333–335
and ancient Greek athletics, 43–45, 49–50 (*see also* Hercules; Milo of Croton)
barriers to adoption as medical treatment, 338–344
behavioral research on social support for, 357–358
benefits of (overview), xv–xvi, xxi–xxv, 220–221, 262–263, 306–307
concepts and definitions of strength, xiv, 19, 24–29, 35–37, 52, 117–120, 135–136, 312–313
gender differences, exercise research on, 114–115, 128, 161–162
gender norms and stereotypes, evolution of, 110–116, 117, 118–119, 126,
130–131, 132–133, 134–136, 137–138, 164–165, 180, 182, 186–191, 210–211, 212
mainstreaming of, xvi, xix, 20–21, 133, 149, 160–162, 174–176, 208–210, 212–216, 336
as medical treatment (*see* American College of Sports Medicine; DeLorme, Thomas; Fiatarone Singh, Maria; Singh, Nalin)
and mental health, xvii–xvix, 123, 251–255, 263, 267–268, 269–274, 332, 337
and mortality risk reduction, 336–338
and muscle hypertrophy, 126–129, 130, 222, 231, 243, 245, 324–325, 348–351
protocols and regimens (*see* DeLorme, Thomas Lanier; periodized training regimens; powerlifting; 3/7 method; Stocking, Charles; Todd, Jan)
safety of (epidemiological evidence), 233–234
as tension, 25, 33, 34, 35–37, 41, 71–72, 176
"Strength Training for Female Athletes: A Position Paper," 161–162
stress, 38–43, 93–94, 103–104, 106
The Stress of Life (Selye), 40, 41
strigil, 8–10, 43, 104–105, 431
stroke rehabilitation, 345
strongwomen. See Sandwina, Katie; Todd, Jan
Suffolk, James, 116, 117
Sukava, Tyson, 57–58
supercompensation, 41–43
Supertraining (Verkhoshansky & Siff), 39–41

tai chi, 280–281, 320
Takarada, Yudai, 348–349
Tang, Yong Q., 265
Taylor, J. Madison, 144–145, 356
tension, strength as, 25, 33, 34, 35–37, 41, 71–72, 176
Tetrads (ancient training programs), 102–104
A Text Book of Physiology (Foster), 143–144

450 INDEX

"Thomas L. DeLorme and the Science of Progressive Resistance Exercise" (Todd, Shurley, and Todd), 212
Thompson, J. Kevin, 251–252
3/7 method, 350–351
Tishler, Dorothy, 233, 234
Tod, David, 254
Todd, Jan
  background and early life, 116–118, 120, 121–122
  on Beaujeaus, 188–191, 194–195
  on Butler, 204
  career overview, xxii–xxiii
  on DeLorme, Thomas, 212–213, 215–216
  early weight training, 111–116, 122–126, 129–133
  as *Guinness Book of World Records* "strongest woman in the world," xxii, 164
  and H.J. Lutcher Stark Center and Archive for Physical Culture and Sports, 209–211
  marriage to and athletic and academic collaboration with Terry Todd, 111–112, 113–116, 122–125, 129–130, 131–132, 134–135, 142, 167–168, 174, 175, 176–177, 180–181, 190, 209–210, 212–213
  on medicine and athletics, tensions between, 212–213, 215–216
  as powerlifter, 112–114, 124–126, 133–136, 137–138, 142–143, 162–163, 165–166, 167–171, 173–174, 253
  on powerlifting and anabolic drugs, 113, 125, 141–142, 166, 172–175
  powerlifting and other lifting records set by, xxii, 112, 115–116, 124–126, 133–134, 137–138, 142–143, 162–163, 167–171, 172–174
  as powerlifting and weight training coach, xxii, 132, 133–134, 135–136, 172, 174, 188–190, 211
  publications by, 130–131, 161–162, 174–175, 176–177, 179–191, 193–196, 199, 201, 202–203, 204, 205, 206, 208, 211, 212–213
  publicity highlights, 137–138, 142–143
  on Rousseau, Jean-Jacques, 186
  on Sandwina, Katie, 163–164, 210–211
  and Schwarzenegger, Arnold, 209, 359
  on science of weight training, 160–162, 212–213, 215–216
  strength shows by, 142–143, 162–163
  on strength training for athletes, 160–162
  training regimens of, 122–124, 129–130, 172, 359–360
  on the Weary Hercules, 208
  on weight training for athletes, 160, 162
  on Windship, George Barker, 199, 201, 202–203, 206
  on Wollstonecraft, Mary, 186–188
  on womanhood and muscular size and strength, xxiii, 112, 117, 119, 130–131, 135–136, 164–165, 167–168, 180, 186–191, 201–211, 212
Todd, Terry
  and Coulter, Ottley, 162, 180–181
  death of, 211
  and H.J. Lutcher Stark Center and Archive for Physical Culture and Sports, 209–211
  marriage to and academic and athletic collaboration with Jan Todd, 111–112, 113–116, 122–125, 129–130, 131–132, 134–135, 142, 167–168, 174, 175, 176–177, 180–181, 190, 209–210, 212–213
  powerlifting championship and records, 114
  publications by, 114;134–135 (*Inside Powerlifting*); 139; 174–175; 176–177; 212–213
  and Schwarzenegger, Arnold, 209
  on the squat, 165–166
tonic muscular contraction, 36–37, 63–64, 71
The Tonight Show, xxiii, 137–138
Torode, Margaret E., 330–332
total joint replacement surgeries, 352
trapezius muscles, 98
A Treatise on Gymnastic Exercises, Or Calisthenics, For the Use of Young Ladies (Beaujeu), 188–192
triceps muscles, xx, 127–128, 190, 315–316
Tsatsouline, Pavel, 34–36, 37–38, 41, 72, 107, 259

Turnen, 183
Twain, Mark, 148–149
type 2 diabetes, 225, 263, 274–279, 281–283, 341–342, 357
2018 Physical Activity Guidelines Advisory Committee Scientific Report, 262; 272–273 (on exercise as treatment for depression); 278–279, 281 (on exercise and type 2 diabetes risk factors); 320 (on falls and fracture risk)

Ultimate Back Fitness and Performance (McGill), 70
Valenzuela, Pedro L., 279–280
Venus, 187–188
Verkhoshansky, Yuri, 39–41
Vernant, Jean-Pierre, 17–18
Vertinsky, Patricia A., 185, 195
Vesalius, Andreas, 196–197
Vesley, Jane de, 115
victory, 16–18, 26, 42–43, 50–52, 80–81, 82, 85, 104–107
A Vindication of the Rights of Woman (Wollstonecraft), 186–188
Vlastos, Gregory, 55

Waddell, Tom, 140
Warren, John Collins, 184, 195, 201–202
Watson, Wendy L., 319
Weary Hercules (statues), 98–102, 208, 209
Wegovy, 278
Weider, Joe, 101, 141, 208
weightlifting. *See* Olympic weightlifting
weight lifting. *See also* powerlifting; strength and strength training principles
    and ancient Greek athletics, 43–45, 49–50
    for athletes, xix, 130–131, 160–162, 164–165 (*see also* sport-specific paradox; Stocking, Charles; Todd, Jan)
    by females, xxiii, 114–115, 130–131, 163–165 (*see also* Todd, Jan)
    mainstreaming of, xvi, xix, 20–21, 133, 149, 160–162, 174–176, 208–210, 212–216, 336
    as medical treatment. *See* DeLorme, Thomas Lanier; Fiatarone Singh, Maria; Singh, Nalin
"Weightlifting in Antiquity: Achievement and Training" (Crowther), 44–45
Welsman, Joanne, 235
WHO (World Health Organization), 273, 275, 305, 319, 333, 334–335, 336
Whorton, James, 145–146
Wilmore, Jack, 114–115, 175
Wilson, Guy, 296–301, 334, 336, 343
Windship, George Barker, 198–203, 204, 206, 207
Wollstonecraft, Mary, 186–188
Woman's Physical Development (magazine), 179
World Health Organization (WHO), 273, 275, 305, 319, 333, 334–335, 336
World Report on Ageing and Health (WHO), 305, 333, 334–335

Yates, Dorian, 250
Yerty, Wilma, 116, 117–118
Young, Archie, 175, 225
Young, David, 80–81
Young People and Physical Activity (Armstrong and Welsman), 235

Zeus, 10–12, 14, 26, 27, 28, 42–43, 95, 325–326

About the Author

A longtime *Vanity Fair* contributing editor, **Michael Joseph Gross** has published investigative reporting, essays, and books about culture, technology, politics, religion, and business. He was raised in rural Illinois and lives in New York City.